Entangled Life

History, Philosophy and Theory of the Life Sciences

Volume 4

For further volumes:
http://www.springer.com/series/8916

Gillian Barker • Eric Desjardins • Trevor Pearce
Editors

Entangled Life

Organism and Environment in the Biological
and Social Sciences

 Springer

Editors
Gillian Barker
Department of Philosophy
Western University
London, Ontario
Canada

Eric Desjardins
Department of Philosophy
Western University
London, Ontario
Canada

Trevor Pearce
Department of Philosophy
University of North Carolina at Charlotte
Charlotte, North Carolina
USA

ISSN 2211-1948   ISSN 2211-1956 (electronic)
ISBN 978-94-007-7066-9   ISBN 978-94-007-7067-6 (eBook)
DOI 10.1007/978-94-007-7067-6
Springer Dordrecht Heidelberg New York London

Library of Congress Control Number: 2013946646

Printed on acid-free paper

Springer is part of Springer Science+Business Media (www.springer.com)

Contents

Introduction: Perspectives on Entangled Life

Gillian Barker, Eric Desjardins, and Trevor Pearce

Abstract Despite burgeoning interest in new and more complex accounts of the organism-environment dyad, biologists and philosophers of biology have paid little attention to the history of these ideas and to their broader deployment in the social sciences and in other disciplines outside biology. Even in biology and philosophy of biology, detailed conceptual models of the organism-environment relationship are still lacking. This volume is designed to fill these lacunae by providing the first multidisciplinary discussion of the topic of organism-environment interaction. It brings together scholars from history, philosophy, psychology, anthropology, medicine, and biology to discuss the common focus of their work: entangled life, or the complex interaction of organisms and environments.

In September 1978, a special issue of *Scientific American* was published, "devoted to the history of life on earth as it is understood in the light of the modern 'synthetic' theory of evolution" (1978, 47). Introduced by the zoologist Ernst Mayr, it comprised a series of articles by prominent scientists showing how that theory made sense of the history of life, from its origins to the emergence of modern human behavior. The final article in the issue, however, stood apart from the others. It offered an extended critique of a notion—adaptation—that was central to the theoretical perspective celebrated by the rest: a notion that had indeed been central to studies of the natural world even before evolution came onto the scene. The idea that the environment sets "problems" that organisms must "solve" was riddled with difficulties, according to geneticist Richard Lewontin (1978, 213). Organisms,

---

G. Barker (✉) • E. Desjardins
Department of Philosophy, Western University, London, ON, Canada
e-mail: gbarker5@uwo.ca; edesjar3@uwo.ca

T. Pearce
Department of Philosophy, University of North Carolina at Charlotte, Charlotte, NC, USA
e-mail: tpearce6@uncc.edu

G. Barker et al. (eds.), *Entangled Life*, History, Philosophy and Theory
of the Life Sciences 4, DOI 10.1007/978-94-007-7067-6_1,
© Springer Science+Business Media Dordrecht 2014

Lewontin insisted, are not passively shaped by the selective forces resulting from changes in environments. Instead, they actively create those changes:

> There is a constant interplay of the organism and the environment, so that although natural selection may be adapting the organism to a particular set of environmental circumstances, the evolution of the organism itself changes those circumstances. (215)

This article closing a special issue devoted to the "modern synthesis" of genetics and natural selection was in fact part of a broad intellectual movement in the late 1970s that began to question certain aspects of that very synthesis—a movement that insisted upon the importance of interaction between organism and environment during ontogeny, or the lifetime of the organism (e.g., Gould 1977; Lewin 1980; Bonner 1982).

Much of the recent interest among biologists in different models of the interaction of organism and environment can be traced back to the new perspectives that emerged in this period. Evolutionary-developmental biology, or "evo-devo," is now a hot topic. Evo-devo has a complex intellectual history going back at least to the nineteenth century, but many historians and practitioners see the modern resurgence of interest in development as a response to the late-1970s critique of the modern synthesis by Lewontin and others (Laubichler 2007; Müller 2007; Wagner 2007; for deeper roots, see Raff and Love 2004; Amundson 2005; and the other chapters in Laubichler and Maienschein 2007). By opening up the black box into which the modern synthesis placed ontogenetic processes, evo-devo explores the interaction of organism and environment at developmental rather than evolutionary timescales.

Lewontin's point about organisms modifying their environments inspired another recent research program in biology even more directly—niche construction. In "Niche-Constructing Phenotypes," the first outline of this approach, John Odling-Smee followed Lewontin in criticizing the modern synthesis for holding "autonomous events in the environment ... to be exclusively responsible for directing the course of evolution down nonrandom paths" (1988, 75). Odling-Smee went on to suggest that the organism-environment relationship—and adaptation itself—involves at least two processes:

> Instead of natural selection's causing organisms to adapt to their environments, ... the constructive activities of phenotypes could cause their environments to become adaptive to themselves. More plausibly, ... the adaptive fit between organisms and their environments could be caused by both of these processes acting together. (77)

This idea of niche construction, and the related notion of ecosystem engineering, opened up new research directions in biology (Odling-Smee et al. 2003; Cuddington et al. 2007), and the resultant models of the relation between organism and environment have been extensively discussed by philosophers (Godfrey-Smith 2000, 2001; Sterelny 2001, 2005; Okasha 2005; Griffiths 2005; Barker 2008; Pearce 2011a).

But despite the burgeoning interest in new and more complex accounts of the organism-environment dyad by biologists and philosophers, little attention has been paid in the resulting discussions to the history of these ideas and to their deployment in disciplines outside biology—especially in the social sciences.

Even in biology and philosophy, there is a lack of detailed conceptual models of the organism-environment relationship. This volume is designed to fill these lacunae by providing the first multidisciplinary discussion of the topic of organism-environment interaction.[1] It brings together scholars from history, philosophy, psychology, anthropology, medicine, and biology to discuss the common focus of their work: *entangled life*, or the complex interaction of organisms and environments.

This multidisciplinary approach is important for at least two reasons. First, it has the potential to reveal historical connections that are not apparent from the perspective of a single modern discipline. For example, when the notion of organism and environment as an interacting system was first articulated in the late nineteenth century, biology, psychology, and philosophy were much less isolated from one another than they are now (and certainly less so than they were in the 1970s). Historical investigation may thus help us recover the set of interdisciplinary problems to which the organism-environment framework was originally applied, and give us new ways of thinking about today's analogous problems. These roots and ramifications of the concept of organism-environment interaction can be traced through various historical periods. In the 1960s, notably, researchers in both psychology and anthropology independently championed "ecological" approaches to their respective sciences: ecological psychology and cultural ecology were both studying humans interacting with their environments, albeit at different levels of organization (Geertz 1963; Gibson 1966; Barker 1968; Rappaport 1968). Histories can connect disciplines, and connecting disciplines can in turn enrich our histories.

Second, bringing researchers from different disciplines together fosters both collaboration and cross-fertilization. As Alan Love has argued, multidisciplinary research is prompted by "complex problem domains that elude scientific explanations arising from specific disciplinary approaches" (2008, 876; cf. Mitchell 2009). When phenomena are complex—and the interaction of organisms and environments surely qualifies—the theories and techniques of individual sciences tend to be inadequate to the challenges of describing, explaining, and intervening on those phenomena. When methods and concepts developed in different disciplinary contexts are combined, however, such difficulties may be met more successfully: a diversity of tools makes problems more tractable. Philosophers have also argued that including a variety of perspectives tends to improve the results of scientific inquiry, since it expands the range of possible interpretations of and approaches to particular problem areas (Wylie 1992; Okruhlik 1994; Longino 2002). (There is reason to suppose that this might be especially true for topics—such as organism-environment interaction—that are deeply interwoven with values and assumptions about human

---

[1]It collects several papers presented in the "Organism-Environment Interaction: Past, Present, and Future" section of the *Integrating Complexity: Environment and History* conference at Western University, 7–10 October 2010. The conference was the off-year workshop of the International Society for the History, Philosophy, and Social Studies of Biology, and was funded by the Rotman Institute of Philosophy and the Social Sciences and Humanities Research Council of Canada. For a brief report of the conference, see Pearce (2011b).

life and human society.) Biologists, social scientists, and philosophers may be able to share insights from their local viewpoints so as to clarify their respective models of organism-environment interaction, and perhaps even develop novel collaborative models.

A final aim of this volume is to show scholars in different disciplines that they really are dealing with similar types of conceptual and empirical problems, despite their apparently divergent goals. Over the last several decades, there has been a quiet revolution across a wide range of fields of study: simplistic understandings of the relation between organism and environment have been increasingly rejected in favor of sophisticated models. Niche construction, evo-devo, nature/nurture, developmental systems, genotype x environment, political ecology, plasticity, feedback effects, affordances—these are among the characteristic concepts of the new approach. But researchers employing these concepts often do not engage with one another's work, and thus do not realize that they are all tackling the same problem: How should we understand organism-environment interaction? This lack of communication is a missed opportunity. The main goal of this volume is thus to convince biologists, philosophers, and social scientists that they are often struggling in the same conceptual thicket even though the foliage they see is different. Identifying the shared object—organism-environment interaction—is the first step to finding a way out.

The volume is divided into three main parts: Historical Perspectives, Contested Models, and Emerging Frameworks. The first part explores the origins of the modern idea of organism-environment interaction in the mid-nineteenth century and its development by later psychologists and anthropologists. In the second part, a variety of controversial models—from mathematical representations of evolution to model organisms in biomedical research—are discussed and reframed in light of recent questions about the interplay between organisms and environment. Finally, the third part investigates several new ideas that have the potential to reshape key aspects of the biological and social sciences.

Today, the idea of organism-environment interaction is ubiquitous. But in the opening chapter, Trevor Pearce shows that this idea, at least in its modern form, dates only to the mid-nineteenth century. It was the philosophers Auguste Comte and Herbert Spencer who first paired the terms 'organism' and 'environment' as part of an account of the nature of life. This dichotomy went on to frame late-nineteenth-century discussion in biology, psychology, and philosophy, specifically the 1890s debates over the causal factors of evolution and the philosophical program of pragmatists like John Dewey.

Christopher Green takes a closer look at a key moment in these 1890s debates: the origins of the idea that environment-induced modifications can pave the way for similar heritable variations—what came to be called the "Baldwin Effect." Green argues that debates about the future of the hundreds of thousands of immigrants who entered the United States each year were an essential part of the context for James Mark Baldwin's much-debated proposal. Arguments over the possibility of improving the lot of these often-destitute immigrants lay in the background of biological debates between neo-Lamarckians and neo-Darwinians over the nature of the organism-environment relationship in evolution.

The next two chapters move to the twentieth century, exploring the history of ecological approaches to psychology and anthropology. Harry Heft links the ecological psychology of James J. Gibson and Roger G. Barker to the radical empiricism of William James and his student Edwin B. Holt. In particular, Holt's notion of action as "out-reaching, outgoing, inquiring, examining, and grasping" laid the groundwork for the modern idea of situated action. Thinking of most behavior as situated helps connect Gibson's "affordances" and Barker's "behavior settings," two important accounts of the relation between organism and environment in human action. The latter account moves beyond consideration of individual organisms in interaction with their individual environments to look at the complex interactions that connect multiple participants, objects, and structures to comprise a functionally-integrated behavior setting, in analogy with the interactions among organisms and abiota that comprise an ecosystem.

As mentioned above, ecological approaches emerged in the 1960s not only in psychology but also in anthropology. After reviewing the origins and development of ecological anthropology, Emily Schultz argues that recent theoretical work by Bruno Latour and others has enriched and extended the traditional anthropological idea that our interaction with environments is invariably culturally mediated. Moreover, this work relates directly to recent discussions in theoretical biology. Schultz suggests that actor-network theory in anthropology and niche construction theory in biology, when combined, form a conceptual framework that can be applied in both fields—especially at the interface of nature and culture.

The second part of the book is focused on contemporary rather than historical questions. The diversity of contemporary issues is reflected in the mix of approaches (and idioms) appearing in this part—two papers engage with formal models in formal terms; two others engage broader conceptual questions about experimental practice and its theoretical connections. In the first half of this part two philosophers analyze the treatment of organism-environment interaction in population genetics models. Bruce Glymour examines the question of whether adaptation should be thought of as adaptation to specific features of the environment or as adaptation to the environment as a whole. He argues that talking about adaptation *to* some environmental feature requires that the feature interactively cause an increase in fitness. Furthermore, such features can be identified only if their causal influence on fitness is measured. Estimates of the strength of selection depend on how these causal processes are modeled.

Marshall Abrams explores different ways of modeling how organisms experience environmental variation. Should we think of organisms in a given region, for example, as sharing a common environment, or as occupying diverse sub-environments? Both representations raise problems for the notion of relative fitness, and the fitness of an organism will come out differently according to the environmental grain we choose. According to Abrams, fitness is a function of probable reproductive success within each sub-environment, weighted according to the probability that the organism is in fact in that sub-environment. He argues that biologists make choices about environmental grain with the intent of capturing the environmental variation that is causally relevant to the population of interest. Given these choices, however, researchers' descriptions of the process of natural selection can be objective.

Jessica Bolker looks to organismal biology to analyze a primary tool of the modern life sciences—the model organism. Bolker argues for the importance of attention to both the biological and epistemological context of such organisms. The former often involves a tension between attempts to standardize and simplify the environments of model organisms and the need to preserve key aspects of organisms' natural environments. The latter depends on whether the organism in question is being used as a surrogate for a different species or as an exemplar of a particular group. Attention to these contexts can help biologists locate deficiencies of current models and develop novel alternatives.

The chapter by Desjardins, Barker, and Madrenas examines the case of human immunology and its inability to translate into clinical outcomes the knowledge obtained from research on the laboratory mouse—a failure that has recently become widely recognized by immunologists. They suggest that in order to achieve clinical success, human immunology will have to depart from the very well established Bernardian reductionist tradition in biomedical research—focusing on finding molecular pathways in animal models in controlled laboratory settings—and instead study humans in their actual environments. This requirement, the authors argue, follows essentially from the fact that the human immune system is such that we cannot sufficiently understand immune responses unless we adopt a research strategy that fully integrates the complex history of interactions between organisms and their environment.

The final part of the book looks at a series of theoretical frameworks for understanding the organism-environment relationship: niche construction, the adaptive landscape, and evo-devo. In the first chapter of this part, Gillian Barker and John Odling-Smee explore the problematical relationship between the conceptions of organism and environment that figure in evolutionary biology and those employed in ecology. They argue that long-standing inconsistencies between the simple idealizations upon which evolutionary and ecological models are based have prevented effective integration of these fields of biological study, despite their obvious interconnections. New perspectives on organism-environment interaction emerging from both disciplines—niche construction and ecosystem engineering—have recently begun to extend these idealizations and bridge the conceptual gap between the two fields. Barker and Odling-Smee argue that further developing these insights to consider the complex effects that organisms have on each other's evolutionary environments as well as their own yields a new theoretical framework—ecological niche construction—that can in turn contribute, along with evolutionary developmental biology, to the emergence of a broad new perspective in biology that takes full account of organism-environment interaction at all levels to integrate evolution, ecology, and development.

Denis Walsh tackles the classic evolutionary metaphor of an adaptive landscape. He begins by criticizing several presuppositions of this metaphor, especially the idea that the topology of the landscape is not affected by whether or not certain points on it are occupied. He proposes instead a new metaphor, the affordance landscape, inspired by Gibson's concept of an affordance—what the environment provides or furnishes to an organism. Walsh argues that the idea of an affordance landscape makes clear that biological form and environmental affordances are co-constituting: i.e., they are reciprocally dependent. On this view, changes in form can result in changes in affordances—movement across the landscape—even without changes in the environment.

Next, Rachael Brown asks why biologists studying behavior have made so little use of the new conceptual framework of evolutionary-developmental biology or "evo-devo." Brown notes that behavioral biologists are missing out, suggesting that the developmental processes emphasized in evo-devo are also important in the evolution of behavior. She draws an important parallel between two non-genetic inheritance channels: the first, chromatin-marking of DNA, is a standard topic in evo-devo, while the second, social learning, is central to studies of behavior. This parallel indicates that behavior—and not just morphology—involves the interplay between development and evolution, and can be understood via the evo-devo framework.

In the final chapter of the volume, Kim Sterelny traces the causes of a series of increases in cooperative behavior across the evolutionary history of the genus *Homo*. He argues that the richness of human cooperative life is due in large part to positive feedback between the natural environment, human populations, and social structures: that is, new forms of cooperation tend to create or promote circumstances that lead to the evolution of yet further cooperative strategies. Sterelny argues that human niche construction—not only modification of the physical environment, but also organization of informational and learning environments for the next generation—has played a central role in the evolution of cooperation.

No volume on so rich and multifarious a theme can address all the issues that merit attention. We cannot hope here to provide a comprehensive overview of the terrain, but more modestly to draw attention to some of its most interesting features as seen from diverse disciplinary perspectives, to introduce readers to some of the explorations already under way, and to indicate the potential for illuminating further work. Some important topics are only touched on in the papers included here; others do not appear at all. Here we briefly indicate some of the many topics that would have been treated in a sufficiently capacious ideal volume on organism-environment interaction. Readers will no doubt think of others—a further indication of the broad importance of these issues.

A range of historical literatures are beginning to trace the origins of organism-environment thinking and its paths in different periods and contexts, from Romantic science to Darwin's own thought; from the American Pragmatists to twentieth-century psychology, psychiatry, and educational theory. The historical papers in this volume give an entree to only some of these discussions. Sterelny and Brown both point toward the need to open up a broader perspective on evolutionary psychology—one that takes full account of organism-environment interaction—but there is much more to explore in this area, notably the contributions of feminist evolutionary psychologists. Several related research programs investigate the broad implications of organism-environment interaction for cognition, under the concepts of embodied cognition, enactivism, situated cognition, and situated knowledge. Heft's paper introduces readers to the roots of ecological psychology; both psychology and philosophy have seen a recent resurgence of interest in approaches that draw on the early ideas that his paper delineates. The notion of niche construction is one of the threads tying this volume together, but there are many extensions of this notion into new areas that we have not captured, including ongoing work on its implications for the concept of adaptation. Green sheds a fascinating new light on the origins of the so-called Baldwin Effect; this idea continues to drive conceptual innovation in biology and philosophy. A particularly fast-growing and exciting family of research programs has grown up around organism-environment interactions that involve regulation, from the genomic to the ecological level. Systems biology, evolutionary developmental biology, and epigenetics are among the programs of biological research emerging in this area; each also has inspired a line of philosophical investigation. Another approach combines elements from biology and the social sciences to explore the ramifications of G x E interactions in behavioral genetics and in psychiatry, among other contexts. And quite diverse literatures are looking at the kinds of complexity that organism-environment interaction gives rise to, and its implications for contingency in processes of biological and social change.

These topics are tremendously diverse, yet the researchers engaging each of them share, with each other and with the authors represented in this volume, a focus on the nature of the relationship between organism and environment and a commitment to unraveling the mysteries of entangled life.

Acknowledgments This collection grew out of a conference held in 2010 at Western University, entitled "Integrating Complexity: Environment and History." The conference was made possible by the generous support of Canada's Social Sciences and Humanities Research Council (SSHRC) and the Rotman Institute of Philosophy, and by the institutional sponsorship and assistance of the International Society for the History, Philosophy and Social Studies of Biology (ISHPSSB). We thank these bodies cordially for their support. Thanks are also due to the many members of Western University's Department of Philosophy who helped to organize and run the conference, and the even larger number of philosophers, historians, and natural and social scientists whose participation made it a success. We wish there had been room in this volume to include many more of the stimulating papers that were presented at the conference. Finally, we are grateful to series co-editor Philippe Huneman and an anonymous reviewer for insightful suggestions, to David Isaac for editorial assistance, and to Springer's Ties Nijssen and Christi Lue for thoughtful editing and a swift and easy publication process.

References

Amundson, Ron. 2005. *The changing role of the embryo in evolutionary thought: Roots of evo-devo*. Cambridge: Cambridge University Press.
Barker, Roger G. 1968. *Ecological psychology: Concepts and methods for studying the environment of human behavior*. Stanford: Stanford University Press.
Barker, Gillian. 2008. Biological levers and extended adaptationism. *Biology and Philosophy* 23: 1–25.
Bonner, J.T. 1982. *Evolution and development: Report of the Dahlem workshop on evolution and development, Berlin 1981, May 10–15*. Berlin: Springer.
Cuddington, Kim, James E. Byers, William G. Wilson, and Alan Hastings. 2007. *Ecosystem engineers: From plants to protists*. Amsterdam: Elsevier.

Geertz, Clifford. 1963. *Agricultural involution: The processes of ecological change in Indonesia.* Berkeley: University of California Press.

Gibson, James J. 1966. *The senses considered as perceptual systems.* Boston: Houghton Mifflin.

Godfrey-Smith, Peter. 2000. Niche construction in biological and philosophical theories. *Behavioral and Brain Sciences* 23: 153–154.

Godfrey-Smith, Peter. 2001. Organism, environment, and dialectics. In *Thinking about evolution: Historical, philosophical, and political perspectives*, ed. Rama S. Singh, Costas B. Krimbas, Diane B. Paul, and John Beatty, 253–266. Cambridge: Cambridge University Press.

Gould, Stephen Jay. 1977. *Ontogeny and phylogeny.* Cambridge: Harvard University Press.

Griffiths, Paul E. 2005. Review of 'Niche construction'. *Biology and Philosophy* 20: 11–20.

Laubichler, Manfred D. 2007. Does history recapitulate itself? Epistemological reflections on the origins of evolutionary developmental biology. In *From embryology to evo-devo: A history of developmental evolution*, ed. Manfred D. Laubichler and Jane Maienschein, 13–33. Cambridge: MIT Press.

Laubichler, Manfred D., and Jane Maienschein. 2007. *From embryology to evo-devo: A history of developmental evolution.* Cambridge: MIT Press.

Lewin, Roger. 1980. Evolutionary theory under fire. *Science* 210: 883–887.

Lewontin, Richard C. 1978. Adaptation. *Scientific American* 239: 213–230.

Longino, Helen E. 2002. *The fate of knowledge.* Princeton: Princeton University Press.

Love, Alan C. 2008. Explaining evolutionary innovations and novelties: Criteria of explanatory adequacy and epistemological prerequisites. *Philosophy of Science* 75: 874–886.

Mayr, Ernst. 1978. Evolution. *Scientific American* 239: 47–55.

Mitchell, Sandra D. 2009. *Unsimple truths: Science, complexity, and policy.* Chicago: University of Chicago Press.

Müller, Gerd. 2007. Six memos for evo-devo. In *From embryology to evo-devo: A history of developmental evolution*, ed. Manfred D. Laubichler and Jane Maienschein, 499–524. Cambridge: MIT Press.

Odling-Smee, F. John. 1988. Niche-constructing phenotypes. In *The role of behavior in evolution*, ed. H.C. Plotkin, 73–132. Cambridge: MIT Press.

Odling-Smee, F. John, Kevin N. Laland, and Marcus W. Feldman. 2003. *Niche construction: The neglected process in evolution.* Princeton: Princeton University Press.

Okasha, Samir. 2005. On niche construction and extended evolutionary theory. *Biology and Philosophy* 20: 1–10.

Okruhlik, Kathleen. 1994. Gender and the biological sciences. In *Biology & society: Reflections on methodology*, ed. Mohan Matthen and R.X. Ware, 21–42. Calgary: University of Calgary Press.

Pearce, Trevor. 2011a. Ecosystem engineering, experiment, and evolution. *Biology and Philosophy* 26: 793–812.

Pearce, Trevor. 2011b. Meeting report: Fourth ISHPSSB off-year workshop. *Biology and Philosophy* 26: 315–316.

Raff, Rudolf A., and Alan C. Love. 2004. Kowalevsky, comparative evolutionary embryology, and the intellectual lineage of evo-devo. *Journal of Experimental Zoology B* 302: 19–34.

Rappaport, Roy A. 1968. *Pigs for the ancestors: Ritual in the ecology of a New Guinea people.* New Haven: Yale University Press.

Sterelny, Kim. 2001. Niche construction, developmental systems, and the extended replicator. In *Cycles of contingency: Developmental systems and evolution*, ed. Susan Oyama, Paul E. Griffiths, and Russell D. Gray, 333–349. Cambridge: MIT Press.

Sterelny, Kim. 2005. Made by each other: Organisms and their environment. *Biology and Philosophy* 20: 21–36.

Wagner, Günter P. 2007. The current state and the future of developmental evolution. In *From embryology to evo-devo: A history of developmental evolution*, ed. Manfred D. Laubichler and Jane Maienschein, 525–545. Cambridge: MIT Press.

Wylie, Alison. 1992. The interplay of evidential constraints and political interests: Recent archaeological research on gender. *American Antiquity* 57: 15–35.

Part I
Historical Perspectives

The Origins and Development of the Idea of Organism-Environment Interaction

Trevor Pearce

Abstract The idea of organism-environment interaction, at least in its modern form, dates only to the mid-nineteenth century. After sketching the origins of the organism-environment dichotomy in the work of Auguste Comte and Herbert Spencer, I will chart its metaphysical and methodological influence on later scientists and philosophers such as Conwy Lloyd Morgan and John Dewey. In biology and psychology, the environment was seen as a causal agent, highlighting questions of organismic variation and plasticity. In philosophy, organism-environment interaction provided a new foundation for ethics, politics, and scientific inquiry. Thinking about organism-environment interaction became indispensable, for it had restructured our view of the biological and social world.

1 Introduction

That creatures are shaped by the world around them is not news. Several centuries before the Common Era, the Hippocratic author of "Airs, Waters, Places" argued that our forms and habits are affected by the climate, the air we breathe, and the water we drink. For example, the inhabitants of Phasis reportedly had the deepest voices known because they breathed "air which is moist and damp and not clean" (Lloyd 1978, 162). As I will show, however, this concrete notion of various external conditions affecting the health and features of living beings was gradually replaced in the second half of the nineteenth century by the abstract idea of an organism's environment. The new dichotomy of organism and environment proved both useful and portable. By the 1890s, it was already operating as an essential framing device in scientific and philosophical arguments. In biology and psychology, the

T. Pearce (✉)
Department of Philosophy, University of North Carolina at Charlotte, Charlotte, NC, USA
e-mail: tpearce6@uncc.edu

G. Barker et al. (eds.), *Entangled Life*, History, Philosophy and Theory of the Life Sciences 4, DOI 10.1007/978-94-007-7067-6_2, © Springer Science+Business Media Dordrecht 2014

environment was seen as a causal agent, highlighting questions of organismic variation and plasticity. In philosophy, organism-environment interaction provided a new foundation for ethics, politics, and scientific inquiry. Thinking about organism-environment interaction became indispensable, for it had restructured our view of the biological and social world.

In the first part of the chapter, I will describe the origins of the idea of organism-environment interaction in the work of Auguste Comte and Herbert Spencer. I will then demonstrate how the idea played a central role in late-nineteenth-century debates over the causal factors of evolution—specifically the controversy over August Weismann's account of heredity and the discovery of the so-called "Baldwin Effect." In the third section, I will follow the idea of organism-environment interaction into philosophy: the pragmatist philosopher John Dewey made the relationship between organism and environment the foundation of his new theories of ethics, education, and scientific inquiry. This chapter and those that follow illustrate how an apparently simple idea—that organisms interact with environments—came to have complicated and lasting consequences, from debates in philosophy and the social sciences to theories of niche construction and human evolution.

2 Origins of an Idea[1]

The English word 'environment' was coined in the late 1820s by the Scottish essayist Thomas Carlyle and popularized in the second half of the century by the philosopher Herbert Spencer. But what is so important about a word? It is not as if earlier thinkers had any trouble discussing the influence of external factors on organisms. For example, Buffon wrote the following in his multi-volume *Natural History*: "The temperature of the climate, the quality of food, and the evils of slavery [i.e., domestication]—these are the three causes of change, alteration, and degeneration in animals" (Buffon 1766, 317). Soon after, French naturalists began to employ umbrella terms for these and other factors, the most influential of which were Jean-Baptiste Lamarck's 'circumstances' and Georges Cuvier's 'conditions of existence.' Lamarck used 'circumstances' to refer to climate, temperature, environing media (water, air), habits, movements, actions, etc. (Lamarck 1801, 13, cf. Lamarck 1809, 1:238). Cuvier's *conditions of existence* was a more formal notion based on the fact that "nothing can exist that does not bring together the conditions that make its existence possible" (Cuvier 1817, 1:6). If terms like 'conditions' and 'circumstances' already existed, why use the word 'environment' in the first place? In this section, I will show that the organism-environment dichotomy emerged from philosophical reflection on the nature of life. Its originator, at least in the English-speaking world, was Spencer.

---

[1] In parts of this section I have drawn on material from Pearce (2010a).

Naturalists in the first third of the nineteenth century, following the work of Carl Linnaeus, Buffon, and Lamarck among others, became more and more interested in the influence of external conditions on organisms. This interest was most pronounced in the proto-ecological writings of Alexander von Humboldt, Augustin de Candolle, and Charles Lyell (see Pearce 2010b, 501–506). The geographical method of Humboldt and Candolle was an attempt to connect specific plants to particular local circumstances. For example, in his "Physical Table of the Equatorial Regions," Humboldt showed how flora vary with altitude, geology, air temperature, the snow line, and the composition and pressure of the atmosphere (Humboldt and Bonpland 1805, 41–42). Candolle, following Humboldt, discussed "the influence of external elements or agents on plants," specifically "the influence of temperature, of light, of water, of the soil, and of the atmosphere" (Candolle 1820, 362). He linked such external influences to Cuvier's notion of conditions of existence: "Specific plants, given their organization, require specific conditions of existence: one cannot live where it does not find a specific quantity of salt water; another where it does not have, at some time of year, some quantity of water or intensity of sunlight, etc." (Ibid., 384). Lyell extended Candolle's work, pointing out that other organisms make up part of the relevant external conditions:

> The stations of different plants and animals depend on a great complication of circumstances,—on an immense variety of relations in the state of the animate and inanimate worlds. Every plant requires a certain climate, soil, and other conditions, and often the aid of many animals, in order to maintain its ground. (Lyell 1832, 140)

Thus naturalists in the early nineteenth century were investigating the influence of external factors—physical and biological—on plants and animals, and employing terms such as 'conditions' and 'circumstances' to refer collectively to such factors.

But though Humboldt and Candolle emphasized the importance of external circumstances, the move to singular terms like '*milieu*' or 'environment'—and to a more explicit organism-environment dyad—was made by philosophers. Spencer's use of the word 'environment' and his emphasis on the organism-environment relationship derived from his reading of the French philosopher Auguste Comte. In the French tradition, the term '*milieu*' (medium) as the counterpart of '*organisme*' was an innovation of the 1830s, although Lamarck had earlier employed the plural '*milieux*' to refer to environing media such as water or air (Canguilhem 1952). In several texts of 1833, for example, the zoologist Étienne Geoffroy Saint-Hilaire linked changes in an organism to changes in its *milieu ambiant*.[2] He claimed that there are two sorts of facts relevant to developing organisms: those belonging to the essence of a type and those involving the intervention of the ambient world. It is

---

[2]Geoffroy (1833a, 88–89n) quotes Blaise Pascal making a related point. However, this is not an accurate quotation but a loose reading of the earlier thinker's well known remark, "I am very afraid that this nature might itself only be a first custom, just as custom is a second nature" (Pascal 1669, 199; Pascal 1991, 208).

the latter that explain why pears from the same orchard are sometimes large and sweet, sometimes small and sour (Geoffroy Saint-Hilaire 1833b, 68–69; see also 1833a, 89n).

Comte went further in the third volume of his *Course of Positive Philosophy*, making the relationship between organism and *milieu* the basis of his conception of life. He attacked Xavier Bichat's claim that life is simply the set of functions that resist death:

> The profound irrationality of [Bichat's] conception consists above all in its complete elimination of one of the two inseparable elements whose harmony necessarily constitutes the general idea of *life*. This idea supposes, indeed, not only a being so organized as to possess the vital state, but also, no less indispensable, some set of external influences that make possible the achievement of that state. Such harmony between the living being and the corresponding *medium* evidently characterizes the fundamental condition of life. (Comte 1838, 288–289, original emphasis)

Comte's notion of life followed that of the naturalist Henri-Marie Ducrotay de Blainville, whose definition of "organized body" (i.e., organism) included "acting on environing external bodies and being affected by these bodies" (Blainville 1822, xxii; see Comte 1838, 295).[3] Comte, however, labeled the two parts of the dichotomy: he insisted that "the idea of life constantly supposes the necessary correlation of two indispensable elements, an appropriate organism and a suitable medium" (Comte 1838, 301). Attaching a footnote to 'medium,' Comte called it a new expression designating "the total ensemble of external circumstances, of any kind, necessary to the existence of each particular organism" (Ibid., 301n). Hence '*milieu*' was introduced as an abstract singular term to replace plural terms such as 'circumstances' or 'conditions of existence' in the context of a new philosophical account of life.[4]

English followers of Comte appropriated his new dichotomy. The author and critic George Henry Lewes, for example, emphasized in a debate over progress in the fossil record that organisms were "the resultant of two factors—Life and

---

[3]For more on the connections between Comte, Blainville, and Lamarck, see Petit (1997) and Braunstein (1997).

[4]Related German concepts and terminology would require a history of their own. Thomas Carlyle seems to have originally coined the word 'environment' to translate the German word '*Umgebung*' (Pearce 2010a, 248). Phrases like "*der Organismus und seine Aussenwelt*" were used in medical writings beginning in the early 1800s: e.g., "the reciprocal determination of the organism and its external world" (Kilian 1802, 150). Philosophically inclined physicians such as Johann Christian Reil and Moritz Naumann also employed this *Organismus-Aussenwelt* dichotomy (Reil 1816, 63; Naumann 1821, 349, 1823, 162). Later in the century German translations used both '*Aussenwelt*' and '*Umgebung*' for Spencer's 'environment' (Spencer 1880, 1:294, 365, 1882, 308, 380). The *Organismus-Umgebung* dyad is apparently absent from German texts prior to the reception of Comte and Spencer. The following is one early usage, before Spencer but after Comte: "form and activity, part and whole, organism and environment are in perfect harmony" (Köstlin 1851, 1:352). Peter Sloterdijk (2005) claims that Jakob von Uexküll (1909) invented the concept of environment, ignoring this rich nineteenth century background.

Circumstance" (Lewes 1851, 996).[5] Lewes's serial summary "Comte's Positive Philosophy" likewise claimed that "*organism* and *medium* are the two correlative ideas of life" (Lewes 1852, 666, original emphasis; cf. Lewes 1853, 167). The word 'environment' was first used in a biological context by the social thinker Harriet Martineau as her preferred translation of Comte's '*milieu.*' Phrases like "the reciprocal action of the organism and its environment" thus appear for the first time in Martineau's translation of Comte's course (Comte 1853, 1:401).

Nevertheless, before Spencer got a hold of it, the word 'environment' was still very rare; he made it a central concept in his popular philosophical accounts of biology and psychology, and by the end of the century it was a common term. Having recently befriended Lewes, Spencer read both Lewes's summary and Martineau's translation of Comte in 1852–1853. Spencer shared Comte's interest in demarcating the living and the non-living, and had previously defined 'life' as "*the co-ordination of actions*" (Spencer 1852, 252, original emphasis). In his later *Principles of Psychology*, however, he adopted Comte's position and Martineau's vocabulary: "the changes or processes displayed by a living body, are specially related to the changes or processes in its environment" (Spencer 1855, 368). This special relation, according to Spencer, is one of correspondence and continuous adjustment:

> The life of the organism will be short or long, low or high, according to the extent to which changes in the environment, are met by corresponding changes in the organism. Allowing a margin for perturbations, the life will continue only while the correspondence continues; the completeness of life will be proportionate to the completeness of the correspondence; and the life will be perfect only when the correspondence is perfect. (Ibid., 376)

This progressive language indicates that Spencer's account of the correspondence between organism and environment was also related to the idea of evolution, for life evolves by improving organism-environment correspondence: as life progresses, said Spencer, this correspondence extends in space and time (i.e., organisms can adapt to external causes less frequently encountered) and increases in speciality, generality, and complexity (Ibid., 394–465). Finally, Spencer declared mind and intelligence merely advanced forms of life; thus he argued that "the manifestations of intelligence are universally found to consist in the establishment of correspondences between relations in the organism and relations in the environment" (Ibid., 483). Spencer's organism-environment dichotomy was thus relevant not only to physiology and zoology but also to psychology, sociology, and ethics, as he attempted to show in later works.

The 1855 edition of Spencer's *Principles of Psychology* was not widely read. But with the publication of the first three parts of his System of Synthetic Philosophy— *First Principles* and *Principles of Biology* in the 1860s and the second edition of the *Psychology* in the early 1870s—his ideas became more and more popular, especially

---

[5]For evidence that Lewes—and not Spencer—wrote this particular article, see Pearce (2010a, 256n17).

in the United States. In 1871, the philosopher-historian John Fiske gave a series of lectures at Harvard on Spencer's evolutionary philosophy that were simultaneously published in *The World*, a New York newspaper (Berman 1961, 79; Nelson 1977; cf. Fiske 1874). The next year, Edward Livingston Youmans founded the magazine *Popular Science Monthly*, which consistently promoted Spencer's views (Spencer 1872; Youmans 1872). By the late 1870s, William James was assigning Spencer's books to his psychology and philosophy classes at Harvard and the young John Dewey was borrowing these same books from his college library in Vermont (James 1988; Feuer 1958). As Spencer's ideas spread, so did his abstract dichotomy of organism and environment. In the next two sections, we will see how the idea of organism-environment interaction framed a series of conceptual discussions in the 1890s—first in biology and then in philosophy.

3 Environment, Plasticity, and Variation

Spencer's *Principles of Psychology* introduced the idea of organism-environment interaction to the English-speaking world. 'Interaction' suggests a mutual influence: the environment affects the organism just as the organism affects the environment. But Spencer talked mostly about just one causal direction: environments modifying organisms. In the fourth section of the chapter, I will show how some philosophers rejected Spencer's account in favor of a more truly interactive view of the organism-environment relationship. But as will become clear in this section, late-nineteenth-century biologists and psychologists focused primarily—as had Spencer—on the environment as an agent of organismal change.

In the late 1880s, Herbert Spencer published a short book entitled *Factors of Organic Evolution*. Spencer emphasized the importance of its topic in the preface, declaring that the question of which casual factors are operative in evolution "demands, beyond all other questions whatever, the attention of scientific men" (Spencer 1887, iv). A few years later, Spencer got his wish: in the 1890s the "factors of evolution" question attracted the attention of a whole variety of scientists and philosophers, becoming the focus of numerous debates, books, and articles. The idea of organism-environment interaction played a key role in these debates, for one of the main points of contention was whether the role of the environment is primarily that of producing or that of preserving variation.

One of the central problems of the factors of evolution debates of the 1890s was the nature and origin of variation. Charles Darwin's first use of the term 'environment'—which appeared only in his last works—shows that the environment was given a kind of causal agency in such discussions:

> In many cases it is most difficult to distinguish between the definite result of changed conditions, and the accumulation through natural selection of indefinite variations which have prove[d] serviceable. If it profited a plant to inhabit a humid instead of an arid station, a fitting change in its constitution might possibly result from the direct action of the environment. (Darwin 1875, 2:281)

This mention of the possible importance of "direct action of the environment" contrasts with Darwin's earlier inclination "to lay very little weight on the direct action of the conditions of life" (Darwin 1859, 134). It is notable that Darwin first speaks of the environment as an important agent in his book *Variation of Animals and Plants under Domestication*: the *Origin of Species* had for the most part placed variation in a black box, whereas *Variation* made it the central theme.

The main player in the debates over the factors of evolution was the German naturalist August Weismann. Darwin, shortly before he died, wrote a prefatory note to a collection of Weismann's early essays. Darwin's words show that the origin of variation was seen as the next big problem in biology:

> Several distinguished naturalists maintain with much confidence that organic beings tend to vary and to rise in the scale, independently of the conditions to which they and their progenitors have been exposed; whilst others maintain that all variation is due to such exposure, though the manner in which the environment acts is as yet quite unknown. At the present time there is hardly any question in biology of more importance than this of the nature and causes of variability. (Weismann 1882, vi)

Variation was an important problem because although most naturalists—even American holdouts—now admitted the fact of evolution, there was much disagreement as to its causes or factors (LeConte 1878, 786–787).[6] For example, the American paleontologist Edward Drinker Cope argued that natural selection is a restrictive but not an originative factor: that is, it rejects variations but does not produce them (Cope 1887, 350–351). Cope was following the Duke of Argyll (among others), who argued that natural selection "gives an explanation, not of the processes by which new Forms first appear, but only of the processes by which, when they have appeared, they become established in the world" (Argyll 1867, 229). Explaining the origin of variation, for Spencer (1887) and Cope (1887), involved determining how the environment could act as a producer of variation and not merely its preserver.

Weismann's essays on heredity, beginning with "On Heredity" in 1883, explicitly attacked the relevance of environment-induced variations to evolution and thus directly contradicted the work of authors such as Spencer, Cope, and Argyll. This new theory of heredity argued that the germ cells that give rise to offspring should "be regarded as something standing opposed to and separate from the entirety of cells composing the body"; a corollary of this claim was that so-called "acquired characters," those caused by the action of the environment during an organism's lifetime, could not be inherited (Weismann 1883, 1885; Moseley 1885, 155). Weismann's theory provoked a storm of criticism, most of which was focused on the problem of variation. George John Romanes for example, following Spencer, argued that mutual co-adaptation of parts within an organism could not be explained

---

[6]For more on this period in the history of biology, see Bowler (1983), (1988), and Richards (1987, 331–503).

by merely "fortuitous variation" and natural selection; it had to rely on a tendency of those parts to vary together, i.e., on "the inherited effects of use and disuse" (Romanes 1887, 406; cf. Spencer 1887, 12–17).[7]

Romanes (1888) coined the term 'Neo-Darwinian' to describe naturalists such as Weismann who "aim at establishing for natural selection a sole and universal sovereignty which was never claimed for it by Darwin himself." There were certainly people whose views approached this sovereignty claim. Alfred Russell Wallace, for example, wrote the following in his book *Darwinism*: "Whatever other causes have been at work, Natural Selection is supreme . . . . The more we study it the more we are convinced of its overpowering importance" (Wallace 1889, 444). Cope (1889) replied by repeating that selection could not be the whole story: "selection cannot explain the *origin* of anything, although it can and does explain survival of something already originated; and evolution consists in the origin of characters, as well as their survival." Argyll (1889) accused the neo-Darwinians of rejecting "any conception which tends to break down the empire of mere fortuity in the phenomena of variation." Nevertheless, Weismann gained many followers, most notably Edward Bagnall Poulton and other Oxford naturalists. As Grant Allen put it a few years later,

> for a year or two after the appearance of Weismann's memoirs, nothing else was heard of in *Nature* and in the scientific societies. Weismannism became the fashionable creed of the day . . . . Young England, as a biologist, swore by the continuity of the germ-plasm, and laughed to scorn the inheritance of the acquired faculty. (Allen 1890, 538)

Naturalists were divided into warring camps: Poulton, in a letter to a friend, actually made a two-column list of individuals arrayed for and against Weismann's view.[8]

The debates over Weismann's theory are usually remembered simply as debates over the inheritance of acquired characters; the problem is that the latter phrase now evokes an easily dismissed Lamarckism, concealing a number of interesting issues. Looking more closely at the relevant texts reveals that the factors debates concerned the importance of organism-environment interaction during ontogeny and its role in evolution, and thus the origin and nature of variation—problems which remain relevant today (Barker 1993; West-Eberhard 2003; Jablonka and Lamb 2005; Laubichler 2010; Schwander and Leimar 2011).

That the relation between organism and environment framed late-nineteenth-century discussions of the factors of evolution is most clearly seen in the work of the three scientists who in 1896 co-discovered what we now refer to as the "Baldwin Effect": Henry Fairfield Osborn, Conwy Lloyd Morgan, and James Mark Baldwin. The Baldwin Effect occurs when environment-induced (and presumably adaptive) ontogenetic variations give groups of organisms time to develop corresponding

---

[7]In their later debate, Weismann capitulated to Spencer on this point, formulating his theory of germinal selection—or selection on elements of the heritable material—as a means of "directing variation" at the organismic level (Weismann 1895, 432). For more on Weismann's germinal selection theory, see Winther (2001).

[8]Poulton to Henry Fairfield Osborn, 31 December 1891: Folder 11, Box 77, General Correspondence, Department of Vertebrate Paleontology Archives, American Museum of Natural History.

phylogenetic variations (Kemp 1896; Baldwin 1896). The importance of this purported "new factor," as Baldwin called it, cannot be understood outside of the context of the factors of evolution debates. (In what follows, I will focus on Osborn and Morgan; Christopher Green discusses Baldwin's contributions in the next chapter.)

At a meeting of the American Society of Naturalists in 1891, Osborn lamented that "after studying Evolution for a century we are in a perfect chaos of opinion as to its factors" (Osborn 1891, 193). In Osborn's framing, the debates over these factors were centrally about the power of the environment to produce variations:

> By the [principle of Lamarck] we diminish the powers of Natural Selection, and increase the powers of Environment; at the same time we greatly simplify the problem of Variation, and render far more complex the problem of Inheritance. By the [principle of Weismann] we throw the entire burden of evolution upon Natural Selection, and eliminate the direct action of Environment; we admit definite laws or causes of Variability, but no definite laws governing the variations of single characters; we greatly simplify the problem of Inheritance. In short, the vulnerable point with the Lamarckians is in solving the problem of Heredity, while their opponents are weakest in solving the problem of variation. (Ibid., 197)

Thus, the followers of Lamarck could take the environment as the primary source of variation, but had difficulty explaining how such variation was inherited, whereas the neo-Darwinians had difficulty accounting for the origin of variation, but no problem explaining how existing variation was passed on.

Employing a distinction between ontogenetic and phylogenetic variation, Osborn was also able to argue that variation in a type of organism following a move to a new environment is not necessarily evidence for the direct action of that environment. The following "crucial experiment" is necessary:

> An organism A, with an environment or habit A, is transferred to environment or habit B, and after one or more generations exhibits variations B; this organism is then retransferred to environment or habit A, and if it still exhibits, even for a single generation, or transitorily, any of the variations B, the experiment is a demonstration of the inheritance of ontogenic variations. (Osborn 1895, 97)

The variations in environment B might be induced by that environment during each successive generation; i.e., the B variations could be merely ontogenetic. But if the B variations persist across generations even when the population has been returned to environment A, then they have become phylogenetic. Osborn is here articulating the important point that a variation induced by a reliable environmental cue each generation mimics a congenital variation.

This point about plasticity and reliable cues was made independently by Morgan during a discussion of several experiments by Poulton: "His experiments neither justify a denial nor involve an assertion of the transmissibility of environmental influence . . . . Can we be sure that there is really a summation of results—that each generation is not affected *de novo* in a similar manner?" He continued: "If each plastic embryo is moulded in turn by similar influence, how can we conclusivly [sic] prove hereditary summation?" (Morgan 1891a, 167). Thus, Morgan agreed with Osborn that ontogenetic plasticity could confound tests of the inheritance of acquired characters: "In experiments to test the question of use-inheritance, the

difficulty is to exclude the effects (1) of selection and (2) of individual plasticity."
The problem was that "extreme plasticity" could indicate that "the influence of
the normal environment is prepotent over the effects of use-inheritance *if* such
occur" (Morgan 1891b, 271–272). Hence both Morgan and Osborn highlighted the
plasticity of organisms and the environment's role as a producer of variation, but
pointed out that such variation was not necessarily heritable.

As Morgan stressed in an essay on Weismann's theories, "all effective variation is
a joint product of the inherent activities of germinal cells and the conditioning effect
of their environment" (Morgan 1893, 30). Osborn agreed, claiming that organic
form is the product of "constitution + the environment" (Dyar 1896, 141). These
ideas laid the groundwork for the Baldwin Effect. Osborn presented his version in
March 1896 before the New York Academy of Sciences:

> During the enormously long period of time in which habits induce ontogenic variations it
> is possible for natural selection to work very slowly and gradually upon predispositions
> to useful correlated variations, and thus what are primarily *ontogenic variations* become
> slowly apparent as *phylogenic variations* or congenital characters of the race. (Ibid., 142)

The idea of "correlated variations" is the key: it seems that Osborn used this phrase
to refer to heritable traits that either mirror or support those traits that had previously
been environmentally induced. The basic point is that plasticity, or ontogenetic
variation in the face of environmental changes, could give organisms time to develop
these correlated congenital variations. The Baldwin Effect was thus a compromise
position between Lamarck and Weismann: it emphasized the role of environment-
induced variation in evolution without depending on the inheritance of acquired
characters. As Osborn put it in a letter to Poulton, "Morgan, Baldwin and myself
have independently arrived at certain conclusions regarding the Lamarckian factor
which will interest you."[9] Osborn argued that this quasi-Lamarckian process was
likely to be important in evolution, "since there is no doubt that the changes of
environment and the habits which it so brings about far outstrip all changes in
constitution" (Dyar 1896, 142).

Like Osborn, Morgan understood the Baldwin Effect as bearing directly on "the
Lamarckian question," and also framed it in terms of the organism-environment
relationship. He outlined the effect in a letter to Poulton dated 12 April 1896, with
'variation' referring to changes "of germinal origin" and 'modification' referring to
changes "of environmental origin":

> Let us suppose that a group of organisms belonging to a plastic species is placed under
> new cond'ns of environment. Those whose innate plasticity is equal to the occasion survive.
> They are modified. Those whose innate plasticity is not equal to the occasion are eliminated.
> Such modification takes place generation after generation but *as such* is not inherited.
> In the meanwhile, however, and concurrently, any congenital variations antagonistic in
> direction to these modifications will tend to thwart them and to render the organism liable to
> elimination; while any congenital variations similar in direction to these modifications will

[9]Osborn to Poulton, 12 June 1896: Folder 11, Box 77, General Correspondence, Department of
Vertebrate Paleontology Archives, American Museum of Natural History.

tend to support them and to favour the individuals in which they occur. (Natural Selection itself will foster variability in given advantageous lines . . . when once initiated.) Thus will arise a congenital *pre-disposition* to the modification in question. The longer the process continues, the more marked will be the predisposition and the greater the tendency for the congenital variations to conform in all respects to the persistent plastic modifications; while the plasticity still continuing in operation, the modifications become yet further adaptive. When relatively perfect adaptation is reached (the conditions remaining uniform) natural selection will slowly yet surely bring the congenital variations up to the level of such adaptation. Thus plastic modification leads, and variation follows: the one paves the way for the other.[10]

In other words, when organisms are plastic, they can adapt to new environmental conditions even without heritable changes; in the longer term, if the conditions persist, more permanent heritable changes that mirror or extend the environment-induced alterations may appear and, via the ordinary action of natural selection, replace the temporary changes.

Morgan's distinction between environment-induced *modification* and congenital *variation* did the same conceptual work as Osborn's division of "ontogenic variation" and "phylogenic variation." These distinctions allowed Morgan and Osborn to tease apart changes caused directly by the environment each generation and inherited changes, and thus to carve out a role for the environment as a producer of variation without endorsing a Lamarckian theory of heredity (although Osborn did later endorse a form of Lamarckism). Traditionally, supporters of Darwin against Spencer had argued that the primary role of the environment in evolution was as "regulator or preserver of . . . variation" (James 1988, 137, cf. James 1880); the work of Morgan and Osborn provided a richer account in which adaptation involved organism-environment interaction both within and across generations. The environment as both producer and preserver of variation was a central part of this new evolutionary story.

4   Organism and Environment in Philosophy

Spencer, despite his influence on the factors of evolution debates, was primarily a philosopher. Given his popularity in America, it is not surprising that philosophers such as William James and John Dewey used Spencer's work as a foil for their own ideas. James was amusing but often unkind in his descriptions of Spencer, whom he associated with the idea that the mind was merely a product of its environment (Godfrey-Smith 1996, 66–99). As he joked in a May 1877 letter to the neurologist James Jackson Putnam, "would *I* were part of [Spencer's] environment! I'd see if his

---

[10]Morgan to Poulton, 12 April 1896: C. Lloyd Morgan letters, Entomological Archives, Hope Entomological Library, Oxford University Museum of Natural History. The quoted points are on a separate sheet enclosed with the letter. Emphasis in original. In the original document, this passage is divided into 11 numbered points (nos. 6–17 of 21 total). I have collapsed them for ease of reading, but have not altered the sentence structure. Cf. Morgan (1896, 316–318).

'intelligence' could establish 'relations' that would 'correspond' to me in any other way than by giving up the ghost before me!"[11] Nevertheless, many philosophers who were critical of Spencer inherited his focus on the organism-environment relationship even as they altered his account of that relationship. In this section, I will argue that the idea of organism-environment interaction formed the basis of John Dewey's pragmatist philosophy.

Dewey was first exposed to Spencer's ideas in college at the University of Vermont, where he borrowed the first volume of Spencer's *Principles of Psychology*—which prominently featured the idea of organism-environment interaction— more often than any other book (Feuer 1958). However, it was not evolutionism but idealism that attracted the young Dewey, and during graduate school he attacked Spencer's evolutionary-empiricist account of knowledge (Dewey 1883).[12] When Dewey began teaching Empirical Psychology at the University of Michigan, he struggled to find a textbook that did not simply adopt Spencer's view that the mind was determined by the environment.[13] In 1884, he used James Sully's *Outlines of Psychology*, which followed Spencer in casting mental life as an adjustment of internal to external relations:

> Through innumerable interactions between the nervous system and the environment the former becomes gradually modified in conformity with the latter. Thus nervous connections are built up in the brain-centres corresponding to external relations. The nervous structures are thus in a manner moulded in agreement to the external order, to the form or structure of the environment. (Sully 1884, 58)

Presumably dissatisfied with Sully's approach, Dewey switched in 1885 to John Clark Murray's *Handbook of Psychology*, which he declared "a great advance on Sully in its philosophical basis."[14] Murray attacked the Spencerian view according to which "man's consciousness is simply the product of the forces in his environment acting on his complicated sensibility, and of that sensibility reacting on the environment" (Murray 1885, 415). Thus it appears that Dewey, in his early career, was critical of Spencer's approach to philosophy and psychology.

Despite this critical stance, Dewey twice taught a class on "The Philosophy of Herbert Spencer" in his early years at Michigan, and Spencer's idea of organism-environment interaction soon began to play a role in Dewey's developing philosophy. The influence of Spencer's ideas is apparent in student lecture notes taken in Dewey's "Speculative Psychology" class of 1887. In one of his lectures,

---

[11] Spencer to James Jackson Putnam, 26 May 1877, in Skrupselis and Berkeley (1992–2004, 4:564, original emphasis). See also James (1878).

[12] Dewey's mentor in graduate school at Johns Hopkins University, the idealist philosopher George Sylvester Morris, was strongly opposed to Spencer's philosophy. He saw it as British empiricism— which for Morris was vulnerable to a variety of standard idealist criticisms—dressed up with new scientific terminology (Morris 1880, 337–388).

[13] For the classes taught by Dewey at the University of Michigan, see the relevant years of the *Calendar of the University of Michigan*. The class textbook is often listed in the calendar.

[14] Dewey to Torrey, 16 February 1886, in Hickman (1999–).

Dewey argued that mind must be an organic unity. Building up to this point, he said that a stone "has no self at all + hence no unity," as it is "wholly dependent upon outside conditions. None of its parts have any necessary relation with one another nor with the world." Moving up the scale, we call a tree an organism because each of its parts "at same time manifests life of whole + at same time contributes to this life." Nevertheless, even a tree is not truly an organism, according to Dewey:

> Material organism not a complete Individual organism for ... [it is] not completely related to all things in the world. Is related to certain things in its environment, those from which it draws its nourishment. But its environment is very limited. It has no direct relation to most things in existence. Higher we go in range of life wider is environment.... If we are to have anything which is completely organic we must have something related to all things however remote or complex. See Spencer's Psyc. Vol I.

The idea that progress in the organic world involves an increase in the number, range, and complexity of organism-environment adjustments is straight out of Spencer's *Principles of Psychology*, as Dewey's citation indicates. But Dewey gave the notion a human-centered twist, arguing that only in our consciousness do we "find a complete organism + hence a true unity or Individual. While there are a great many things in world Indifferent to a material organism there is nothing which is not either actually or potentially in relation to Intelligence. Environment of mind is coextensive with Universe."[15]

The basic problem of knowledge, according to Dewey's idealist account of it in these Speculative Psychology lectures, is the tension between this potentially universal character of consciousness and its inability to realize this potential in practice. We continually overcome this tension by a process of adjustment—of stimulus and response. Environment provides the stimulus: "Man's intelligence dependent for its content upon its surroundings. A mind shut off from contact with the world remains a blank." Prompted by its sensations, "mind must respond to the stimulus and construct something out of this material." Dewey here returned to Spencer's idea of organism-environment interaction, placing it in the context of his idealist account of knowledge:

> Response of mind brings out + makes real for human intelligence relations which are already real for Universal intelligence. This Response includes
>
> 1 – A wider + wider environment
> 2 – A higher development of reacting self.
>
> i.e. range of anyone's world narrowly depends on extent to which it can react to stimuli. World of lowest Organism is simply few inches of surrounding temperature + food. Higher animals will include to certain extent environment of sights + sounds + also certain number of remembered images. Since man has power of reacting in an indefinite number of ways, no limit can be put to his environment. i.e. merely being surrounded by a world does not

---

[15]Dewey, Speculative Psychology, Lecture 6 (16 March 1887), Box 2, Edwin C. Goddard Papers, Bentley Historical Library, University of Michigan. Cf. Spencer (1880, 1:294). I have replaced abbreviations such as 'Iv.' and 'Uv.' with the terms for which they stand. A copy of these notes is held at the Center for Dewey Studies, Southern Illinois University—Carbondale.

constitute having a world. To have a world must be also power of selecting + responding to things in the surroundings. See Spencer's Princ. of Psyc. Vol. 1 pp 291–305.[16]

Thus although Dewey employed Spencer's idea of organism-environment inter-action, he differed from Spencer in two key respects: first, in the idealist notion of a universal intelligence or consciousness implied by the universal *potential* of our own more limited consciousness; and second, in the emphasis on the mind's active "power of selecting + responding" to the environment. Thus Dewey, inspired by but critical of Spencer, was already developing his own account of organism-environment interaction in the late 1880s.

As I have demonstrated elsewhere, when Dewey began teaching courses on ethics, politics, and Hegel's philosophy in the 1890s, he also started connecting the organism-environment relationship to a dialectical account of adjustment or adap-tation (Pearce forthcoming). This account was derived in part from the work of the philosopher Samuel Alexander, who was at the time attempting to combine German philosophy and evolutionary ethics (see Dewey 1894, 885). Alexander worried that Spencer and his followers often seemed to assume "that the environment is itself something fixed and permanent, according to which, as he gradually discovers its character, [the individual] must arrange his conduct." Instead, argued Alexander, "adaptation can only be understood as a joint action of the individual and his environment, in which both sides are adjusted to each other. What the environment is depends upon the character or the qualities of the individual, for it is only in so far as it responds to him that it can affect him at all" (Alexander 1889, 271). Dewey, in his book *Outlines of a Critical Theory of Ethics*, adopted Alexander's notion of adjustment/adaptation:

> Even a plant must do something more than adjust itself *to* a fixed environment; it must assert itself *against* its surroundings, subordinating them and transforming them into material and nutriment; and, on the surface of things, it is evident that *transformation* of existing circumstances is moral duty rather than mere reproduction of them. The environment must be plastic to the ends of the agent. (Dewey 1891, 115, original emphasis)[17]

There are two routes to adaptation, a change in the organism or a change in the environment, and the latter may be more important to understanding human behav-ior and ethics. Thus Dewey differed from Spencer in emphasizing the importance of construction and reconstruction—i.e., modifications of the environment by the organism—in the (co)-adaptation of organism and environment (see Godfrey-Smith 1996, 131–165).

Dewey's conception of organism-environment interaction, which solidified in the 1890s, became the cornerstone of his philosophy. William James, reviewing the

---

[16]Dewey, Speculative Psychology, Lectures 10/11 (13/15 April 1887). Dewey is here citing Spencer's chapters "Life and Mind as Correspondence," "The Correspondence as Direct but Heterogeneous," and the opening of "The Correspondence as Extending in Space" (Spencer 1880, 1:291–305).

[17]In the preface to this book, Dewey lists Alexander's *Moral Order and Progress* among those books to which he is "especially indebted" (1891, vii).

approach of Dewey's "Chicago School" of philosophy and psychology, noted the importance of the conception:

> Like Spencer, . . . Dewey makes biology and psychology continuous. 'Life,' or 'experience,' is the fundamental conception; and whether you take it physically or mentally, it involves an adjustment between terms. Dewey's favorite word is 'situation.' A situation implies at least two factors, each of which is both an independent variable and a function of the other variable. Call them E (environment) and O (organism) for simplicity's sake. They interact and develop each other without end; for each action of E upon O changes O, whose reaction in turn upon E changes E, so that E's new action upon O gets different, eliciting a new reaction, and so on indefinitely. The situation gets perpetually 'reconstructed,' to use another of Professor Dewey's favorite words, and this reconstruction is the process of which all reality consists. (James 1904, 2)[18]

This basic idea, that experience and inquiry fundamentally involve a mutual adjustment of organism and environment—or transformation/reconstruction of a situation—in response to a concrete problem, would reappear in various guises and contexts for the rest of Dewey's career.

Dewey's famous works on education, metaphysics, aesthetics, and scientific inquiry all depend on the notion of organism-environment interaction. A complete overview is not possible in this short chapter, but the following series of examples gives a sense of how important the organism-environment relationship is in Dewey's philosophical work. In the early pages of *Democracy and Education*—after outlining the meaning of 'environment' and the importance of the social environment—he declares, "we never educate directly, but indirectly by means of the environment" (Dewey 1916, 22). In the Body-Mind chapter of *Experience and Nature* he writes, contra Spencer, "what the organism actually does [in adjusting/adapting] is to act so as to change its relationship to the environment" (Dewey 1925, 283). In *Art as Experience*, describing the reconstructive work of experience as the site of the aesthetic, he says, "attainment of a period of equilibrium is at the same time the initiation of a new relation to the environment, one that brings with it potency of new adjustments to be made through struggle" (Dewey 1934, 17). Finally, in *Logic: The Theory of Inquiry*, he grounds the central idea of an unsettled or problematic situation (which prompts inquiry) in the notion of a "state of imbalance in organic-environmental interactions" (Dewey 1938, 106). The conception of organism-environment interaction that he developed in the 1890s, related to but also critical of Spencer's version, was foundational for Dewey's mature philosophical work. In the pragmatist philosophy of Dewey and James, organism-environment interaction became fully interactive.

---

[18]This passage describing Dewey's biological approach to philosophy foreshadows the "dialectical biology" of Richard Lewontin, who famously argued that $dO/dt = f(O,E)$ and $dE/dt = f(O,E)$. See Levins and Lewontin (1985, 104–105) and Godfrey-Smith (2001). For more on the Dewey-Lewontin connection, see Pearce (forthcoming).

5 Conclusion

The organism-environment dyad, so prominent in turn-of-the-century scientific and philosophical debates, was also invoked throughout the twentieth century. Both Spencer and Dewey were influential on continuing discussions of the proper role of government and ongoing arguments about the best way to educate children. Both were also important figures in the developing social sciences—anthropology, sociology, and psychology. Dewey's essay on "The Reflex Arc," for example, is often seen as a founding document of functionalist psychology; it even mentions the Spencer-Weismann controversy in a footnote, illustrating the kinship between biological and philosophical discussions at the time (Dewey 1896, 360n2). By the mid-twentieth century, ecology—originally defined as the science of organism-environment relations—had become a key notion for social scientists who wanted to focus on human-environment or culture-environment interactions. Echoes of the 1890s debates described in this chapter can be heard in those of the 1950s, '60s, and '70s (Heft, this volume; Schultz, this volume).

Today, organism-environment talk is more common than ever before. Variation and plasticity are once again major topics in the biological sciences (West-Eberhard 2003; Carroll 2005), and philosophers are increasingly attending to the fact that organisms modify their biological and social environments (Pearce 2011; Barker and Odling-Smee, this volume; Sterelny, this volume). Late-nineteenth-century thinkers such as Dewey and Morgan sometimes seem as if they could have been writing yesterday. Thus looking back at the history of the notion of organism-environment interaction, we also look forward—to a century in which we continue to build with old tools made new.

Acknowledgments I am grateful to Gillian Barker and Eric Desjardins for comments on this chapter. I also received helpful feedback on earlier versions from audiences at the University of North Carolina at Charlotte and at the "Romanticism & Evolution" conference at Western University. Research for the chapter was made possible by generous funding from the Rotman Institute of Philosophy, the Social Sciences and Humanities Research Council of Canada, and the Andrew W. Mellon Foundation. Finally, thanks to the Center for Dewey Studies at SIU-Carbondale, the Bentley Historical Library at the University of Michigan, the Oxford University Museum of Natural History, and the American Museum of Natural History for providing access to archival materials and permission to reproduce some of those materials here.

References

Alexander, Samuel. 1889. *Moral order and progress: An analysis of ethical conceptions*. London: Trübner.
Allen, Grant. 1890. The new theory of heredity. *Review of Reviews* 1 (June): 537–538.
Argyll, George Campbell, Duke of. 1867. *The reign of law*. London: Alexander Strahan.
Argyll, George Campbell, Duke of. 1889. Acquired characters and congenital variation. *Nature* 41 (December 26): 173–174.

Baldwin, James Mark. 1896. A new factor in evolution. *American Naturalist* 30 (June–July): 441–451, 536–553.

Barker, Gillian A. 1993. Models of biological change: Implications of three recent cases of 'Lamarckian' change. *Perspectives in Ethology* 10: 229–248.

Berman, Milton. 1961. *John Fiske: The evolution of a popularizer*. Cambridge: Harvard University Press.

Blainville, Henri-Marie Ducrotay de. 1822. *De l'organisation des animaux, ou principes d'anatomie comparée*, vol. 1. Paris: F.G. Levrault.

Bowler, Peter J. 1983. *The eclipse of Darwinism: Anti-Darwinian theories in the decades around 1900*. Baltimore: Johns Hopkins University Press.

Bowler, Peter J. 1988. *The non-Darwinian revolution: Reinterpreting a historical myth*. Baltimore: Johns Hopkins University Press.

Braunstein, Jean-François. 1997. Le concept de milieu, de Lamarck à Comte et aux positivismes. In *Jean-Baptiste Lamarck, 1744–1829*, ed. Goulven Laurent, 557–571. Paris: CTHS.

Buffon, Georges-Louis Leclerc, Comte de. 1766. De la dégénération des animaux. In *Histoire naturalle, générale et particulière, avec la description du cabinet du roi*, vol. 14, 311–374. Paris: De l'imprimerie royale.

Candolle, Augustin Pyrame de. 1820. Géographie botanique. In *Dictionnaire des sciences naturelles*, vol. 18, ed. Frédéric Cuvier, 359–436. Strasbourg: F.G. Levrault.

Canguilhem, Georges. 1952. Le vivant et son milieu. In *La connaissance de la vie*, 160–193. Paris: Hachette.

Carroll, Sean B. 2005. *Endless forms most beautiful: The new science of evo devo and the making of the animal kingdom*. New York: Norton.

Comte, Auguste. 1838. *La philosophie chimique et la philosophie biologique*, Cours de philosophie positive, vol. 3. Paris: Bachelier.

Comte, Auguste. 1853. *The positive philosophy of August Comte*. 2 vols. Trans. Harriet Martineau. London: John Chapman.

Cope, Edward Drinker. 1887. *The origin of the fittest: Essays on evolution*. New York: D. Appleton.

Cope, Edward Drinker. 1889. Lamarck *versus* Weismann. *Nature* 41 (November 28): 79.

Cuvier, Georges. 1817. *Le règne animal distribué d'après son organisation, pour servir de base à l'histoire naturelle des animaux et d'introduction à l'anatomie comparée*. 4 vols. Paris: Deterville.

Darwin, Charles. 1859. *On the origin of species by means of natural selection, or the preservation of favoured races in the struggle for life*. London: John Murray.

Darwin, Charles. 1875. *The variation of animals and plants under domestication*. 2 vols. 2nd ed. London: John Murray.

Dewey, John. 1883. Knowledge and the relativity of feeling. *Journal of Speculative Philosophy* 17 (January): 56–70.

Dewey, John. 1891. *Outlines of a critical theory of ethics*. Ann Arbor: Inland Press.

Dewey, John. 1894. Moral philosophy. In *Johnson's universal cyclopaedia: A new edition*, vol. 5, ed. Charles Kendall Adams, 880–885. New York: A.J. Johnson.

Dewey, John. 1896. The reflex arc concept in psychology. *Psychological Review* 3 (July): 357–370.

Dewey, John. 1916. *Democracy and education*. New York: Macmillan.

Dewey, John. 1925. *Experience and nature*. Chicago: Open Court.

Dewey, John. 1934. *Art as experience*. New York: Milton, Balch.

Dewey, John. 1938. *Logic: The theory of inquiry*. New York: Henry Holt.

Dyar, Harrison G. 1896. Stated meeting [March 9]. *Transactions of the New York Academy of Sciences* 15: 137–143.

Feuer, Lewis S. 1958. John Dewey's reading at college. *Journal of the History of Ideas* 19: 415–421.

Fiske, John. 1874. *Outlines of cosmic philosophy, based on the doctrine of evolution, with criticisms on the positive philosophy*. 2 vols. London: Macmillan.

Geoffroy Saint-Hilaire, Étienne. 1833a. Considérations sur des ossemens fossiles, la plupart inconnus, trouvés et observés dans les bassins de l'Auvergne. *Revue encyclopédique* 59: 76–95.

Geoffroy Saint-Hilaire, Étienne. 1833b. Le degré d'influence du monde ambiant pour modifier les formes animales; question intéressant l'origine des espèces téléosauriennes et successivement celle des animaux de l'époque actuelle. *Mémoires de l'Académie Royale des Sciences de l'Institut de France* 12: 63–92.

Godfrey-Smith, Peter. 1996. *Complexity and the function of mind in nature*. Cambridge: Cambridge University Press.

Godfrey-Smith, Peter. 2001. Organism, environment, and dialectics. In *Thinking about evolution: Historical, philosophical, and political perspectives*, ed. Rama S. Singh, Costas B. Krimbas, Diane B. Paul, and John Beatty, 253–266. Cambridge: Cambridge University Press.

Hickman, Larry A. (ed.). 1999–. *The correspondence of John Dewey, 1871–1953*. Charlottesville: InteLex.

Humboldt, Alexander von., and Aimé Bonpland. 1805. *Essai sur la géographie des plantes; accompagné d'un tableau physique des régions équinoxiales*. Paris: Levrault, Schoell.

Jablonka, Eva, and Marion J. Lamb. 2005. *Evolution in four dimensions: Genetic, epigenetic, behavioral, and symbolic variation in the history of life*. Cambridge: MIT Press.

James, William. 1878. Remarks on Spencer's definition of mind as correspondence. *Journal of Speculative Philosophy* 12 (January): 1–18.

James, William. 1880. Great men, great thoughts, and the environment. *Atlantic Monthly* 46(October): 441–459.

James, William. 1904. The Chicago school. *Psychological Bulletin* 1: 1–5.

James, William. 1988. *Manuscript lectures*. Cambridge: Harvard University Press.

Kemp, James F. 1896. Extra meeting of the biological section. *Transactions of the New York Academy of Sciences* 15: 123.

Kilian, Conrad Joseph. 1802. *Entwurf eines Systems der gesammten Medizin*. Jena: Friedrich Frommann.

Köstlin, Otto. 1851. *Gott in der Natur: Die Erscheinungen und Gesetze der Natur im Sinne der Bridgewaterbücher als Werke Gottes*. 2 vols. Stuttgart: Paul Reff.

Lamarck, Jean-Baptiste. 1801. *Système des animaux sans vertèbres, ou tableau général des classes, des ordres et des genres de ces animaux*. Paris: Deterville.

Lamarck, Jean-Baptiste. 1809. *Philosophie zoologique, ou exposition des considérations relatives à l'histoire naturelle des animaux*. 2 vols. Paris: Dentu.

Laubichler, Manfred D. 2010. Evolutionary developmental biology offers a significant challenge to the neo-Darwinian paradigm. In *Contemporary debates in philosophy of biology*, ed. Francisco J. Ayala and Robert Arp, 199–212. Chichester: Wiley-Blackwell.

LeConte, Joseph. 1878. Man's place in nature. *Princeton Review* 55 (November): 776–803.

Levins, Richard, and Richard C. Lewontin. 1985. *The dialectical biologist*. Cambridge: Harvard University Press.

Lewes, George Henry. 1851. Lyell and Owen on development. *The Leader* 2 (October 18): 996–997.

Lewes, George Henry. 1852. Comte's positive philosophy. Part XIV – The science of life. *The Leader* 3 (July 10): 665–666.

Lewes, George Henry. 1853. *Comte's philosophy of the sciences: Being an exposition of the principles of the Cours de philosophie positive of Auguste Comte*. London: Henry G. Bohn.

Lloyd, G.E.R. (ed.). 1978. *Hippocratic writings*. New York: Penguin.

Lyell, Charles. 1832. *Principles of geology, being an attempt to explain the former changes of the earth's surface, by reference to causes now in operation*. vol. 2, London: John Murray.

Morgan, C. Lloyd. 1891a. *Animal life and intelligence*. Boston: Ginn.

Morgan, C. Lloyd. 1891b. The nature and origin of variations. *Proceedings of the Bristol Naturalists' Society* 6: 249–273.

Morgan, C. Lloyd. 1893. Dr. Weismann on heredity and progress. *The Monist* 4 (October): 20–30.

Morgan, C. Lloyd. 1896. *Habit and instinct*. London: E. Arnold.

Morris, George Sylvester. 1880. *British thought and thinkers: Introductory studies, critical, biographical and philosophical*. Chicago: S.C. Griggs.

Moseley, Henry N. 1885. The continuity of the germ-plasma considered as the basis of a theory of heredity. *Nature* 33 (December 17): 154–157.

Murray, John Clark. 1885. *A handbook of psychology*. Montreal: Dawson Brothers.

Naumann, Moritz. 1821. Versuch zur Wiederlegung einiger Behauptungen in Kiefers System der Medicin. *Isis von Oken* 8: 344–351.

Naumann, Moritz. 1823. *Ueber die Grenzen zwischen Philosophie und Naturwissenschaften*. Leipzig: Adolph Wienbrack.

Nelson, Clinton Eugene. 1977. John Fiske's Harvard lectures: A case study of philosophical lectures. Ph.D. thesis, University of Iowa.

Osborn, Henry F. 1891. Are acquired variations inherited? *American Naturalist* 25: 191–216.

Osborn, Henry F. 1895. The hereditary mechanism and the search for the unknown factors of evolution. In *Biological lectures delivered at the Marine Biological Laboratory of Wood's Holl in the summer session of 1894*, 79–100. Boston: Ginn.

Pascal, Blaise. 1669. *Pensées de M. Pascal sur la religion et sur quelques autres sujets*. Paris: Guillaume Desprez.

Pascal, Blaise. 1991. *Pensées*. Paris: Bordas.

Pearce, Trevor. 2010a. From 'circumstances' to 'environment': Herbert Spencer and the origins of the idea of organism–environment interaction. *Studies in History and Philosophy of Biological and Biomedical Sciences* 41: 241–252.

Pearce, Trevor. 2010b. "A great complication of circumstances" – Darwin and the economy of nature. *Journal of the History of Biology* 43: 493–528.

Pearce, Trevor. 2011. Ecosystem engineering, experiment, and evolution. *Biology and Philosophy* 26: 793–812.

Pearce, Trevor. Forthcoming. The dialectical biologist, circa 1890: John Dewey and the Oxford Hegelians. *Journal of the History of Philosophy*.

Petit, Annie. 1997. L'héritage de Lamarck dans la philosophie positive d'Auguste Comte. In *Jean-Baptiste Lamarck, 1744–1829*, ed. Goulven Laurent, 543–556. Paris: CTHS.

Reil, Johann Christian. 1816. *Entwurf einer allgemeinen Pathologie*. 3 vols. Halle: Curt.

Richards, Robert J. 1987. *Darwin and the emergence of evolutionary theories of mind and behavior*. Chicago: University of Chicago Press.

Romanes, George J. 1887. The factors of organic evolution. *Nature* 36 (August 25): 401–407.

Romanes, George J. 1888. Lamarckism *versus* Darwinism. *Nature* 38 (August 30): 413.

Schwander, Tanja, and Olof Leimar. 2011. Genes as leaders and followers in evolution. *Trends in Ecology & Evolution* 26: 143–151.

Skrupskelis, Ignas K., and Elizabeth M. Berkeley (eds.). 1992–2004. *The correspondence of William James*. Charlottesville: University Press of Virginia.

Sloterdijk, Peter. 2005. Atmospheric politics. In *Making things public: Atmospheres of democracy*, ed. Bruno Latour and Peter Weibel, 944–951. Cambridge: MIT Press.

Spencer, Herbert. 1852. A theory of population, deduced from the general law of animal fertility. *Westminster and Foreign Quarterly Review* 57 (April): 468–501.

Spencer, Herbert. 1855. *The principles of psychology*. London: Longman, Brown, Green, and Longmans.

Spencer, Herbert. 1872. The study of sociology. *Popular Science Monthly* 1 (May): 1–17.

Spencer, Herbert. 1880. *The principles of psychology*. 2 vols. 3rd ed. New York: D. Appleton.

Spencer, Herbert. 1882. *Die Principien der Psychologie*. 3rd ed. Vol. 1. Trans. Benjamin Vetter. Stuttgart: E. Schweizerbart.

Spencer, Herbert. 1887. *The factors of organic evolution*. London: Williams and Norgate.

Sully, James. 1884. *Outlines of psychology*. New York: D. Appleton.

Uexküll, Jakob von. 1909. *Umwelt und Innenwelt der Tiere*. Berlin: Julius Springer.

Wallace, Alfred Russell. 1889. *Darwinism: An exposition of the theory of natural selection with some of its applications*. London: Macmillan.

Weismann, August. 1882. *Studies in the theory of descent*. 2 vols. Trans. Raphael Meldola. London: Sampson Low, Marston, Searle, & Rivington.

Weismann, August. 1883. *Ueber die Vererbung*. Jena: G. Fischer.

Weismann, August. 1885. *Die Continuität des Keimplasma's als Grundlage einer Theorie der Vererbung*. Jena: G. Fischer.

Weismann, August. 1895. Heredity once more. *Contemporary Review* 68 (September): 420–456.

West-Eberhard, Mary Jane. 2003. *Developmental plasticity and evolution*. Oxford: Oxford University Press.

Winther, Rasmus G. 2001. August Weismann on germ-plasm variation. *Journal of the History of Biology* 34: 517–555.

Youmans, Edward Livingston. 1872. Editor's table. *Popular Science Monthly* 1 (May): 113–117.

James Mark Baldwin, the Baldwin Effect, Organic Selection, and the American "Immigrant Crisis" at the Turn of the Twentieth Century

Christopher D. Green

Abstract The "Baldwin Effect," named after the turn-of-the-twentieth-century American psychologist James Mark Baldwin, has experienced a revival over the last few decades, driven primarily by some cognitive scientists who think it might be able to solve problems related to the evolution of consciousness. Baldwin's own interests when he developed the theory, which he called "organic selection," were somewhat different from those of modern cognitivists, and his social context was enormously different. This chapter aims to recover the social challenges of Baldwin's time and explore how they might have been related to his proposal. Chief among these challenges was the widespread perception in the United States that the massive immigrant slums in New York and other cities posed a kind of existential threat to the American way of life. This perception, in turn, led to a number of radical and disturbing eugenic proposals for meeting the "immigrant problem." It is suggested here that, although Baldwin did not address the immigrant issue directly, it was in his mind as he developed his theory of "organic selection," and also that it offered a way out of the crisis that many Americans thought they then faced.

1 Introduction

In June 1953, George Gaylord Simpson published an article in the journal *Evolution* about a phenomenon that was, up to that time, known as "organic selection." He renamed it the "Baldwin effect" because he thought that the phrase "organic selection" had come to be used in too many different ways, and he wanted to focus on a single one—the one he associated with James Mark Baldwin. He characterized the "Baldwin effect" thus:

C.D. Green (✉)
Department of Psychology, York University, 4700 Keele Street, Toronto, ON, Canada
e-mail: christo@yorku.ca

G. Barker et al. (eds.), *Entangled Life*, History, Philosophy and Theory of the Life Sciences 4, DOI 10.1007/978-94-007-7067-6_3,
© Springer Science+Business Media Dordrecht 2014

> The effect may be analyzed as involving three distinct (but partly simultaneous) steps: (1) Individual organisms interact with the environment in such a way as systematically to produce in them behavioral, physiological, or structural modifications that are not hereditary as such but that are advantageous for survival, i.e., are adaptive for the individuals having them. (2) There occur in the population genetic factors producing hereditary characteristics similar to the individual modifications referred to in (1), or having the same sorts of adaptive advantages. (3) The genetic factors of (2) are favored by natural selection and tend to spread in the population over the course of generations. The net result is that adaptation originally individual and non-hereditary becomes hereditary. (Simpson 1953, 112)

He went on to conclude that, although possible, it was unlikely that just the correct confluence of events occurred often enough for the "Baldwin effect" to be an important factor in evolution.

Just six months later, however, Conrad Waddington (1953b) demurred, very nearly declaring that his theory of canalization (Waddington 1942) and the associated phenomenon of "genetic assimilation" (which had been announced in the same issue as Simpson's article; Waddington 1953a) are the mechanisms by which organic selection operates.

Although there had been previous mentions of "organic selection" in the evolutionary literature prior to Simpson's and Waddington's (most notably in Julian Huxley's monumental book, *Evolution: The Modern Synthesis*, 1942), the two 1953 articles are the primary basis of the modern interest in the Baldwin effect, as reflected by the fact that Simpson's phrase—"the Baldwin Effect"—is the one used nearly universally to denote it today.

Curiously, however, when we "moderns" go back to what is supposed to be the *locus classicus* of the "Baldwin effect"—James Mark Baldwin's 1896 article "A New Factor in Evolution"—we find a paper that is much more patchy and obscure than we might expect. Indeed the paper is intentionally patchy. As Baldwin himself said in the opening, it is mostly a compilation of passages from articles and one book he had previously published. It uses a vocabulary of technical terms with which most of us are unfamiliar. It seems unclear whether "organic selection"—the phrase that Simpson redubbed the "Baldwin effect"—refers to a process that goes on only in the learning of single individuals (sometimes in social interactions with other individuals) or whether it refers to the process by which such learning becomes congenital in future generations. Perhaps most frustrating, only the barest sketch of a mechanism is clearly outlined: *viz.*, that learning serves as a "bridge" to instincts that are required by environmental change but have not yet had time to evolve through natural selection. This seems easily refutable: if an animal has learned to do the things that will keep it alive in a changing environment, then it has simultaneously relieved the selection pressure that bore upon it to change its range of congenital instincts. (One can, of course, start speculating at this point about the additional resources that might hypothetically be required to learn and hold in memory new behavioral routines, compared to their (again hypothetical) instinctive equivalents, but in doing so one passes into the realm of shoring-up a flagging research program ad hoc.)

The intellectual reward that one gets from a first reading of Baldwin's "New Factor" article, when approached in this way, is often so meager that some have accused Baldwin of a variety of nefarious motives. Perhaps he was a closet Lamarckian. Perhaps he was trying to steal an idea from Conwy Lloyd Morgan and Henry Fairfield Osborn. Perhaps he illicitly attributed to his earlier writings ideas that only appeared in his later writings in order to snatch priority for himself. Perhaps he was an amateur, an academic unknown foolishly trying to play in the "Big Time" of evolutionary theory (see, e.g., Griffiths 2003).

Since the late 1980s, there has been a great deal of debate about whether the "Baldwin effect" is actually possible, what would count as evidence for a "Baldwin effect," and whether it might have played an important role in evolution, particularly in the evolution of humans (e.g., Hinton and Nolan 1987; Maynard Smith 1987; Dennett 1995; Deacon 1997; Weber and Depew 2003). I do not propose to add to that debate here.

Instead, I want to go back to Baldwin's time. I want to show that the issues that motivated Baldwin and his colleagues to propose and elaborate organic selection in the 1890s have been lost, and are sometimes misconstrued today. Few have come to Baldwin trying to find out what he was interested in, what he was concerned about, and what pressures—intellectual, social, and political—he had to contend with. Most come expecting to be able to scoop up a sizeable chunk of 'gold' and immediately spend it here in the modern world (or, alternatively, scoop up what appears to others to be 'gold,' and then show it to be naught but pyrite).

I aim to reconstruct the historical context in which Baldwin worked in order to show that Simpson's and Waddington's concerns were not well aligned with Baldwin's and that, as a result, the significance of organic selection has been distorted in many modern discussions of the "Baldwin effect." Baldwin's main interest was not in showing how learned behaviors could become congenital, though he did presume that the "bridging" principle would do the trick. "Organic selection" was intended to *supplement* (his term) natural selection by showing that many behaviors that Lamarckians presumed to have become congenital were, instead, inherited socially and "evolve" by way of a process that is analogous to natural selection but that does not involve the congenital "germ line" except in a very general and indirect way.

2   Some Background on Baldwin

James Mark Baldwin was born in the capital of South Carolina, Columbia, in 1861. It was the first year of the Civil War. He was the third of five children in the family of the prosperous businessman Cyrus Hull Baldwin and his wife Lydia Eunice Ford. Although Mark, as he was always known, was born in the South, his family had deep roots in Connecticut, having first settled there in the 1630s. After moving to South Carolina, Cyrus was known to buy slaves just so that he could set them free.

He headed north during the war to avoid being drafted into the Confederate army. The rest of the family, however, stayed behind for most of the war. Only after Columbia was burned in February 1865, following Sherman's "March to the Sea," was the Baldwin family moved north by the Union Army. The whole family returned to Columbia after the war, and Cyrus held a variety of appointments in the military government during Reconstruction, including mayor of Columbia (see Baldwin 1926, chap. 1). In 1878 Baldwin was sent to New Jersey for collegiate preparation. In 1881, bucking his family's Yale tradition, he opted to enter the College of New Jersey (not re-dubbed Princeton University until 1896). His initial intention had been to study for the ministry but philosophy caught his interest, and he was soon training under the prominent Scottish "common sense" philosopher, James McCosh, who was also president of the College. Although a Presbyterian minister, McCosh was open to evolutionary theory, and had, in some of his writings, worked to find an accommodation between it and Christianity (see, e.g., McCosh 1890).

Wilhelm Wundt had founded the world's first experimental psychology research laboratory in Leipzig just two years earlier and courses in Wundt's brand of "physiological psychology" were beginning to appear in American colleges, including the College of New Jersey. One of the young instructors from whom Baldwin learned the "new psychology" in 1883–1884 was Henry Fairfield Osborn,[1] with whom he would later co-develop the theory of organic selection (see Pearce, this volume). Upon finishing his undergraduate degree, Baldwin won a scholarship to study in Germany for a year. He spent some of his time in Leipzig hearing Wundt's lectures and serving as a subject in some of his students' experiments. Baldwin was most intrigued at the time, however, by the work of Spinoza, which he studied under Friedrich Paulsen in Berlin. Returning to Princeton in 1885, Baldwin wanted to write his dissertation on the thought of the Jewish idealist, but McCosh insisted that he write a refutation of materialism instead. He completed his doctorate the following year and took his first significant academic appointment at Lake Forest College, near Chicago, in 1887. After two tumultuous years at what he later recalled as a "narrow and mercantile" institution (at one point he had tendered his resignation; Baldwin 1926, 40), Baldwin escaped to a professorship in Metaphysics and Logic at the University of Toronto. There he founded his first experimental psychology laboratory (the first in the British Empire, he claimed), completed a two-volume textbook of psychology, and began conducting research on child development, using his two daughters as subjects. In 1893, after five years in Canada, he was called back to the College of New Jersey as a professor of philosophy and psychology, where he founded the school's first psychology laboratory, co-founded the *Psychological Review* with James McKeen Cattell of Columbia University, and began to work seriously on the relationship between mental development and evolution.

---

[1] For a recent account of Osborn's psychological research while at Princeton, see Young (2012) (and for more context, Young 2009).

3 Evolutionary Theory and American Psychology

Although it is not widely recognized today, evolutionary theory lay at the heart of the most influential movement in American psychology of the late nineteenth century, and Baldwin was laying claim to that already well-established tradition. As far back as 1870, the independent Cambridge, Massachusetts scholar, Chauncey Wright, had hypothesized that "our knowledges and rational beliefs result, *truly and literally*, from the survival of the fittest among our original and spontaneous beliefs" (Wright 1870, 301). Darwin himself saw Wright's work and cited it in his *Descent of Man* (Darwin 1871). A year later Darwin republished, at his own expense, another of Wright's (1871) articles on evolution. Finally he invited Wright to his home at Downe, where he commissioned Wright to compose an article on the evolution of consciousness that appeared the following year (Wright 1873). Now this would be of little significance to the history of psychology were it not for the fact that Wright was simultaneously heading a discussion group that called itself the "Metaphysical Club" and which included the young Charles Sanders Peirce and William James among its members.

By 1875 Wright was dead, at the age of only 45, but James was already carrying Wright's Darwinian message forward in his physiological psychology course at Harvard, and in his review ([James] 1875) of Wundt's 1874 textbook, *Grundzüge der physiologeschen Psychologie*. In 1878 James began publishing the material that would make up his landmark *Principles of Psychology*, which finally appeared in 1890. In this work, among other things, he steered American psychology away from the project that occupied the Wundt lab—that of distilling pure apperception from the rest of consciousness—and, instead, attempted to examine the person's (and the animal's) interaction with its environment more holistically: What evolutionary purpose might consciousness serve? What are emotions and what role do they play in life? How are habits acquired and maintained, and what are their functions? Although Baldwin was educated in the 1880s under McCosh, his whole generation was influenced by William James's effort to "re-found" scientific psychology on a broadly evolutionary basis.

It is important to keep in mind, however, that not all evolutionists were strict natural selectionists. Even Darwin himself had steadily ceded ground, over the course of the six editions of the *Origin of Species*, to those who believed that the fossil record could only be explained by allowing that characters acquired during the lifetime of an organism somehow become congenital and are transmitted to offspring: these were the so-called neo-Lamarckians. Darwin's hand-picked intellectual successor, the Canadian-born George John Romanes, gave over to neo-Lamarckism completely in the realm of instinct (Romanes 1877, 1882, 1888). Edward Drinker Cope, perhaps the leading American paleontologist of the era, was a staunch neo-Lamarckian (Cope 1887), and trained his star protégé in the theory—none other than Henry Fairfield Osborn, Baldwin's former teacher and the future co-developer of "organic selection."

The theory of natural selection was losing ground rapidly until the publication of the work of the German zoologist August Weismann on heredity and evolution.

Weismann declared that only natural selection accounts for evolution, and to make his point he chopped the tails off of several successive generations of mice to show that they became congenitally no shorter as a result. The experiment was a bit of a caricature, convincing almost no one, but it re-energized Darwinian theory, and a whole generation of young "neo-Darwinians" appeared, partly as a result. More seriously, Weismann put forward a new theory of heredity: that the germ cells that give rise to offspring are wholly isolated from the somatic cells and, thus, cannot transmit to the offspring any changes that occurred during the lifetimes of the parents (Weismann 1885, 1892). Between 1893 and 1895 Weismann engaged in a bitter debate with Herbert Spencer on the pages of *Contemporary Review*, a debate in which Romanes participated and, perhaps surprisingly, sided with Weismann on the question of the inadequacy of Spencer's critique, though remaining a Lamarckian himself (Weismann 1893, 1894; Spencer 1893a, b, c; Romanes 1893a, b, c). There can be no doubt that Baldwin was aware of all of this and was assessing where best to throw in his own lot (see Pearce 2010, chap. 2).

4 The Social Context

Although the formal debate about natural selection was carried on mostly in terms of arcane biological questions (e.g., how does the rest of a stag's body adjust to accommodate the increasing size of its antlers?; how can the transmission of acquired characters account for neuter organisms such as worker bees and soldier ants?), behind all of this loomed the most pressing social question of the age: What was to become of the wave of destitute immigrants then pouring into New York, Boston, Chicago, and other American cities, straining not only the resources, but also the ethical imaginations, of the American people?

The debate over what to do with, to, or about the legions of the impoverished, who were literally stuffed into the slums of America's cities, dated back to *before* the flood of a million Irish Catholic refugees, and perhaps a million more German Catholics, who came to America in the 1840s and 1850s. There was another wave of African-American migrants from the South to northern cities in the aftermath of the Civil War as well. But the so-called "new" immigration—mostly of southern and eastern Europeans: Italians, Jews, Russians, Poles, Greeks—starting around 1880 and the unprecedented crowding and poverty it created led to a renewed sense of crisis, and even of catastrophe. The numbers *remain* truly staggering. Hundreds of thousands of immigrants entered New York City alone *each year* between 1880 and 1920, peaking at over one million in 1907 alone (this at a time when entire population of Manhattan was 2.3 million).[2] Between a quarter and a third of these

---

[2] Exact figures vary somewhat from source to source. Two particularly accessible and reliable sources are the Ellis Island Foundation's own timeline (http://bit.ly/3fYIo6) and the Fordham University website on New York City History (http://bit.ly/zJpFnb).

Fig. 1 "Lodgers in a crowded Bayard Street tenement" (Photograph by Jacob Riis, ca. 1890, from the collections of the Museum of the City of New York (90.13.1.158). Reproduced with permission. A sketch based on this photograph was included in Riis (1890, 69).)

stayed in New York City to live out their lives. By the first decade of the twentieth century, 40 % of New York City's residents were immigrants. In the public schools, more than 70 % of the students had at least one foreign-born parent.[3] As the New York newspaperman Edwin C. Hill put it, "every 4 years, New York [City] adds to itself a city the size of Boston or St Louis.... It is the whirlpool of the races."[4]

Numbers so vast as these made it easy to regard the newcomers as a kind of horde—a deindividuated mass that had to be dealt with on a mass scale. However, the photographs published in Jacob Riis's book, *How the Other Half Lives* (1890), of the filth and misery of the urban slums—photographs which were only made possible by the invention of the magnesium flash just three years earlier—served to humanize and individualize those who had previously been seen by much of the public merely as a problematic swarm and infestation (e.g., Figs. 1 and 2). The book was a sensation, going through eleven editions in just the first five years after its initial publication. The photographs spurred a widespread movement to

---

[3]See Camille Avena's essay "Progressive education in New York City" on the Fordham University website on New York City history (http://bit.ly/oyGTs2).

[4]Cited in Ric Burns' (1999) "New York: A Documentary Film," Episode 4, 34:00 http://www.pbs.org/wnet/newyork/.

Fig. 2 "Yard in Jersey Street" (Photograph by Jacob Riis, ca. 1897, from the collections of the Museum of the City of New York (90.13.1.102). Reproduced with permission.)

do something to alleviate the suffering of these people by improving the quality of their housing, the sanitation of their neighborhoods, and the conditions of their employment. The movement was led by New York's Civil Service Commissioner: one Theodore Roosevelt.

Although Riis's work led to better conditions in the notorious Five Points slum and tenement ghettoes elsewhere in the city, it would be a mistake to think that Riis or most of his followers believed that mere situational poverty was the new immigrants' only problem. Like much of the American population, he believed southern- and eastern-European immigrants to be mentally and morally inferior to those of northern-European ancestry. They were seen as being unable to govern themselves. And this, note, was from the "progressive" end of the political spectrum. So, the question faced by American social thinkers, such as Baldwin, was whether anything could be done to improve the newcomers' lot on a permanent basis, or whether these newcomers and their descendants would, forever more, be dependent on the good graces of their presumed "betters."

In this matter, the neo-Lamarckians offered a much more optimistic and "progressive" vision of the future than did the neo-Darwinians. If acquired characteristics could be transmitted to offspring, then it might be possible to raise the mental and moral status of lowly immigrants through education, and those improvements would be passed on to future generations congenitally. And then they could be improved further, which would in turn be passed on again, and so on, until the descendants were indistinguishable from Americans of northern-European stock. If it were not the case, however, that improvements to one generation would be passed on to the next, as the neo-Darwinians held, then efforts to educate poor, indigent immigrants would, at best, be only a very partial and very temporary solution. Even if some could be taught—through strenuous remedial training—to fend for themselves in the hurly-burly of modern urban life, the next generation would simply fall back to where the parents began, and the whole expensive, labor-intensive process would have to begin again, on into the future indefinitely. One can see how people with so dark a vision of the future might feel themselves driven toward eugenic schemes to limit the number of offspring that such people might produce, as a means of limiting the ultimately unsustainable burden that would fall to society at large, generation after generation, should this dreary cycle not be broken. David Starr Jordan, the newly-installed president of Stanford University, for instance, observed that neo-Lamarckians "who see the key to the elevation of the human race in the direct inheritance of the results of education, training, and ethical living" viewed neo-Darwinism as the "Gospel of Despair" (Jordan 1892, 244).[5] Bolstered by widespread disapproval of the newcomers' "alien" ways and a palpable fear of how they were transforming America's cities, the pressure to do something—even something radical—became nearly irresistible.

5 Baldwin's Contribution

Baldwin, however, offered a third option—a way in which neo-Lamarckism and the lingering problem of its actual mechanism could be abandoned but one could still hold out hope that future generations could benefit from the achievements of previous generations. At its most basic level, it was little more than learning by imitation, which would hardly have been a revolutionary proposition, but when combined with the process of natural selection—though pitched at the social level rather than at the biological level—it held new promise for both the world of science and for the world of social policy.

The idea is sketched in Baldwin's 1896 article, "A New Factor in Evolution," which, as noted above, is usually taken to be the *locus classicus* of the "Baldwin effect." But, as was also already mentioned, that article is little more than a series of previously-published passages, stitched together so roughly that it is often difficult to make out exactly what Baldwin is claiming and—just as important—what he is not.

---

[5]Jordan himself did not take this to be a "just criticism," however, because, essentially, whatever is true is true, whether it lead to despair or no. He also noted that Osborn was predicting the rapid decline of Weismann's influence in 1892. Of course, it was Osborn's Lamarckism that rapidly declined, and Osborn himself became a eugenicist before long.

In order to understand clearly what Baldwin was after, one has to look at the three books he wrote on the topic between 1895 and 1902: *Mental Development in the Child and the Race* (1895c), *Social and Ethical Interpretations in Mental Development* (1897c),[6] and *Development and Evolution* (1902). These books formed a series on the topic of mental development taken, respectively, from the psychological, the social, and the biological perspectives.

It is true that Baldwin believed that an organism that is able to learn strategies for dealing with a changing environment would give itself evolutionary "breathing space"—the time required for natural selection to gradually produce an instinct to handle the environmental change. The learned behavior was to serve as a "bridge" between where the animal starts congenitally and where it has to get to in order to continue to survive and reproduce in the new environment.

There are two things to be said about this. First, it is not clear that natural selection would, in fact, produce the relevant instinct once the animal had solved the problem through learning. Having learned to solve the problem, the selection pressure is removed, unless the process of learning is somehow so arduous that random variations that are even slightly in the "right" direction ("partial" instincts, Baldwin called them) actually provide selective advantage for organisms by making the learning required of future generations slightly less arduous. But, of course, if the learning process is so arduous that the species is effectively teetering on the brink of extinction for an extended period of time, it is a wonder that the species survives the hundreds or thousands of generations that would be needed to get a "full instinct" in place. Where the question of the evolution of the congenital germ line did exercise Baldwin, it was not in the matter of developing new instincts but in the evolution of ever-greater mental and behavioral plasticity. Greater plasticity allows for a greater range of learning capacity, which, in turn, allows for more rapid and effective responses to whatever challenges the environment brings forth.

Second, as has been pointed out by many Baldwin-bashers, the mechanism he sketched by which learning might eventually become congenital is not spelled out in any detail—hypothetical examples are few and actual biological examples are practically non-existent in his published writings. It is possible, however, that the reason for this is that the "congenitalization" of learned responses was not the primary aim of Baldwin's work (contrary to what almost everyone today is led to believe they will find in Baldwin). He thought he had a little mechanism to handle what he regarded as the *preliminary* problem of explaining away phenomena that seemed to support the neo-Lamarckian position. And two other prominent evolutionists, on opposite sides of the debate, no less (Henry Fairfield Osborn and Conwy Lloyd Morgan), seemed to agree with him about that. But what Baldwin saw

---

[6]It is easy to become distracted by the inclusion of the term "ethical" in the title of the second book, but the matter is explained in Baldwin's autobiography (Baldwin 1926, 66–67). Essentially he added the term to the title at the last minute in order to make it more appealing to the Danish Royal Academy of Sciences awards committee. Originally the book was subtitled just "A Study in Social Psychology."

as the real prize was an application of the structure of natural selection to mental and social processes that would render moot the whole argument about whether adjustments in congenital instincts were the products of natural selection or of the transmission of acquired characters. Accounting for new instincts was not Baldwin's main goal.

Learning from the environment, and then transmitting that learning to new generations through a process he dubbed "social heredity" was where the action really was for Baldwin. The process was thought to go like this: at root, the young organism learns, through trial and error, to solve challenges posed by the environment (e.g., moving oneself around, picking up objects, etc.). This process has exactly the same structure as natural selection: an action is attempted in the effort to solve a particular challenge; multiple variants of this action are tried out (Baldwin called this the "circular reaction"—each unsuccessful attempt calls out for an approximate repetition of the action). Those variations that are unsuccessful are "selected out." Those that are successful, or that approximate success better than previous efforts, are retained and become the basis of future variations. Eventually a successful action pattern "evolves," so to speak, to address the particular environmental challenge at hand. The successful variation is stored as a "habit," to be used again in the future. It is important to note that, thus far, none of this is really original to Baldwin. It is, rather, a case of Baldwin adopting the application of natural selection to the mental and behavioral realms that had been put forward earlier by Chauncey Wright and William James.

Baldwin's real insight occurred when he noted that, if every single organism had to go through this entire process from scratch, learning every non-instinctive behavior in its repertoire through trial and error, not many would survive the process. They just wouldn't be able to learn everything they needed to know in the time available before a lethal environmental challenge presented itself. But Baldwin noted that children learn a great deal from imitating their parents (and other members of their immediate group). In itself, this was nothing new, but Baldwin was able to integrate it into a process that he called "social heredity," which also borrowed the structure of natural selection, but this time pitched at the social level of analysis: Most initial attempts at imitation (a kind of behavioral analog to "reproduction") will fail, and will quickly be selected "out." But, with repeated and varied attempts at imitation, the child will quickly discover a variation of the model that is successful at meeting the particular challenge s/he faces (e.g., getting a piece of food into the mouth), and this will be retained or selected "in." With further imitations and variations, that minimally successful version will be "perfected"—made as efficient as the model on which it is based. This version will be stored as a "habit." *Or*, some random variation will actually prove to be better (more "fit") than the model being imitated, and *that* version will be stored as a habit. And, as that "evolved" version is, in turn, imitated by the next generation (where a behavioral "generation" is defined by who *learns* from whom, rather than by who is borne of whom), it will gradually become the "standard" version of the behavior in that particular population. Individuals who use it will have a selective advantage over those who do not.

Now, will this learned habit somehow descend into the germ line and become congenital? This is the question that exercises pretty well everyone who considers the "Baldwin effect" today, but it does not seem to have much exercised Baldwin himself. As I said before, he thought that there was probably a non-Lamarckian mechanism by which it might do so. But he did not believe that this would be a wholly good thing for the growing organism, even if it happened through natural selection, because he feared that too many rigid instincts would block up the organism's ability to learn new things. He believed that the process of evolution worked more effectively if psychological and behavioral plasticity were kept at a maximum. Indeed, he believed that what had made humans superior to other animals was that their evolutionary history had given them a high degree of mental and behavioral plasticity rather than an extensive assortment of special-purpose instincts. As Baldwin put the matter,

> In the animals, the social transmission seems to be mainly useful as enabling a species to get instincts slowly in determinate directions, by keeping off the operation of natural selection. Social Heredity is then the lesser factor; it serves Biological Heredity. But in man, the reverse. Social transmission is the important factor, and the congenital equipment of instincts is actually broken up in order to allow the plasticity which the human being's social learning requires him to have .... The [human] child is the animal which inherits the smallest number of congenital co-ordinations, but he is the one that learns the greatest number. (Baldwin 1896a, 539–540)

Baldwin also thought that the process of social heredity might well have tricked investigators into thinking that they were witnessing a Lamarckian transmission of acquired characteristics when in fact they were witnessing the process of social heredity—what appeared to be a young individual having congenitally acquired a behavior that its parents had only learned through an effortful process of trial and error, was actually a simple act of imitation. (And this was elaborated by the possibility of new variations resulting in new versions that were even more "fit" than those the parents had originally learned.)

Baldwin repeatedly described this process as a "supplement to natural selection." *This* was the "new factor"—a kind of natural selection that operated in the psychological and social realms, rather than in the biological. Unfortunately, Baldwin applied the phrase "organic selection" rather indifferently to various parts of the larger process. Sometimes it included only the social and psychological parts of it. Later he turned to the phrase "functional selection" to denote just this part. Sometimes he used "organic selection" to cover the whole thing, including the possibility of learned behaviors "bridging" into new instincts. Because this was the version that most interested Simpson and the evolutionists who came after, this full version is the one most people take to be "organic selection," or the Baldwin Effect, today.

By the time of his 1902 book, *Development and Evolution*, however, Baldwin had come to the conclusion that a hard distinction between the physical, the psychological, and the social was untenable. Evolution, he declared, operates "psychophysically": "the organic and the mental are welded in the process of evolution" (Baldwin 1902, 29).

In an even later book, published in 1909 in honor of the centennial of Darwin's birth, Baldwin produced what is probably the most lucid and sophisticated description of his position. Unfortunately, it is one that is almost never cited today. Baldwin noted in 1909 that there is

> at every stage of growth, *a combination of congenital characters with acquired* [learned] *modifications*; natural selection would fall in each case upon *this joint or correlated result*; and the organisms showing the most effective combinations would survive. Variation plus modification, the joint product actually present at the time the struggle comes on, *this is what selection proceeds upon*, and not, as strict neo-Darwinism or Weismannism supposes, upon the congenital variations alone. (Baldwin 1909, 18)

That is to say, for example, having a long narrow beak doesn't do one any good unless one knows how to jam it into a crack in a tree in order to extract edible insects. Once one starts jamming it into trees, however, there arises a selective advantage for a longer, thinner beak. It is the *combination* of complementary physical and behavioral attributes that natural selection acts on. Having the physical basis of a solution to an ecological challenge without knowing how to use it is no better than a baby having a pencil.

So, one might well ask, what does any of this have to do with the crisis over the mass immigration to the United States of millions of people who were supposedly mentally and morally inferior? Baldwin's proposal blazed a trail out of the dilemma. Before, one either had to declare for neo-Lamarckism, and hold firm to the belief that the effects of education would be carried forward in future generations by the transmission of acquired characters, or one had to declare for neo-Darwinism, and thereby accept the dark implication that education, to the degree it was even possible, would have to be repeated with each successive generation indefinitely into the future; or, even more darkly, that we could not allow there to be many future generations of these putatively inferior people. Baldwin cut through this problem because his theory implied that if we educated a single generation, some of the fruits of that effort would be carried forward by the process of social heredity, just as it presumably already had in those of Western European descent from the time of their distant ancestors. Moreover, the very state of being better educated, and of being surrounded by others who were better educated, would change the kinds of environmental challenges that immigrants had to face and, by that very fact, change the course of biological evolution for those who had arrived here, in the modern cities of America, from very different environments that had posed a very different set of challenges.

I do not wish to imply that Baldwin was a thoroughgoing racial egalitarian. It is clear that he perceived differences in the mental and moral status of the races. He even wrote approvingly of Galton's program of positive eugenics from time to time. But his theory permitted a much wider potential for *social* amelioration of those differences than did many of the standard social views of the day. What made Baldwin different was that he did not believe the congenital and the social to be wholly distinct categories. He recognized that they shaded into each other and influenced each other to a much greater degree than was generally recognized.

So, if Baldwin's theory had major implications for the immigration crisis that America faced at the turn of the twentieth century, why didn't he say so explicitly, and why is he not now known as the man whose ideas solved it for us? There is any number of reasons. First, his arrest in a police raid on a Baltimore brothel in the fall of 1908 led to his being more or less erased from history. Although scandal was avoided in the months immediately following the incident, political enemies threatened to turn it into a public affair early the following year, when Baldwin was nominated to sit on Baltimore's school commission. Baldwin was forced to resign his professorship at Johns Hopkins. He was also forced to resign his presidency of the International Congress of Psychology, which was to have been held in 1910. He and his family left the country, moving first to Mexico and later to France, where he lived out most of the rest of his life. There, he became close friends with the famous psychiatrist, Pierre Janet, who likely passed some of Baldwin's ideas about development on to a later student of his: Jean Piaget, probably the pre-eminent child psychologist of the twentieth century.[7] During World War I, Baldwin wrote prolifically urging the US to come to France's defense, which it ultimately did.[8] In 1929, however, when E. G. Boring wrote the most influential history of psychology textbook of the twentieth century, he consigned Baldwin to a minor role in the discipline's past, and Baldwin was mostly forgotten by both psychologists and evolutionists until his "effect" was revived by G. G. Simpson, nearly 20 years after Baldwin's death in 1934.

Second, public policy statements were not Baldwin's *métier*.[9] During the mid-1890s, he wrote a number of articles for *The Inland Educator: A Journal for the Progressive Teacher*, which was edited by Francis Staler, an old Princeton school chum of Baldwin's who had since taken a position at Indiana State Normal School (Baldwin 1895a, b, 1896b, c, 1897a, b). These mostly focused on the problems of how to teach children who had what we would today call different "learning styles"—children who are more "sensory" vs. those who are more "motor"—and on the relationship between imitation and invention in children. Although these ideas held important implications for the issues posed by immigration in an era in which some races were widely regarded as being naturally more impulsive

---

[7]See Wozniak (2009) for a detailed account of the professional disaster that befell Baldwin in the wake of his arrest and resignation.

[8]There is little reason to believe that Baldwin's writings had much to do with that decision. Baldwin and Woodrow Wilson had despised each other when, respectively, professor and president at Princeton. Indeed, Wilson's high-handedness is one of the reasons Baldwin cites for having left Princeton for Hopkins in 1905. It was the "Zimmerman telegram" from Germany, urging Mexico to attack the US in exchange for a return of New Mexico and Arizona after the war that prompted Wilson to act in 1917. Incidentally, a ship on which Baldwin and his family were travelling across the English Channel, the *Sussex*, was torpedoed and sunk during WWI. One of his daughters was seriously wounded in the attack. A young Wilder Penfield, who would later become a leading neuroscientist in Montréal, was on the same ship, and became acquainted with the Baldwin family there.

[9]Though, see Baldwin's (1902, 144–148) comments on "Social Progress."

and less reflective than others, Baldwin never addressed the matter directly. In his autobiography, describing his episode in Mexico, he commented dismissively that "the more or less barbarous hordes, bound up in century-old customs or cults... must work gradually into the new freedom which rests on self-government and social continence."[10] But one finds very little explicit mention of contemporary social dilemmas in Baldwin's academic writings until, long out of the American academy, he took up the cause of US entry into World War I. It seems to have been just how he preferred to conduct himself as a scientist.

John Dewey's educational proposals for the schools were not far removed from the implications of Baldwin's ideas, though steeped much more weakly in evolutionary theory. And, unlike Baldwin, Dewey actually set up his famous laboratory school right in the heart a city that was notoriously suffering from the immigration crisis: Chicago. The model of Dewey's school was imitated (more or less) far and wide and, as a result, Dewey became known as the man who changed the course of American education. In short, Dewey was much more active on this front, and he continued as a prominent public intellectual for nearly half a century after Baldwin had been removed from the scene.[11]

To conclude, the faults and lacunae that modern philosophers and biologists find in Baldwin's writings are partly real, but are also partly the result of their attempting to find answers to modern questions in historical texts. If one approaches Baldwin's work with the issues that Baldwin faced in mind, many of the complications and obscurities fade away. Baldwin was more interested in transcending the neo-Lamarckian/neo-Darwinian dispute of his era than in adjudicating it. It is primarily in his later work—especially *Development and Evolution* (1902) and *Darwin and the Humanities* (1909)—which is now read by almost no one, that he brought the social, psychological, and biological aspects of his theory together in an interesting and sophisticated way—a way that, in this era of reductionist "evolutionary psychology," might still have a thing or two to teach us today.

References

Baldwin, J. Mark. 1895a. Differences in children, from the teacher's point of view [Part 1]. *The Inland Educator* 1: 5–10.
Baldwin, J. Mark. 1895b. Differences in pupils from the teacher's point of view [Part 2]. *The Inland Educator* 1: 269–273.

---

[10]Thanks to Robert Wozniak for directing me to this passage. He believes that by 'superstition,' Baldwin probably meant Catholicism and, if so, then he probably felt the same way about the Irish and German immigrants of the mid-nineteenth century, and the Italian immigrants of the late nineteenth as well.

[11]Dewey (1898) published a review of Baldwin's *Social and Ethical Interpretations*, but it did not touch on the present issues either, focusing mainly on what Dewey took to be a certain theoretical confusion in Baldwin's view between the process of developing a conscious self and the specific contents of that consciousness.

Baldwin, J. Mark. 1895c. *Mental development in the child and the race.* New York: Macmillan.

Baldwin, J. Mark. 1896a. A new factor in evolution. *American Naturalist* 30: 441–451. 536–553.

Baldwin, J. Mark. 1896b. Differences in pupils from the teacher's point of view [Part 3]. *The Inland Educator* 2: 126–129.

Baldwin, J. Mark. 1896c. Differences in pupils from the teacher's point of view [Part 4]. *The Inland Educator* 3: 232–235.

Baldwin, J. Mark. 1897a. Invention vs. imitation in children [Part 1]. *The Inland Educator* 4: 267–271.

Baldwin, J. Mark. 1897b. Invention vs. imitation in children [Part 2]. *The Inland Educator* 5: 58–63.

Baldwin, J. Mark. 1897c. *Social and ethical interpretations in mental development.* New York: Macmillan.

Baldwin, J. Mark. 1902. *Development and evolution.* New York: Macmillan.

Baldwin, J. Mark. 1909. *Darwin and the humanities.* Baltimore: Review.

Baldwin, J. Mark. 1926. *Between two wars, 1861–1921: Being memories, opinions and letters received.* Boston: Stratford.

Cope, E.D. 1887. *The origin of the fittest: Essays on evolution.* New York: Appleton.

Darwin, Charles. 1871. *The descent of man, and selection in relation to sex,* 1st ed. London: John Murray.

Deacon, Terrence W. 1997. *The symbolic species.* New York: W. W. Norton.

Dennett, Daniel C. 1995. *Darwin's dangerous idea.* New York: Simon & Schuster.

Dewey, John. 1898. [Review of] Social and ethical interpretations of mental development. *Philosophical Review* 7: 398–409.

Griffiths, P.E. 2003. Beyond the Baldwin effect: James Mark Baldwin's "social heredity," epigenetic inheritance, and niche-construction. In *Evolution and learning: The Baldwin effect reconsidered,* ed. Bruce H. Weber and David J. Depew, 193–215. Cambridge, MA: MIT Press.

Hinton, Geoffrey E., and Steven J. Nolan. 1987. How learning can guide evolution. *Complex Systems* 1: 495–502.

Huxley, Julian. 1942. *Evolution: The modern synthesis.* London: Allen and Unwin.

[James, William]. 1875. [Review of] Grundzüge der physiologischen Psychologie. *North American Review* 121: 195–201.

Jordan, David Starr. 1892. The present battle-ground of evolution. *The Dial* 13 (October 16): 242–244.

Maynard Smith, John. 1987. Natural selection: When learning guides evolution. *Nature* 329: 761–762.

McCosh, James. 1890. *The religious aspect of evolution.* New York: Charles Scribner's Sons.

Pearce, Trevor. 2010. "A perfect chaos": Organism-environment interaction and the causal factors of evolution. Ph.D. dissertation, University of Chicago.

Riis, Jacob A. 1890. *How the other half lives: Studies among the tenements of New York.* New York: Charles Scribner's Sons.

Romanes, George J. 1877. *The scientific evidences of organic evolution.* London: Macmillan.

Romanes, George J. 1882. *Animal intelligence.* London: Macmillan.

Romanes, George J. 1888. *Mental evolution in man.* London: Kegan Paul.

Romanes, George J. 1893a. Mr. Herbert Spencer on "natural selection". *Contemporary Review* 63: 499–517.

Romanes, George J. 1893b. The Spencer-Weismann controversy. *Contemporary Review* 64: 50–53.

Romanes, George J. 1893c. A note on panmixia. *Contemporary Review* 64: 611–612.

Simpson, George Gaylord. 1953. The Baldwin effect. *Evolution* 7: 110–117.

Spencer, Herbert. 1893a. The inadequacy of "natural selection". *Contemporary Review* 63: 153–166, 439–456.

Spencer, Herbert. 1893b. Professor Weismann's theories. *Contemporary Review* 63: 743–760.

Spencer, Herbert. 1893c. Rejoinder to Professor Weismann. *Contemporary Review* 64: 893–912.

Waddington, Conrad. 1942. Canalization of development and the inheritance of acquired characters. *Nature* 150: 563–565.

Waddington, Conrad. 1953a. Genetic assimilation of an acquired character. *Evolution* 7: 118–126.

Waddington, Conrad. 1953b. "The Baldwin effect," "genetic assimilation," and homeostasis. *Evolution* 7: 386–387.

Weber, Bruce H., and David J. Depew (eds.). 2003. *Evolution and learning: The Baldwin effect reconsidered*. Cambridge, MA: MIT Press.

Weismann, August. 1885. *Die Kontinuität des Keimplasmas als Grundlage einer Theorie der Vererbung* [The continuity of the germ-plasm as the principle of a theory of heredity]. Jena: Gustav Fischer.

Weismann, August. 1892. *Das Keimplasma: Eine Theorie der Vererbung* [The germ-plasm: A theory of heredity]. Jena: Gustav Fisher.

Weismann, August. 1893. The all-sufficiency of natural selection. *Contemporary Review* 64: 309–338, 596–610.

Weismann, August. 1894. *The effect of external influences upon development*. London: Henry Frowde, Amen Corner. (Originally delivered as the Romanes Lecture for 1894 at Oxford.).

Wozniak, Robert H. 2009. James Mark Baldwin, professional disaster, and the European connection. *Rassegna di Psicologia* 26: 111–128.

Wright, Chauncey. 1870. Limits of natural selection. *North American Review* 111 (October): 282–311.

Wright, Chauncey. 1871. The genesis of species. *North American Review* 113 (July): 63–103.

Wright, Chauncey. 1873. Evolution of self-consciousness. *North American Review* 116 (April): 245–310.

Young, Jacy L. 2009. Evolution, education, and eugenics: Organic selection in progressive era America. M.A. thesis, York University, Toronto.

Young, Jacy L. 2012. The biologist as psychologist: Henry Fairfield Osborn's early mental ability investigations. *Journal of the History of the Behavioral Sciences* 48: 197–217.

The Tension Between the Psychological and Ecological Sciences: Making Psychology More Ecological

Harry Heft

Abstract In spite of the fact that psychology has been committed to an evolutionary framework for over a century, ecological approaches to psychology, first proposed several decades ago, continue to be marginalized within the discipline. Considering the shared lineage of evolutionary and ecological thinking, this situation seems paradoxical, and, indeed, it reflects an underlying tension between the psychological and ecological sciences. The basis for this tension can be traced historically to psychology's early embrace of Herbert Spencer's evolutionary view of environment-mind correspondence, which is incompatible with the dynamic, relational character of ecosystems thinking. In this respect, William James criticized Spencer for failing to recognize the active and selective character of thought and action, which for James, is the hallmark of psychological processes. From this starting point, James's psychology and philosophy of radical empiricism offers a relational and dynamic approach that is more in keeping with ecological thinking, particularly as these ideas were extended by James's student, E. B. Holt, in his treatment of purposive, situated behavior. James Gibson's ecological approach to perceiving builds, in part, on these bodies of work, and his concept of affordances locates meaning in perceiver-environment relations, that is, in situated action. Further, the ecological approach of Roger Barker, with its concept of behavior setting, offers an opportunity to bring sociocultural processes to bear on situated action. It is seen that socially normative actions are situated in behavior settings and have the character of being both regulated and flexible, dual properties that are examined through a consideration of Hayek's analysis of purposive action. Collectively, these contributions advance an approach to psychology that is coordinative with the perspective of the ecological sciences.

H. Heft (✉)
Department of Psychology, Denison University, Granville, OH, USA
e-mail: heft@denison.edu

G. Barker et al. (eds.), *Entangled Life*, History, Philosophy and Theory
of the Life Sciences 4, DOI 10.1007/978-94-007-7067-6_4,
© Springer Science+Business Media Dordrecht 2014

1   Introduction

The historical relationship between psychology, on the one hand, and Darwinian evolutionary thinking and its allied field of ecology, on the other, has never been straightforward. In spite of the immediate attention given by much of the scientific community to Darwin's writings on natural selection, his work had very little impact on the initial formulation of experimental psychology in the 1870s and 1880s. By the end of the nineteenth century, psychology began to adopt a functionalist stance, broadly embracing an evolutionary perspective. It took another six decades, however, before ecological approaches to psychology were proposed, and even then, these frameworks were marginalized in the discipline from the outset—a pattern that continues to the present day. On its face, at least, this disjunction appears somewhat paradoxical. How can it be that the value of an evolutionary framework for scientific psychology ceased to be in question over a hundred years ago, while ecological approaches to psychology continue to have little impact on the field? Perhaps the answer rests with the conceptualization of organismal evolution that was embraced from the outset.

The present chapter will begin by exploring this tension between evolutionary and ecological thinking in psychology from an historical and conceptual perspective. In the course of doing so, I will integrate the work of several twentieth-century thinkers with the goal of explicating some of the necessary foundations for an ecological approach to psychology. This analysis will be rooted philosophically in William James's radical empiricism. From there, selected features of the mostly forgotten radical empiricist writings of James's student, Edwin Bissell Holt, will be examined in order to prepare the ground for a consideration of the nature of *situated action*, which will be shown to be a central idea for an ecological psychology. The ecological approach of Holt's student James J. Gibson will then be briefly considered, as will the independently developed ecological psychology of Roger G. Barker. Their concepts of affordances and behavior settings, respectively, point to two eco-psychological structures of the everyday environment that are essential to the formulation of an ecological psychology. Finally, I will examine behavior settings from the perspective of Holt's writings in order to advance an understanding of the relationship between situated action and social practice.

The exploration of these issues will go a long way toward redressing what will be seen to be psychology's century-long, half-hearted embrace of evolutionary thinking, and its on-going neglect of an ecological perspective.

2   Some Historical Considerations

Over the last decades of the nineteenth century, the first generation of experimental psychologists, such as Wilhelm Wundt, G.E. Müller, and Oswald Külpe, took the contents of consciousness and their laws of operation to be the primary subject matter for the new discipline. For this reason, functional questions that an evolutionary perspective would prompt, such as why mental operations had the particular character that they had, were deemed to be of minor importance.[1] Evolutionary considerations also had little immediate interest for those early physiologists and psychologists engaged in psychophysical measures of sensory processes or reaction time studies. Moreover, with their sole focus on the adult human mind, they also held at arm's length the comparative work of the day on animal behavior. In this light, it is unsurprising that there was also no apparent interest in evolutionary theory's offshoot, *ecology*, which was proposed in the 1860s by Darwin's early supporter Ernst Haeckel as "the science of the mutual relationships of organisms to one another" (Richards 2008, 27).

Psychology's indifference to evolutionary thinking began to change in the 1890s with the contributions of William James, James Mark Baldwin, John Dewey, and the subsequent emergence at the turn of the century of American functionalism. But this initial burst of functionalist attention to evolutionary thinking was short-circuited. James passed from the intellectual scene by 1910, and functionalism quickly fell out of favor among growing numbers of behaviorists who responded to John Watson's (1913) call for a positivistic, mechanistic, and reductionist psychology. James's psychology was lacking in all of these qualities. Baldwin was drummed out of American psychology around the same time (see Green this volume); and Dewey, having never engaged in laboratory experimentation, could be safely ignored. By 1920, even though species evolution was in place as a background assumption of psychology, much work in the field was directed toward developing an adequate learning theory with little regard to questions of a broader functionalist nature.

The term 'ecology' and its cognates seems to have appeared initially in psychology through the writings of Kurt Lewin (1943) on "psychological ecology" and Egon Brunswik (1955) concerning the "ecological validity" of research designs. While these uses conveyed the general connotation of the term "ecology," they were intended to serve a more limited purpose within the context of Lewin's and Brunswik's respective theoretical stances. For a broader use of "ecology" and proposals for an ecological approach, we must look somewhat later to the independently developed programs of the perceptual psychologist James J. Gibson (1966, 1979), the child/social psychologist Roger G. Barker (1968), and the developmental psychologist Urie Bronfenbrenner (1979). And yet none of these programs were embraced by mainstream psychology at the time they were formulated, and they remain on the margins of psychology today. Even with the resurgence of evolutionarily grounded thinking in psychology beginning in the 1980s, which was spurred both by sociobiology and by Chomskyian arguments for innate universal

---

[1] In his writings of the 1860s, Wundt gave considerable attention to Darwin's ideas, but by the 1890s Wundt no longer accorded them much weight when it came to mental processes (see Richards 1980, 6–61).

grammar, ecological approaches to psychology have received scant attention across diverse areas of psychology.[2]

How was it possible for psychology to assimilate evolutionary thinking early on, and by mid-century to reinvigorate its interest in such matters, while maintaining a century-long indifference to an ecological perspective? To answer this question, we need to turn back to the nature of psychology's early commitments to evolutionary thinking.

3   Spencer and James on Mind/World Relations

Although it might be assumed from today's vantage point that Charles Darwin's works were the primary basis for psychology's early embrace of an evolutionary outlook, this is not the case. Instead, it was Herbert Spencer's treatment of evolution, and not Darwin's, that had the greatest early impact on psychology. The prominent nineteenth-century Scottish philosopher/psychologist Alexander Bain, whose writings formed the backbone of many early psychology curricula, referred to Spencer in a letter to him as "the philosopher of the doctrine of Development, notwithstanding that Darwin has supplied a most important link in that chain" (quoted in Richards 1987, 244). Spencer had written copiously about evolution, publishing even before Darwin. His writings were mostly philosophical, indeed metaphysical in nature, rather than being rooted in fieldwork as were Darwin's. Nor had Spencer formulated the mechanism of natural selection. But his work was more influential within psychology for two reasons. First, although Spencer assiduously distanced himself from religious views, he proposed that evolution operates progressively toward "higher" adapted forms. Not only did this perspective serve a social purpose in seeming to justify social inequities of the day, but more critical for our aims here, his evolutionary progressivism fit comfortably with widely held religious views of salvation, held especially in the United States, than did Darwin's. Natural selection may have helped to explain species' adaptation to a changing environment, but there was little justification for viewing environmental change as being in any way progressive. Darwin's vision of undirected, contingent evolution was very much out of step with dominant cultural commitments at the time. Second, a related set of existing commitments were also at play. While Spencer's framework stretched received views in new and exciting ways, it did not challenge the meta-theoretical commitment to the matter/mind distinction that dominated nineteenth- and twentieth-century Anglo-American psychology. In contrast, Darwin's theory in the hands of functionalists such as James, Baldwin, and Dewey did just that.

---

[2]Evolutionary psychologists do claim James as a forbearer, but their use of James is highly selective. With some justification, they see their position anticipated by James' chapter on instinct in *The Principles of Psychology,* but they fail to evaluate this chapter within the corpus of James's writings. Failing to do so, they miss the central thrust of his perspective.

Anglo-American psychology's commitments to a dualistic distinction between the environment and the organism, and its early alignment with the associationistic British tradition, thus made Spencer's framework a relatively easy fit (Young 1970; Pearce 2010; Pearce this volume). Apart from the particulars of his sweeping analyses, Spencer's vision in the General Synthesis portion of his *Principles of Psychology* (1855) may be summarized as follows: "the life of every organism is a continuous adaptation of its inner actions to outer actions." Applied to mind, Spencer proposed that the workings of evolution result in mental structures that correspond to environmental conditions. This formulation offered a powerful template for thinking about the relationship between organism and environment, and importantly, with its reference to inner and outer relations, one that was in keeping with the prevailing mind-body/world dualism of psychological theory.

One might argue that after the rise of behaviorism and its commitment to positivism, such dualisms were out of play. However, although behaviorists rejected the mentalism of the "older psychology," they often tacitly preserved the long-held dualism between a physical world and a mental domain in their embrace of the stimulus and response dichotomy (Heidbreder 1933). As Dewey (1896) presciently pointed out, "conceptions of the nature of sensation and of action derived from the nominally displaced psychology are still in control. .... The older-dualism of body and soul finds a distinct echo in the current dualism of stimulus and response" (37). Behaviorism preserved a focus on the correspondence between environment and behavior. Therefore, while evolutionary thinking *might* have transformed psychology through Darwin's influence (as we will see), instead it was mostly assimilated into existing frameworks, with the new functional discourse cloaking long-held dualistic thinking. Most of the received theoretical assumptions of the discipline were left untouched but for superficial "functional" flourishes that we still see in much contemporary psychology.

4 William James as Psychology's Early Darwinian

Although Spencer's celebrity has dimmed since his death in 1903, the opposite may be said about William James. His contributions to psychology are cited today in nearly every corner of the discipline as being formative influences. And yet, there is something deeply ironic in this turn of events. Although James deserves much credit for directing psychology toward functionalist thinking, the dualistic Spencerian form that psychology took—and that is still apparent in much of the discipline—was not at all what James had intended. As early as 1878 James roundly criticized Spencer's approach of viewing organismic adaptation as a growing *correspondence* of inner to outer structures (James 1878). In *The Principles of Psychology* (1890), he expressed appreciation for Spencer's naturalism, in contrast to the "brass instrument" experimentalists, but described Spencer's correspondence stance as "vagueness incarnate" (1:19); and in his critique of automaton theory (Chapter 5) James identified some of its shortcomings. Specifically, James criticized any notion of a passive mind shaped mechanistically by the press of environmental influences, thereby rejecting correspondence theories. Mental life, a functional process of animate beings, is *directed activity* by a thinking and feeling organism. In James's hands, the nineteenth-century emergence of the life sciences as championed by Darwin offered the opportunity for a transformation of long held thinking: what philosopher Marjorie Grene described as "a revolution of life against dead [mechanistic] nature and of understanding as against the calculi of logical machines" (Grene 1974, 3).

The Spencerian conception of "mind as correspondence to nature" treats mind as a container-like domain partitioned from the natural world, and yet whose structure corresponds, to varying degrees, with the "outside" world. In contrast, James (1904) famously argued that mind was grounded in on-going, brain-based processes, and that consciousness is not a thing or a container, but instead is a *function*. The distinctive functional quality of a complex mind/brain for James was its selectivity: *"The pursuance of future ends and the choice of means for their attainment are thus the mark and the criterion of the presence of mentality* in a phenomenon" (James 1890, 1:21). It is on this very point that James's approach diverges sharply from Spencer's, and, I must add, from the many contemporary psychologies that unknowingly take a Spencerian, correspondence form. In his 1878 essay, "Remarks on Spencer's definition of mind as correspondence," James concluded,

> I, for my part, cannot escape the consideration, forced upon me at every turn, that the knower is not simply a mirror floating with no foot-hold anywhere, and passively reflecting an order that he comes upon and finds simply existing. The knower is an actor, and co-efficient of the truth on one side, whilst on the other he registers the truth which he helps to create. ... In other words, there belongs to mind, from its birth upward, a spontaneity, a vote. It is in the game, and not a mere looker-on. (James 1878, 17)

Along with its selective nature, then, James's account of mind emphasized its active character ("the stream of thought") in contrast to the passive 'mirror of nature' as proposed by Spencer. This emphasis on mind's active nature and James' conceptualization of "new modes of thought and conceptual innovations [springing] up in mind as spontaneous mental variations" (Richards 1987, 427) are consonant with a Darwinian view of natural processes teeming with *activity, variation and possibility*. Process and the production of variation are at the heart of Darwinian thinking, but not Spencerian correspondence thinking. And for James, when it comes to initiating action, the complex brain is "an instrument of possibilities, but of no certainties" (James 1890, 1:144).

> The dilemma in regard to the nervous system seems, in short, to be of the following kind. We may construct one which will react infallibly and certainly, but it will then be capable of reacting to a very few changes in the environment—it will fail to be adapted to all the rest. We may, on the other hand, construct a nervous system potentially adapted to respond to an infinite variety of minute features in the situation; but its fallibility will then be as great as its elaboration. We can never be sure that its equilibrium will be upset in the appropriate direction. (Ibid., 1:143)

The contingent view of evolution from a Darwinian perspective is echoed in James's account of action and thought.

How then, James asks, are we to explain the fact that actions taken by animals with complex brains are often appropriate to circumstances, which are themselves changing, and that on-going immediate human experience is typically focused rather than being overwhelmed by a flood of thoughts? His hypothesis is that "superadded" to complex brains is an "organ added for the sake of steering a nervous system grown too complex to regulate itself" (147). The function of this "organ"—what we would refer to today as executive functions associated with the prefrontal cortex— is to provide a means for *selecting* actions and thoughts (Donald 2001). Most of the time, complex organisms are not pushed around as mere automata, but they selectively engage their immediate experience which is marked by variation and possibility. And on the heels of conscious selection, the accrual of habitual action then follows (James 1890, chap. 4).

Having introduced selection as a hallmark of consciousness, James then must address an additional question: what is the basis for selection at the level of individual experience? James is clear that selection is guided by the *interests* of the individual. And these are not "interests" merely in the sense of a dispassionate, utilitarian calculus. Rather, interests coincide with *feelings*. Selection is based on what is felt to be significant for us, for good or ill. In this respect, James's account of cognition differs from the path followed by much of later cognitive psychology, which up until quite recently has neglected the place of feelings in cognition (Damasio 1999). For James, objects of thought are suffused with feeling. And it is essential that they are, otherwise on what basis do we select one among the many? Complex organisms are not "mere lookers-on," or in Dewey's words, mere "spectators." Mind is always "in the game," and what the individual cares about matters in ways that are entirely absent from Spencer's account, as from much of contemporary cognitive science. If psychologists hope to understand the dynamic character of organism-environment processes, they cannot neglect this fact, which in turn demands that they keep the selective and committed character of organismic functioning at the forefront of any analysis. Keeping the selective nature of functioning central compels us to reformulate how we conceptualize the *environing context* of thinking and action, and in turn, both thinking and acting themselves.

These reasons for his rejection of Spencer's traditional stance distance James's thinking from the received views and thus, they help to explain why James's writings had less of an impact on early psychology than is often supposed by contemporary psychologists—and indeed why the full effect of James's perspective has yet to be felt broadly in psychology, in spite of widespread assertions to the contrary (Heft 2001). These claims will likely come as a surprise to many contemporary psychologists for whom James is an icon and Spencer a relative unknown. One source of the problem here is the widespread neglect of James's later writings, especially his proposal for *radical empiricism* which often is wrongly dismissed as merely philosophical (e.g., Leahey 2000; Mandler 2007; Hergenhan 1988; for exceptions, see Viney 1989; Crosby and Viney 1993; Heft 2001). As we will see, James's earlier psychological writings, viewed from the perspective of his radical empiricism and its thorough rejection of a dualistic view, remove any ambiguity as to where his theoretical commitments lay (Perry 1935).

I submit that psychology's embrace of a Spencerian evolutionary stance, which blends evolutionary thought with British Empiricist associationism (Young 1970), instead of an intentional approach along the lines of James's radical empiricism (see below), has made it impossible to move beyond dualistic thinking that establishes a sharp divide between the environment and psychological functioning. And saddled with that dichotomy, psychology has been unable to adopt the kind of dynamic, relational thinking that James advocated, that pervades Darwin's writings, and that has come to characterize the *ecological sciences*.

In addition, most modern psychological theory, particularly in the areas of perception and cognition and those subfields that draw on these areas (e.g., social psychology), is rooted in modes of thinking derived from mechanistic Newtonian physical science, the foundation of seventeenth- and eighteenth-century British Empiricism. But in the past two centuries, theoretical biology has gradually taken on its modern, dynamic form; and physics has long left behind a reliance on a Newtonian framework to embrace dynamic models. For the most part, however, psychology has remained rooted in seventeenth-century mechanistic thinking originally developed for the analysis of the inanimate, although important signs of change are afoot (e.g., Kelso 1995). For a discipline such as psychology that is concerned with the study of the animate world, a mechanistic commitment would seem to be an impediment to its development. The life sciences, and more specifically, the ecological sciences, are surely a more suitable home for psychology (e.g., Reed 1995; Herrman 1998). Psychology's long-standing commitment to mechanistic thinking, and much that it entails, helps to explain psychology's indifference to ecological thinking—to which we now turn.

5   The Hallmarks of an Ecological Science

Ecological science is concerned with dynamic, interdependent processes constituting natural systems—*ecosystems* for short—that are comprised of living and nonliving things. They are *dynamic* in the sense that ecosystem processes are continually in flux even as they function to maintain the stability of the system as a whole: that is, they are *quasi-stable systems*. Ecosystems are *self-organizing*, in the sense that their dynamic structure arises from the on-going interrelationships among their constituents. The quality of *interdependence* that is characteristic of ecosystems means that any living constituent of the system is viable only as a participant in that network of reciprocal, interdependent processes over time. From an ecosystems perspective, it is apparent that living things function neither in isolation as self-contained units nor as individual passive entities that are solely shaped by "external" influences. Rather, living things viewed from an ecosystems perspective are *active participants* in a network of reciprocal influences. This network of ecosystem interdependencies is in place as a result of both the individual and the joint *histories* of its constituent processes. Owing to their partial co-evolution, ecosystem constituents are functionally interconnected, rather than separate and autonomous. Finally, from an ecological perspective, natural systems tend to be structured as *nested hierarchies*, at successively more macro and micro levels of organization.

Ecological approaches to psychology are relatively rare because, as was already noted, many foundational ideas in psychology are grounded in long-standing, mechanistic and dualistic commitments that run counter to such a view. John Dewey (1920) repeatedly took psychology to task for adopting a Cartesian, "spectator" approach to knowing, whereby psychological processes stand apart from the environment. Instead, he advocated for a transactional psychology built on the idea of the functional, interdependent, and on-going reciprocity of environment and individual processes. A dichotomous way of thinking about environment and persons has persisted in various forms through much of experimental psychology's history, most notably in the dichotomies between stimulus and response, and between input and output. This is one of the reasons why computer models have been so readily assimilated into psychological theory. This framework promotes uni-directional, linear causal thinking, usually proceeding from environment to organism, rather than the systemic, reciprocal, mutually sustaining relationships characteristic of the ecological sciences.

Ecosystems thinking cannot be assimilated into a mechanistic, dualistic theoretical perspective. What then would a psychology look like that is informed by an ecological perspective? One way to begin to reconstruct an ecological approach to psychology is by turning again to William James's view of selective functioning, this time in the context of his radical empiricism. From there we will examine the work of one of his last students, Edwin Bissell Holt (1873–1946), who was an ardent proponent of James's radical empiricism and a graduate school mentor of James Gibson.

6 Radical Empiricism

6.1 The Field of Immediate Experience

William James (1904) proposed that the study of psychology begin with the flux of immediate awareness ("pure experience") that is neither objective or subjective. "Pure experience for me antedates the [objective or subjective] distinction ... Its determinations are all retrospective, drawn from what it develops into" (James to Warner Fite, 3 April 1906; cited in Perry 1935, 2:351). Pure experience is characterized by a relatively undifferentiated field of possible relations, a multiplicity of "sensible natures" (James 1912, 15), "a *that* which is not yet any definite *what*, tho' ready to be all sorts of whats" (46). The metaphysics of radical empiricism assumes from the outset a network of relations; and although James did not explicitly describe it as such, this view has the character of a field theory. These relations are poised to be differentiated *selectively* by one portion of the field, which will eventually be realized as the knower. Selective action is a psychological manifestation of a defining property of animate beings, namely, *agency*. Immediate experience, then, is a pre-reflective awareness of a field of relations; and the distinction between self and world, as well as the particulars of the world, stem *a posteriori* from selective engagement with that field.

This beginning contrasts both with Cartesian rationalist thinking which starts with self-awareness of an isolated thinker and then traces the logical grounds for subjective experience of an external world, and with the traditional British Empiricist approach which begins with particulars in mind (e.g., things, impressions, ideas) assumed to correspond partially to particulars in the world. In the latter tradition, an approximation of the order of the external world is established in the mind in a "bottom-up" fashion through laws of association grounded in the contiguities of experience.

In James's view, immediate experience consists of things and their relations; "the relations between things, conjunctive, as well as disjunctive … [are] as much a matter of direct particular experience, neither more so or less so, than the things themselves" (James 1904, 173). What is critical in James's account is that relations are experienced rather than imposed on experience *a posteriori* as a result of exposure to contiguous spatial and temporal patterns. The semi-structured, "quasi-chaos" of James's world of immediate experience is reminiscent of Darwin's "entangled bank" with its network of relations and structure available to be discovered by the naturalist. With this rich account of immediate experience, functioning adaptively necessitates selection and discovery of latent structure in the world, and as a result the development of an adaptive interconnectedness of knower and known, rather than the joining together of disparate atomistic parts of experience, and the wholesale construction of a mental model of the world that exists apart from the world itself.

6.2   Holt's Molar Behaviorism

The selective nature of acting and thinking was developed further by E. B. Holt, who succeeded in bringing these Jamesian proposals into the arena of behaviorist thought. Like other early behaviorists, Holt distanced himself from those experimental psychologists who took the *content* of consciousness to be their central concern. Following his mentor, Holt took consciousness as a function of an active organism, as something an organism does in an environing context. Like other bodily functions, such as breathing, digesting, or walking, consciousness was a *process* rooted in the biological operations of the body in an environment. Hence, in invoking consciousness, he was not positing the sort of non-material "spooky" stuff that haunted reductive behaviorists such as Watson. Consciousness was a function or activity of a complex biological organism, one among many biological processes viewed from the standpoint of the organism as a whole functioning in its environment.

Holt further diverged from the Watsonian brand of behaviorism that viewed behavior as being both *reducible* to component reflexes and *triggered* by environmental influences. In the place of this reductive and passive view of the organism's response to environmental stimuli, Holt's behaviorism was both non-reductive (molar) and purposive, with the integrated actions of the organism as his starting point. The positivist Holt transposed James's stream of thought, things and their relations, into a flow of integrated, directed responses.

6.3 Adient Responses

In the spirit of positivist behaviorism, Holt felt that the goal of any natural science is to identity lawful and specifiable relationships in its domain of study. In the case of psychology, predictive relationships linking environment and action were to be sought. However, unlike the mechanistic approach of most behaviorists who looked for linear, causal relationships along the lines of stimuli evoking responses, Holt followed James and Dewey in calling for the study of functional relationships characterized by organisms systematically directing their actions toward some referent object. In a Jamesian vein, Holt urged that we take an individual's behavior, not as a reaction to the prodding of a stimulus, but fundamentally as being directed toward a stimulus. That is, he distinguished behavior (action) from mere reflexive responses, with the hallmark of action being its character of *adience* (Holt 1931). By 'adience,' Holt means that behavior or, better, action, is by definition an "out-reaching, outgoing, inquiring, examining, and grasping" response, not a mere reaction to impinging stimuli (Holt 1931, 41). Moreover, because action is an out-reaching, it has a referent, an "object" toward which it is directed, and, crucially—and this point cannot be over-emphasized—*the referent of directed action is enfolded within action itself.*

Such behavioral "out-reaching" is not to be understood reductively as the activity of individual, separate muscle "twitches," or single reflexive actions, but as a directed *integration of responses* by an organism considered as a *whole* biological entity.[3] In Holt's view, it is only when we consider such integrated, adient action that we are truly led from the study of biological questions to the study of the *psychological* functions.

It is important to see the underlying Jamesian relational metaphysics that Holt is intending here. Although in *The Principles of Psychology* (1890) James equivocated concerning his position on mind-body (or mind-world) dualism, by the late 1890s— the time when Holt was his student—James had thoroughly rejected dualism. James's philosophy of radical empiricism was intended to provide a *relational* meta-theoretical alternative to dualistic models. Holt, later followed by his student James J. Gibson, continued to work in this vein (Heft 2001).

---

[3]Later in the century, Merleau-Ponty (1963) refers to such integrative, holistic, directed actions as "the body's projects."

The passive view of the organism that accompanied many behaviorist models has, for the most part, faded from view; but its dualistic baggage remains. For decades now, cognitive scientists have proposed a multitude of active mental processing models, all the while retaining a mind-world dichotomy. Many of these models tend to have the character of top-down processes that control action and thought, rather than more passive, bottom-up associative models. In spite of this noteworthy shift from a passive to an active view of mind, such models preserve a dichotomy in thinking, and as such continue to impede our understanding of how actions and environmental contingencies are interwoven.

To see how this is so, let us consider an influential example of top-down control— namely, a *mental script*. These considerations will also prepare the way for a comparison between a dualistic approach to thought and action, and a relational, ecological view to be taken up later in this chapter. A script is a mental represen- tation (schema) consisting of "a predetermined, stereotyped *sequence of actions* that defines a well-known situation" (Schank and Abelson 1977, 41; emphasis added). Although script schemas were posited principally as serving language processing, they are also assumed to control action. They have been defined in an authoritative reference as "the subclass of schemata that are used to account for generic stereotyped sequences of action" (Brewer 1999, 729; see also Schank 1999, chap. 10). Taken as such, scripts specify the appropriate sequence of actions that individuals follow in particular situations—a prescribed set of decision rules that control action. While their activation is initially triggered when the individual finds herself in a certain setting (e.g., a restaurant), the subsequent actions are controlled ("top-down") by the script. Knowledge of this sort that guides situation-specific action is self-contained "within" the operational structure of the mental schema, and the decision-rules *pre-exist* the individual's entering the setting. It can be seen that although mind is taken to be active, a dichotomy between environment (the situation) and mind (the script) is retained. The character of the environment does little more than trigger the pre-established operations of the script, and perhaps switch it off if actions go awry.

Adient responses, as Holt conceives them, differ from mental schemas such as scripts. As already noted, functional properties of the object toward which an integrated action is directed are themselves *constituents* of that action (Holt 1915, 55). Action, seen from this point of view, has a relational character. Research on infant reaching and grasping over recent decades provides support for this view. Grasping is not pre-figured, but by five months reaches are adjusted in the course of action (von Hofsten 1991; Butterworth et al. 1997); and they involve more than "micro-movements" of the hand, involving a *prospective* organization of the whole body (Rochat and Senders 1991), e.g., preparatory adjustment of the abdominal muscles and the back (Gibson and Pick 2000). These findings are lovely realizations of what Holt meant by behavior being adient. Action is *on-going* (akin to James's stream of thought), not the "switching on" of a pre-existing set of linked actions (Dewey 1896). Action is continuously adjusted *in the course of activity* with the referents of action among its constituents. In the flow of experience (James's starting point), features of the environment toward which action is directed are enfolded within the action, as our examples above illustrate, blurring the boundary between world and mind. The focus shifts accordingly from an environment and behavior dichotomy to an integrated environment-action function within an eco-system-like field of possibilities.

6.4   The Place of Meaning

One of the problems arising from environment-mind dualism is how to account for the apparently *meaningful* character of the environment. To take a comparatively simple case, features of the environment from the point of view of human (first-person) experience are functionally meaningful ("the sofa looks to be a good place to sit"). But for a theorist working from an environment-behavior (mind) dichotomy, where is meaning to be located? There is no place for meaning among the physical and mathematical descriptors typically employed by psychologists to characterize environmental properties. For this reason, meaning is usually considered to be a subjective quality imposed on the environment by individual minds. This point of view introduces no end of philosophical problems (James 1904; Rorty 1979; Putnam 1999), not the least of which is the postulation of so many private, intra-subjective domains with little obvious means to connect to the environment or to each other. Faced with this circumstance, adaptive functioning seems nearly miraculous.

If, however, features of the environment are enfolded in the course of action, then it might be possible to locate meaning in the dynamic of the environment-individual relationship (Johnson 2007). This possibility comports with the concept of *affordance* as developed by James Gibson (1979).[4] Affordances are the functional possibilities of the environment *for an individual*, and for this reason, they must be specified relationally. An object affords grasping for an organism that has a prehensile appendage, such as a hand or a tail, in those cases where the diameter of the object is smaller than the span of the appendage. The *meaning* of "being-graspable" is a joint property of the environmental feature and the organism's possibilities for action. The concept of affordance has generated a substantial body of empirical research (see Heft 2001; Heft and Richardson, in press). This is familiar ground to anyone acquainted with Gibson's ecological approach.[5]

Affordances, in principle, can be specified with reference to stimulus information scaled relative to an individual's action capabilities (Warren 1984). A detailed explanation of Gibson's idea of information is beyond the focus of this chapter (see Gibson 1966, 1979); but suffice it to say that, e.g., in the case of vision, information is carried by an array of reflected light which is structured by the surface properties of the environmental layout and its features and is most readily revealed through

---

[4]See also William James's (1905) essay "The Place of Affectional Facts in a World of Pure Experience."

[5]For an application of the concept of affordance to fitness landscapes, see Walsh (this volume).

actions of the perceiver. In short, a perceiver is embedded in a field of potential information (structure)—an instantiation of James's characterization of the field of immediate experience.

To summarize, we have tried to lay out some of the groundwork for an ecological approach to psychology. We began with a Jamesian field of potential structure (immediate experience), and proposed that mental functions operate selectively to reveal some of that structure—there is a selective awareness of structure. Translating this account into less mentalistic language, this structure is carried in the array of potential stimulus information (e.g., Gibson's ambient array), and the act of information pick-up has, in Holt's terminology, an adient quality, which means that the referent of directed action (that which the actor is engaging) is enfolded within action itself. Action, in other words, is not an isolated response that operates independently of context, but it is structured with reference to context. It has an intrinsic relational character that is characteristic of any facet of an interdependent (ecological) system.

This framework raises a new question that has received little attention, and it is one that will eventually allow us to draw Barker's ecological program into the discussion. If we are to anchor an ecological approach to the claim that action is selectively directed to some features of the environment, with the character of the referent a constituent of the action, what is the possible *scope* of directed action? Simply put, how "far out" beyond the skin of the organism, both distally and temporally, can adient action be directed in order to account for the meaningfulness of human experience?

7 The Scope of Directed Action

7.1 Recession of the Stimulus

The ecosystems perspective locates organismic functioning in a wide network of natural phenomena. How can we apply this stance to psychological functioning? A long-standing impediment to doing so is the heritage of a mainstay of psychological discourse, the term 'stimulus.'

Before 'stimulus' was stretched imprecisely to apply to *anything* outside of the organism that has a causal influence on it, the term appropriately referred to that which immediately initiated neural activity at receptor surfaces (Gibson 1960). For example, a sudden withdrawal of my arm could be traced to something that just pricked me, that stimulus initiating neural activity from receptors in my skin. From a "stimulus" perspective, the specific *source* of this sensation is inconsequential because my resulting movement is the same, whether the prick be due to a pencil point, a syringe, the tip of a knife, or what have you. But consider cases when the individual experiences a sharp prick on the arm and resists withdrawing it, such as when the individual is receiving an injection from a physician or an adolescent male is demonstrating his toughness. Here, the source of the skin prick and the context (i.e., its *meaning*) are essential for understanding the action from a psychological standpoint. This action (resistance to withdrawal) is only understandable if it is viewed with reference to a source in context that is more distal than stimulation at the receptor surface. Considered psychologically (but not neurophysiologically) the skin prick alone recedes in its functional significance. This is Holt's important and yet much neglected concept of the "*recession of the stimulus.*" The response is not merely to the skin prick but to this sensation in context, or to use more modern terminology, it is a *situated action*. Moreover, the response is not merely a simple reflexive movement of a limb, but a result of crosscurrents of integrated, adient actions specific to, indeed "into," the situation.

> As the number of component reflexes involved in the response increases [i.e., when we consider integrated responses], the immediate stimulus itself recedes further and further from view as a significant factor. (Holt 1915, 76–77)

This point has significance from an evolutionary standpoint because it comports with the view that *selection* (in the sense of natural selection) must be understood with respect to integrated action by the whole organism, rather than piecemeal stimulus–response sequences. The focus of *integrated* action, which always extends beyond the body boundary, generates behavior that may have adaptive significance for the individual. Although it may be adaptive to withdraw from any stimulus that causes pain, it may also be adaptive in the long run to inhibit such responses by attending to more extended meaningful structures in the situation. In short, to understand selection pressures at work in relation to animate life from a psychological perspective—and human life, in particular—we need to be attuned to the "scale" of the environment that is relevant to the functioning of the whole organism at a particular time.

7.2   Situated Action

Let us clarify further Holt's idea of adience, and in doing so explore some of its wider implications. The focus of an adient (integrated) action is some feature of the environment beyond the body surface (i.e., the sensory receptors). He offers the following example:

> [T]he hen has got a retinal image of a hawk and she is clucking to her brood—shoot the hawk and [/or] remove the brood and she stops clucking, for she is reacting to neither one nor the other, but to a situation in which both are involved. (Holt 1915, 161)

Whereas the presence of the hawk, taken alone, might trigger a simple response by the hen, such as fleeing or freezing, in the presence of her brood it results in an alarm call. The reaction to the predator is conditional on other situational features of the environment that are also present. Its functional significance is established within a field of features. Integrated action is adient with respect to a set of relationships, including the perceiver. Or more directly put, action is *situated*.

The previous example was a comparatively simple case of situated action. Let us expand our consideration of what might "count" as part of a situation in the case of human action. In the context of a discussion about the need for researchers to determine the referent for action, Holt illustrates anecdotally how the focus of action may go beyond the immediacy of "present" circumstances:

> [T]he man is walking past my window; no, I am wrong, it is not past my window that he is walking; it is *to* the theater; or am I wrong again? Perhaps the man is a journalist, and not the theater, nor yet the play, but the 'society write-up' it is to which the creature's movements are adjusted; further investigation is needed. (Holt 1915, 161–162)

This example obviously introduces far more complexities than the hen-brood-hawk case, not only because the situation here is more extended temporally, but also because the psychological significance of the situation is embedded within a sociocultural framework. Indeed, these two factors are interrelated; and the ecological contexts for human action would appear to be different from other species in these respects. Human intentional action can have as its focus some end point that is temporally quite remote because action can be symbolically mediated. For some species, the end point of actions can also be temporally remote (e.g., long-distance migrations), but there the similarity to many symbolically-mediated human actions ends. Most human action is adient with respect to symbolic and culturally specific meanings at varying levels of scale and time. Holt did not extend his analysis of behavior in sociocultural contexts beyond this single, hypothetical example.

In the remainder of this chapter, and for the purposes of contributing to an ecological psychology, we will explore the idea of situated action within a sociocultural context from an ecological standpoint. The central idea here is that action taken at the scale of the whole organism is best understood in relation to a field of functional possibilities. In the human case, many of these functional possibilities are embedded within a network of sociocultural significance.

Gibson's concept of affordance described above would permit such an analysis, as my previous writings have described in various ways (see Heft 1989, 1990, 2001, 2007). For that reason, we will consider only one example here to illustrate the claim. In the ecological psychology literature, a commonplace example of an affordance is a sit-on-able-surface, or simply stated, a seat. A horizontal surface is perceived as affording sitting-on if approximately at knee-height relative to the perceiver. In circumstances when the individual desires a respite from standing, he may utilize this affordance. Now consider a case of an individual walking around a museum and encountering a display of furnishings from the eighteenth century. In spite of the fact that the visitor may be weary from hours of walking, a chair on display will not be perceived as a place to sit, even if it meets the material and relational criteria of a sit-on-able-surface. The reason why the chair would not be perceived as a place to sit is plain. In human experience, objects are never encountered independently of a sociocultural context. Here the relational character of affordances extends (in this case) beyond the body-scaled attributes of the object in question to include a wider set of contextual relations. The action is only understandable when the reference for action is taken in context, and in the case of human action, this context encompasses a field of sociocultural practices.

Admittedly, in this example reference to a context of sociocultural practices involves a good deal of hand-waving. In an effort to be more concrete, let us return to Holt's notion of "the recession of the stimulus." Situated action occurs with reference to affordances that are themselves embedded in a sociocultural context. In this respect, action would be adient with respect to wider set of *environmental structures* arising from socially-derived practices within which affordances are embedded. If we are to maintain an ecological approach, such structures should be specifiable in the contexts of everyday action, and not merely intra-subjective mental states. Moreover, action will be adient with respect to these more encompassing (higher-order) structures. But what are these ecological structures?

Barker's neglected ecological program in psychology will help us to make a first run at this question, as well as providing further groundwork for an ecological approach to psychology.

7.3   Behavior Settings

In the mid-1940s, the child psychologist Roger G. Barker established a field research station in a small Kansas town for the purposes of studying "human behavior and its environment *in situ*" (Barker 1968, 1). Although such a step is altogether commonplace in field biology, with its long historical ties to natural history, it was highly unusual in psychology at mid-century, and remarkably, it remains so today.

Barker's research team observed a large number of children individually as they went about their activities over the course of a day. Behavior in everyday settings has some degree of orderliness to it. Generally, children (and adults) act in ways that are appropriate to the settings that they enter, and violations from normative practices are relatively rare. What might account for this pattern of activity? The two usual candidates are causal influences in the environment, such as directives from others, and intra-personal factors, such as traits or personality. In keeping with behaviorist thinking in American psychology, Barker anticipated that most of the behaviors of the children would follow, in a stimulus–response fashion, from the antecedent actions ("social inputs") of others. To his surprise, these inputs were found to be unreliable predictors. Although a sizeable proportion of behaviors followed immediately from "social inputs" (40 % being a rough estimate overall), much of the time they did not. When instead Barker considered intra-personal attributes as predictors of behavior *in situ*, he found that they were not very helpful either. The data pushed him to recognize the need for a higher level of analysis than either the environment or the person.

> We found, in short, that we could predict some aspects of children's behavior more ade-
> quately from knowledge of the behavior characteristics of the drugstores, arithmetic classes,
> and basketball games they inhabited than from knowledge of the behavior tendencies of
> particular children. (Ibid., 4)

In other words, the best predictor of child's overall pattern of behavior was "where" she was—the setting in which she was a participant. Knowing "where" she was provided better grounds for predicting her behavior than identifying antecedent actions of other individuals or knowing about her individually.

The data indicated, in short, that a child's actions at any given time occur within extra-individual dynamic structures that Barker called *behavior settings*. Behavior settings are constituted by the joint actions of their participants in conjunction with the physical milieu (affordances) of the setting (see Barker 1968; Schoggen 1989). For example, the behavior setting of a rehearsal of the school orchestra would be constituted by the actions of the various members of the group, including the music teacher, as well as the instruments, music stands, sheet music, chairs, etc. Behavior settings are dynamic, quasi-stable patterns of action and milieu that have a specifiable geographical location, as well as temporal boundaries (starting and ending points). The actions of individuals as participants in a behavior setting, as a matter of course, contribute to its constitution, and their actions are situated with respect to their place in the setting. Barker and his colleagues went on to study the properties of behavior settings in a variety of ways, resulting in a remarkable analysis of these "eco-behavioral" dynamic structures that accounted for a great deal of the order we see in everyday social action (e.g., Barker and Wright 1955; Barker and Gump 1964; Barker 1968, 1978; Barker and Schoggen 1973).

From a Holtian perspective, the findings that a child's actions were only weakly related to immediate "social inputs" would suggest that the proper environmental referent for the action had not yet been identified—a conclusion Barker reached as well. In Holt's terminology, proper analysis requires that we recognize a "recession of the stimulus" when considering increasingly complex actions. Barker claimed that the referent for action was the behavior setting itself, or perhaps more accurately stated, the *opportunities and affordances available within the context of a behavior setting*. To borrow an example from Barker's data, a young girl (Maud) and her mother and brother were observed in a drugstore. Approximately 60 % of Maud's individual acts were not prompted by anything her mother, brother, or workers in the store said or otherwise did. Indeed, many of her actions ran counter to their directives. But at the same time, these same actions were uniformly consistent with what was normative for that setting (see Barker 1968, 146–151). As a participant/customer in a drug store, Maud browsed through the magazines, sat on a stool at the soda fountain, etc., without being prompted by her mother to do so, while refraining from running through the store or transgressing in other ways, even in the absence of admonitions not to do so. That is, while the child engaged in a range of activities with seeming independence of immediate social input, for the most part those activities were consistent with actions *normatively appropriate* to drugstores *qua* behavior settings. The set of actions was adient with respect to the behavior setting, if not always responsive to social inputs.

In this instance, as with Holt's example of the journalist, action was structured over an extended duration as an individual engaged particular affordances within a behavior setting. Indeed, the reason why behavior settings are reliable predictors of behavior is because *behavior settings are constituted, in part, by activities of their participants*. Individual actions are structured with respect to the social practices that define the setting. Returning to our example, the drugstore functioned as a drugstore because its participants, including Maud, acted in ways that made the very functioning of the behavior setting possible. To see why this is so, consider those very rare instances when an individual behaves in non-normative ways. If Maud began running and yelling through the drugstore, it would cease to function qua drugstore until corrective measures were taken, such as removing her from the setting. Such disruptive actions would be seen as such because of their non-normativity. But what is less obvious, and often overlooked, is the fact that situationally normative actions contribute to the very existence of a behavior setting in the first place. We can also state this point in terms of Holt's idea of adience: *situationally normative action is structured by its referent*. And in this case the referent is a higher-order (extra-individual) structure of the environment.

We can see with this example that Barker ultimately shifted from what was, in effect, a causal approach to environment-individual relations to a more Jamesian approach. Environmental factors did not "cause" or evoke behavior; instead, individuals selectively engaged particular settings ("mind is in the game"), and their actions were constrained by virtue of their participation in behavior settings. This observation opens up possibilities for understanding individual psychological processes in context, leading to the question raised earlier: How can we begin to understand psychological processes from the standpoint of situated action and thinking?

7.4 Scripts or Situated Action?

Barker insisted that the principles accounting for behavior setting dynamics were quite different from those that account for individual actions. What led him to hold this view? Behavior settings are higher-order (extra-individual) complex systems—or in Barker's terminology "eco-behavioral entities"—and "the reality and the nature of behavior settings as eco-behavioral entities *do not reside in psychological processes* of the inhabitants, but in the circuitry that interconnects behavior settings, inhabitants, and other behavior setting components" (Barker 1968, 174; emphasis added). A discussion of behavior settings as complex systems is beyond the scope of this chapter.

But what can be said about the "psychological processes of the inhabitants"? Barker did not offer much in the way of an account of psychological (individual) processes in the context of behavior settings. But one of Barker's last students, Alan Wicker (1987), proposed that individuals possess script-like knowledge of the type described earlier in this chapter. Wicker writes, "in some ways, the script appears to be a rediscovery of behavior setting programs from a quite different approach than Barker's" (624). In his invaluable update of Barker's seminal book, his long-time collaborator Phil Schoggen (1989) also cited scripts (among other constructs) as having much in common with the behavior setting analysis. However, Schoggen is less ready than Wicker to embrace scripts, because of their nearly exclusive focus on mental structures without due attention to the environment as well. That criticism is apt. He further suggests that "script analysis would be greatly strengthened by including a more systematic representation of the objective physical and social world, such as that provided by the study of behavior settings" (321). The "more systematic representation ... such as that provided by ... behavior settings" that Schoggen calls for would surely involve more than giving greater attention to environmental features in a script analysis, as I take Wicker to be suggesting. In script-like (schema) models, action is controlled by intra-psychological processes in a top-down manner, from cognition to action. Environmental features are treated as supplemental factors that cue or trigger cognition and action rather than being integral and partially constitutive of these processes. With this view, a sharp dichotomy between environment and mind is assumed. In contrast, we have proposed following Holt's suggestion that psychological processes should viewed in relation to their referent—situated action (adient responses) includes environmental features as constituents of action.

Several of the inadequacies of schema-like, intra-psychological models are revealed when we consider the dynamic character of an individual's on-going adjustments *during the course of participating* in behavior settings. First, a script, like any program, is a set of pre-established decision rules that guide action, and as such, once triggered, scripts seem to "run-off" without any obvious means of fine-tuning and adjusting actions "on-line" over time. For this reason, a script is not very flexible. As Schank (1999) acknowledged, "one disadvantage of the script-based method is its lack of usability in similar but nonidentical situations. Reliance on scripts inhibits learning from experience" (12). Scripts are of value only in familiar and highly routinized situations. Like mental representations generally, they are lacking in generative possibilities.

But actions in situations are typically more flexibly dynamic than scripts permit, with choices needing to be made "on the fly" with the interweaving of situational events and one's own actions. Indeed, actions in settings seem more improvisational than rule-governed. The actions of the young girl in the drugstore, for example, are not planned in any rigid manner, but improvised as environmental features (e.g., the magazine stand) are encountered *seriatim* and as others act in the setting. And yet her actions are regulated nonetheless. They are *constrained* by her *practical* understanding of the normative limits in this specific behavior setting. The challenge then for psychological theory is how to conceptualize psychological functioning in such a way that simultaneously captures both the *structured* and the *flexible* character of situated action and thought. Scripts, as they have been conceptualized in cognitive science, will not fill the bill.

Second, a script theorist might argue that a sequence of prescribed actions can be terminated if conditions change, and a new script can take its place. But such a system would be unworkable, because it would call for a seemingly endless number of scripts, "layer upon layer of increasingly minute plans exhaustively controlling our every move" (Leudar and Costall 1996, 160). To draw an analogy in the area of language development, it was similar problems with associationistic explanations of language learning that made a generative linguistic theory so appealing for many psychologists.

What might be an alternative to a script-driven explanation of individual actions in a setting? Let us turn once more to E. B. Holt's concepts of adience and integrated action, seeking a better grip on the nature of situated action.

7.5 *"Specification by Superposition"*

Adience, as we have seen, refers to the essential quality of action which is an "out-reaching, outgoing, inquiring, and examining, and grasping" response (Holt 1931, 41). It is not an aimless out-reaching, but a directed and sustained act such that the properties of the referent partially structure the action. For this reason, action has "a limited locus of freedom" (218)—it is subject to action constraints, with some activity possibly varying within those constraints. In other words, there are degrees of freedom within constraints. Holt illustrates this point with the example of a small child grasping a ball: "one will often notice the little fingers bending and unbending while *retaining the contact* between the tip of each finger ... and the ball" (215). There is exploratory movement within constraints: "In general, every sustained adient reflex ... defines a limited *locus of freedom* for random movement" (218, emphasis added).

Still this example is a bit too simple because it fails to capture either the complexity or the dynamic character of what Holt had in mind. As for the complexity, voluntary actions rarely are simple motor responses, but instead involve an integration of multiple action tendencies. If so, then the "locus of freedom" would be established by these converging action tendencies.

> If, then, several adient reflexes are *simultaneously (and continuously) operative* [emphasis added], the actual locus of freedom is reduced to that more limited range that may be *common* to the several loci defined by the several sustained reflexes. The freedom of the organism is more sharply defined and limited by every additional response that is actively maintained. (218)

As for its dynamic character, on-going adjustments bring different combinations of action tendencies to bear on the environment over time. There are overlapping patterns of action shifting kaleidoscopically in attunement with on-going changes in circumstances. For example, as one reaches for an object there is a continual fine-tuning of the reach with respect to the object's position relative to the body and continual fine-tuning of the grasp as the object shape is assessed.

Perhaps we can shed further light on this conceptualization by considering a psychological model proposed by Friedrich A. Hayek (1969).[6] Hayek follows a different path then that taken in much of cognitive science. He rejects the view that perceptual experience and action are composed, "bottom-up," of particulars or component units (sensations and micro-movements, respectively). Nor does he advocate "top-down" models (e.g., schemas) that impose structure from "above" on particulars. Instead, he envisions activity as fundamentally a process of classification or pattern recognition. It is critical to see that for Hayek classifications of patterns are very broad or "abstract." Here I will focus solely on Hayek's account of action.[7] The abstract classifications can be thought of as very broad action tendencies. They are lacking in specificity, instead being inclinations to act in a certain broad manner and, as a matter of course, they exclude other ways of acting. It is the superposition of numerous broad action categories that would give rise to structured action, with some degrees of freedom.

> A disposition to act will be directed towards a particular pattern of movement only in the abstract sense of pattern, and the execution of the movement will take one of many different possible concrete forms *adjusted to the situation* taken into account by the joint effect of many other dispositions *existing at the moment*. (Hayek 1969, 314–315, emphases added)

In other words, the particulars of action and experience are "the product of a super-imposition of many 'classifications'" (310). The conceptualization is quite similar to Holt's proposal of the "locus of freedom" in behaving.

I hope it is apparent how this model of action is sensitive to the very issues that the script schema neglects—namely, how on-going actions can operate in an ad hoc, and yet regulated fashion over the course of a sequence of actions. Rather than a prescribed program (e.g., a script) being triggered by environmental input, resulting in an ensuing commitment to a sequence of actions, actions are produced through the overlapping of classes of action. Because the specificity of the action is attributable to the joint convergence of abstract classes, the particulars of action can be adjusted "on the fly" with the inclusion of additional abstract classes and the elimination of others. Hayek referred to this structurally flexible process as "specification by superposition" (322).

Let us speculate about how this model might be applied to the data reported earlier from Barker (1968). Observations of Maud in the drugstore showed that her actions could only be weakly predicted by antecedent social "inputs," such as instructions by her mother. It would appear that social inputs are not reliably triggering action. Alternatively, one could suppose that Maud is simply executing a series of rule-governed ("top-down") actions, perhaps a sequence of mini-scripts specific to a series of activities within the broader drugstore script. *Post*

---

[6]Hayek is far better known for his work in economics, which has drawn much public attention and notoriety of late. His interest in psychology began early in his career, and the principal link between his economic and psychological writings is a concern with the nature of complex systems (Weimer 1982; Beck 2009).

[7]For his account of perceiving, see Hayek (1952).

hoc it might seem that she is following a pre-established plan, and no doubt, individuals do enter places with a measure of pre-meditation. However, assuming the operation of prescribed action eliminates the measure of spontaneity that seems always to characterize action-in-context. Possibilities for spontaneity (creativity) are preserved if we view actions has being governed by a dynamically changing set of overlapping constraints than rather than by prescribed rules.

> But the rules of which we are speaking ... will often merely determine or limit the range of possibilities within which choice is made consciously. By eliminating certain kinds of action altogether and providing certain routine ways of achieving the object, they merely restrict the alternatives on which a conscious choice is required. ... [T]he rules which guide an individual's action are better seen as determining what he will not do rather than what he will do. (Hayek 1967, 56–57)

To approach Maud's actions from the point of view of a script would seem to eliminate what was *spontaneous* about her actions in the setting. If instead, we consider Maud's actions in terms of dynamically overlapping "abstract" constraints, then "specification of action through superposition" would constrain action, presumably along (tacitly) normative lines, while leaving available degrees of freedom for individual actions. This account would apply even to the most ritualized actions where some variation is unavoidable.

To reiterate the ecological focus here, these action tendencies are embedded and situated in the sense that they are adient with respect to environmental structures. Actions are directed toward environmental features, and in the process they are structured by them. For this reason they are relational in nature. In other words, actions are anticipatory, and flexibly so, and some of the information that directs action resides in the environment relationally considered. Affordances and behavior settings are among such sources of information considered relative to an individual, and they specify action possibilities and constraints. For this reason, knowledge does not need to be carried only "in the head," where psychological tradition has exclusively placed it, because the situation itself—the ecological context—is a rich source of information. As Suchman (1987) has pointed out, "the situation for action is thus an inexhaustible resource" (see also Clark 1998).

From this stance, it can be seen that when psychologists overlook affordance possibilities in the setting, the challenge of explaining action is greatly exacerbated.

> The enormous problems of specification that arise in cognitive science's theorizing about intelligible action have less to do with action than with the [ill-conceived] project of substituting definite procedures for vague plans, and representation of the situation, for *action's actual circumstances*. (Suchman 1987, 46, emphasis added)

And because the operations of any setting are in flux to some degree, action-in-context always involves a degree of vagueness and uncertainty. The assumption of a mental representation has tended to assume definite procedures (programs), which in reality are not especially adaptive "on the ground" of changing circumstances, even while it overlooks the environment itself as a rich source of information.

The relational concepts of affordances and behavior settings are invaluable as a means of anchoring what is too often only loosely referred to as situated action. In the process, they are instrumental for articulating an ecological approach to psychology.

8   Conclusion

An uneasy tension has existed between the psychological and the ecological sciences, and it continues to this day. This tension is attributable to that fact that early in its history psychology's embrace of species evolution followed Herbert Spencer's vision of the environment-organism correspondence rather than Darwin's view of a dynamic, reciprocal accommodation. Psychology's early Darwinian, William James, took Spencer to task for failing to appreciate that organisms are active, selectively engaging their surround, rather than being passively shaped by circumstances. It is ironic that although much of contemporary psychology venerates the contributions of William James, while Spencer has become a forgotten figure, it is the approach of Herbert Spencer that still holds a grip on psychological theory, much like the grin of the Cheshire cat that remains after its body has faded from view.[8] Spencer's correspondence approach to environment-organism relations succeeded in carrying the dualism of psychology's early foundations into its era of evolutionary thinking. In doing so, it has sustained a tension between psychology and the ecological sciences.

Ecological thinking can be recovered in psychology by returning to the path James initially blazed and that was developed by a line of successors, including E. B. Holt and J. J. Gibson. We saw that James's philosophy of radical empiricism, with its basic assumption that psychological experience begins in a field of possibilities, comports with ecological thinking. So does his active view of mind participating in the on-going events of the world. Mind is a participant in a field of relations in the way that an individual organism is a participating constituent of an ecosystem. James's student Holt, in turn, developed the idea that enfolded within action are properties of action's referent (a quality of adience), and that the referent for integrated action is located distally from the body surface (the "recession of the stimulus"). When we begin by noting that animate life is characterized by organisms engaging their surround—that is, that action is situated—the psychologically significant features of the environment that complement the individual's action and thought, such as affordances and behavior settings, come to the forefront. At this stage, the notion of situated action remains rather vague. However, an examination of the individual's role in constituting, and also in adjusting to, the dynamics of a behavior setting as conceived by Barker—guided by some insights from the writings of Holt and Hayek—sheds new light into the character of situated action. These

---

[8]Thanks to Gillian Barker who suggested this allusion.

ideas serve as a foundation for ecological psychology, and in doing so can begin to resolve the tension that exists at present between the psychological and ecological sciences.

Acknowledgments I am very grateful to Jonathan Barker and to the editors of this volume for their helpful comments on drafts of this chapter. I also thank Rob Wozniak for helpful conversations on nineteenth-century psychology.

References

Barker, Roger G. 1968. *Ecological psychology: Concepts and methods for studying the environment of human behavior*. Stanford: Stanford University Press.

Barker, Roger G. 1978. *Habitats, environments, and human behavior*. San Francisco: Jossey-Bass.

Barker, Roger G., and Herbert F. Wright. 1955. *Midwest and its children: The psychological ecology of an American town*. Hamden: Archon Books.

Barker, Roger G., and Paul V. Gump. 1964. *Big school, small school: High school size and student behavior*. Stanford: Stanford University Press.

Barker, Roger G., and Phil Schoggen. 1973. *Qualities of community life: Methods of measuring environment and behavior applied to an American and an English town*. San Francisco: Jossey-Bass.

Beck, Naomi. 2009. In search of the proper scientific approach: Hayek's views on biology, methodology, and the nature of economics. *Science in Context* 22: 567–585.

Brewer, William F. 1999. Schemata. In *The MIT encyclopedia of the cognitive sciences*, ed. Robert A. Wilson and Frank C. Keil, 729–730. Cambridge: Cambridge University Press.

Bronfenbrenner, Urie. 1979. *The ecology of human development: Experiments by nature and design*. Cambridge, MA: Harvard University Press.

Brunswik, Egon. 1955. Representative design and probabilistic theory in functional psychology. *Psychological Review* 62: 193–217. (Reprinted in *The essential Brunswik*, eds. Kenneth R. Hammond and Thomas R. Stewart, 135–156. New York: Oxford University Press.)

Butterworth, G., E. Verweij, and B. Hopkins. 1997. The development of prehension in infants: Halverson revisited. *British Journal of Developmental Psychology* 15: 223–236.

Clark, Andy. 1998. *Being there: Putting brain, body, and world together again*. Cambridge, MA: MIT Press.

Crosby, Donald A., and Wayne Viney. 1993. Toward a psychology that is radically empirical: Recapturing the vision of William James. In *Reinterpreting the legacy of William James*, ed. Margaret E. Donnelly, 101–111. Washington, DC: American Psychological Association.

Damasio, Antonio. 1999. *The feeling of what happens: Body and emotion in the making of consciousness*. Orlando: Harcourt Press.

Dewey, John. 1896. The reflex arc concept in psychology. *Psychological Review* 3: 357–370. (Reprinted in *The philosophy of John Dewey*, ed. John J. McDermott. Chicago: University of Chicago Press.)

Dewey, John. 1920. *Reconstruction in philosophy*. New York: Henry Holt.

Donald, Merlin. 2001. *A mind so rare: The evolution of human consciousness*. New York: W.W. Norton.

Gibson, James J. 1960. The concept of the stimulus in psychology. *American Psychologist* 16: 694–703.

Gibson, James J. 1966. *The senses considered as perceptual systems*. Boston: Houghton Mifflin.

Gibson, James J. 1979. *The ecological approach to visual perception*. Boston: Houghton Mifflin.

Gibson, Eleanor J., and Anne D. Pick. 2000. *An ecological approach to perceptual learning and development*. New York: Oxford University Press.

Grene, Marjorie. 1974. *The knower and the known*. Berkeley: University of California Press.

Hayek, Friedrich August. 1952. *The sensory order*. Chicago: University of Chicago Press.

Hayek, Friedrich August. 1967. Rules, perception, and intelligibility. In *Studies in philosophy, politics, and economics*, ed. Friedrich August Hayek, 43–65. Chicago: University of Chicago Press.

Hayek, Friedrich August. 1969. The primacy of the abstract. In *Beyond reductionism: New perspectives in the life sciences*, ed. Arthur Koestler and John Raymond Smythies, 309–323. Chicago: University of Chicago Press.

Heft, Harry. 1989. Affordances and the body: An intentional analysis of Gibson's ecological approach to visual perception. *Journal for the Theory of Social Behavior* 19: 1–30.

Heft, Harry. 1990. Perceiving affordances in context: A reply to Chow. *Journal for the Theory of Social Behavior* 20: 277–284.

Heft, Harry. 2001. *Ecological psychology in context: James Gibson, Roger Barker, and the legacy of William James's radical empiricism*. Mahwah: Lawrence Erlbaum.

Heft, Harry. 2007. The social constitution of perceiver-environment reciprocity. *Ecological Psychology* 19: 85–105.

Heft, Harry and Michael Richardson. In press. Ecological psychology. In *Oxford bibliographies in psychology*, ed. Dana S. Dunn. New York: Oxford University Press.

Heidbreder, Edna. 1933. *Seven psychologies*. New York: Appleton-Century.

Hergenhan, Baldwin Ross. 1988. *An introduction to the history of psychology*, 6th ed. Belmont: Wadsworth.

Herrman, Heinz. 1998. *From biology to sociopolitics: Conceptual continuity in complex systems*. New Haven: Yale University Press.

Holt, Edwin Bissell. 1915. *The Freudian wish and its place in ethics*. New York: Henry Holt.

Holt, Edwin Bissell. 1931. *Animal drive and the learning process: An essay toward radical empiricism*, vol. 1. New York: Henry Holt.

James, William. 1878. Remarks on Spencer's definition of mind as correspondence. *Journal of Speculative Philosophy* 12: 1–18.

James, William. [1890] 1981. *The principles of psychology*, 3 vols. Cambridge, MA: Harvard University Press.

James, William. 1904. Does consciousness exist? *Journal of Philosophy, Psychology and Scientific Methods* 1: 477–491.

James, William. 1905. The place of affectional facts in a world of pure experience. *Journal of Philosophy, Psychology and Scientific Methods* 2: 281–287.

James, William. [1912] 1976. *Essays in radical empiricism*. Cambridge, MA: Harvard University Press.

Johnson, Mark. 2007. *The meaning of the body: Aesthetics of human understanding*. Chicago: University of Chicago Press.

Kelso, J.A. Scott. 1995. *Dynamic patterns: The self-organization of brain and behavior*. Cambridge, MA: MIT Press.

Leahey, Thomas Hardy. 2000. *A history of psychology: Main currents in psychological thought*, 5th ed. New York: Prentice Hall.

Leudar, Ivan, and Alan Costall. 1996. Situating action IV: Planning as situated action. *Ecological Psychology* 8: 153–170.

Lewin, Kurt. 1943. Psychological ecology. (Reprinted in *Field theory in social science: Selected theoretical papers*, ed. Dorwin Cartwright. New York: Harper.)

Mandler, George. 2007. *A history of modern experimental psychology: From James and Wundt to cognitive science*. Cambridge, MA: MIT Press.

Merleau-Ponty, Maurice. 1963. *The phenomenology of perception*. London: Routledge & Kegan Paul.

Pearce, Trevor. 2010. From 'circumstances' to 'environment': Herbert Spencer and the origins of the idea of organism-environment interaction. *Studies in History and Philosophy of Biological and Biomedical Sciences* 43: 241–252.

Perry, Ralph Barton. 1935. *The thought and character of William James*. 2 vols. Boston: Little Brown.

Putnam, Hilary. 1999. *The threefold cord: Mind, body, and world*. New York: Columbia University Press.

Reed, Edward S. 1995. *Encountering the world: Toward an ecological psychology*. New York: Oxford University Press.

Richards, Robert J. 1980. Wundt's early theories of unconscious inference and cognitive evolution in their relation to Darwinian biopsychology. In *Wundt studies*, ed. Wolfgang G. Bringmann and Ryan D. Tweney, 42–70. Toronto: Hogrefe.

Richards, Robert J. 1987. *Darwin and the emergence of evolutionary theories of mind and behavior*. Chicago: University of Chicago Press.

Richards, Robert J. 2008. *The tragic sense of life: Ernst Haeckel and the struggle over evolutionary thought*. Chicago: University of Chicago Press.

Rochat, Philippe, and Stefan J. Senders. 1991. Active touch in infancy: Action systems in development. In *Newborn attention*, ed. Michael J. Weiss and Philip R. Zelazo, 412–442. Norwood: Ablex.

Rorty, Richard. 1979. *Philosophy and the mirror of nature*. Princeton: Princeton University Press.

Schank, Roger C. 1999. *Dynamic memory revisited*. New York: Cambridge University Press.

Schank, Roger C., and Robert P. Abelson. 1977. *Scripts, plans, goals, and understanding*. Hillsdale: Lawrence Erlbaum.

Schoggen, Phil. 1989. *Behavior settings: A revision and extension of Roger G. Barker's ecological psychology*. Stanford: Stanford University Press.

Spencer, Herbert. 1855. *The principles of psychology*. London: Longman, Brown, Green, and Longmans.

Suchman, Lucy A. 1987. *Plans and situated action: The problem of human-machine communication*. Cambridge: Cambridge University Press.

Viney, Wayne. 1989. The cyclops and the twelve-eyed toad: William James and the unity-disunity problem in psychology. *American Psychologist* 44: 1261–1265.

von Hofsten, Claes. 1991. Structuring of early reaching movements: A longitudinal study. *Journal of Motor Behavior* 23: 280–292.

Warren, William H. 1984. Perceiving affordances: Visual guidance of stair climbing. *Journal of Experimental Psychology (Human Perception and Performance)* 10: 683–703.

Watson, John B. 1913. Psychology as the behaviorist views it. *Psychological Review* 20: 158–177.

Weimer, Walter B. 1982. Hayek's approach to the problems of complex phenomena: An introduction to the theoretical psychology of the sensory order. In *Cognition and the symbolic processes*, vol. 2, ed. Walter B. Weimer and David Stuart Palermo, 241–286. Hillsdale: Lawrence Erlbaum.

Wicker, Allan W. 1987. Behavior settings reconsidered: Temporal stages, resources, internal dynamics, context. In *Handbook of environmental psychology*, vol. 1, ed. Daniel Stokols and Irwin Altman, 613–654. New York: Wiley.

Young, Robert M. 1970. *Mind, brain, and adaptation in the nineteenth century*. Oxford: Oxford University Press.

New Perspectives on Organism-Environment Interactions in Anthropology

Emily A. Schultz

Abstract Anthropologists contend that the organism-environment connections responsible for human evolution are indirect—mediated by culture. This chapter reviews influential twentieth-century anthropological interpretations of the cultural mediation of human adaptations to environments, arguing that ethnography and other qualitative forms of analysis reveal important phenomena overlooked by quantitative analysts committed to methodological individualism. It highlights work by post-positivist anthropologists, who describe relations among human and non-human organisms, cultural forms, and features of environments as "natural-cultural" networks, an approach reminiscent of developmental systems theory and niche construction. Evolutionary theorists have much to gain by incorporating these sophisticated, contemporary post-positivist anthropological understandings of culture into their models of human-environment connections.

1 Introduction

In early twentieth-century North America, eugenicists were claiming to be able to sort and rank human populations in terms of biological "race," arguing that such biological "races" were the direct products of past natural selection. Anthropologist Franz Boas and his students challenged such claims, not by denying that the human species had evolved by natural selection, but by arguing that the organism-environment connections that that produced distinct human adaptations were *indirect*, mediated by *culture*; that is, by learned beliefs and behaviors acquired by human beings as members of particular social groups. Because culturally-mediated human connections to the environment were shared, newborn human individuals

E.A. Schultz (✉)
Department of Sociology and Anthropology, St. Cloud State University, St. Cloud, MN, USA
e-mail: eschultz@stcloudstate.edu

G. Barker et al. (eds.), *Entangled Life*, History, Philosophy and Theory
of the Life Sciences 4, DOI 10.1007/978-94-007-7067-6_5,
© Springer Science+Business Media Dordrecht 2014

were spared having to invent new mediated connections to the environment on their own, from scratch. Cultural mediation of human-environment connections appeared to be ancient, associated with a long period of post-natal dependency, during which human children learned from their elders the skills and knowledge (including language) required for successful survival and reproduction. In sum, human organisms were conceived as highly generalized, behaviorally plastic social organisms whose adaptive connections to their environments varied from population to population, because each population's connections to its environment were mediated by particular sets of cultural knowledge and skills, passed down from one generation to the next. Among other things, emphasis on the centrality of culturally mediated human connections to environments allowed anthropologists to explain human diversity without recourse to the concept of biological "race."

Since Boas's day, definitions of culture, and demonstrations of how culture mediates human adaptations to environments, have varied across subfields of anthropology, and have not been without contention (Abu-Lughod 1991; Trouillot 2002). Nevertheless, I argue here that contemporary post-positivist anthropological theorizing about culture can refine and strengthen understandings in theoretical biology about the roles culture may play in mediating human-environment connections. Post-positivist perspectives can be found in the work of anthropologists working across the subfields of biological anthropology, cultural anthropology, linguistic anthropology, and archaeology, and they are well-established in such growing specialties as environmental anthropology and the anthropology of science, technology, and medicine (both of which regularly incorporate insights derived from a newer subfield, applied anthropology). To be sure, post-positivist anthropology remains controversial among those anthropologists who continue to believe that positivist science is science *tout court*, and it is ignored by theorists of cultural evolution who cast their discussions of the cultural mediation of human-environment connections in terms of gene-culture coevolution (e.g. Boyd and Richerson 1985; Richerson and Boyd 2005; Durham 1991). At the same time, most anthropologists who adopt post-positivist perspectives often describe their diachronic analyses of culturally mediated human-environment connections as "historical" rather than "evolutionary," and they rarely engage with selectionist and adaptationist forms of explanation. But it is also true that adaptationist and selectionist arguments offer few resources for illuminating the messy contingencies shaping the kinds of organism-environment entanglements of interest to post-positivist anthropologists.

In my view, serious scholarly discussions of organism-environment connections can no longer afford to ignore the post-positivist anthropological contributions reviewed in this essay, for at least three reasons: (1) this work demonstrates the breadth and sophistication of contemporary post-positivist anthropological analysis, and exposes the speciousness of allegations that rejecting positivism means rejecting science, or that criticizing selectionism means rejecting evolution; (2) this work highlights valuable insights gained from ethnography and other qualitative forms of analysis, thereby exposing the limitations of social science perspectives that favor quantitative analysis and commitments to methodological individualism; and (3) this work displays surprising affinities with insights from "dissident" traditions in theoretical biology, such as developmental systems theory and niche construction, which should be further developed.

My argument is set out in three sections. Section 2 begins with Julian Steward's "classic" mid-twentieth-century account of the cultural mediation of organism-environment connections, recalls the criticism it generated, and describes successor approaches like political ecology that became well established in the late twentieth century. By highlighting struggles to better account for the patterns Steward attempted to capture in his distinction between the core and periphery of human cultural adaptations, this section shows how some anthropologists came to abandon the assumption that either "cultures" or "environments" could be unproblematically conceived as separate, self-contained entities. In addition, it shows how ethnographic work in colonial and post-colonial settings led some anthropologists to draw attention both to the capacities of human beings to rework cultural mediations in drastically altered environments, and to the importance of contextualizing these processes within fields of power.

Under conditions of post-Cold-War globalization, however, some anthropologists have found that political ecology cannot fully account for the remarkable ways people everywhere now mediate connections to contemporary environments, mixing and matching cultural objects and practices inherited from the past with cultural objects and practices imported from elsewhere. Section 3 showcases some of the innovative post-positivist anthropological research that attempts to make sense of these processes. It also shows an imbalance in interdisciplinary exchanges between theoretical biology and anthropology, for this and other relevant anthropological work has been largely ignored by theoretical biologists who write about culture. I describe significant anthropological research inspired by thinkers like Charles Sanders Peirce, Bruno Latour, and Gilles Deleuze and Félix Guattari, which can help theoretical biologists provide more nuanced accounts of the cultural mediation of human-environment connections. Finally, in Sect. 4, I show one way of more directly connecting this newer anthropological work to current work in theoretical biology. It turns out that key features of actor network theory, developed by Bruno Latour and his colleagues, bear a strong family resemblance to key features of niche construction, developed by John Odling-Smee and his colleagues. Drawing on a recent ethnography, I show how conceiving of constructed niches as actor-networks can provide a path that permits insights from cutting-edge post-positivist archaeology and cultural anthropology to enter into current discussions of organism-environment connections in theoretical biology, where they are badly needed. This move requires abandoning dualistic "nature-nurture" thinking for new perspectives that conceive of relations among human organisms, cultures, and environments in terms of "natural-cultural" networks. But it also promises to incorporate a more refined understanding of culture into theoretical biology, where it is long overdue.

2   Cultural Ecology and Its Progeny

Any anthropological discussion of relations between human organisms and their environments must begin with Julian Steward because, as ecological anthropologist Emilio Moran observes, "Steward delimited, more than anyone before him, the field of human/environment interactions" (Moran 1990, 10). Indeed, Steward's *Theory of Culture Change: The Methodology of Multilinear Evolution* (1955) embodies both the achievements and the difficulties that continue to challenge anthropological thinking about culture, ecology, and evolution. This volume contains the fruit of Steward's prewar ethnographic comparisons and theoretical innovation; his critiques of L. H. Morgan's (1963) and E. B. Tylor's (1958) nineteenth-century unilinear evolutionary schemes (and their twentieth-century descendant, the "universal" evolution of Leslie White (1949) and V. Gordon Childe (1951)); the lessons learned from Boasian "cultural relativist" ethnography; and the limitations of British functional anthropology. His multilinear evolutionary approach is meant to avoid the pitfalls of these alternatives, without abandoning scientific cause-and-effect explanations. Steward struggles mightily, however, to bind all these threads together:

> Whereas [Morgan, Tylor, Childe, and White] have sought to formulate cultural development in terms of universal stages, my objective is to seek causes of culture change. Since 'evolution' still strongly connotes the nineteenth-century view, I hesitate to use it but find no better term.

Chapter 2 of Steward's book develops a method for recognizing the ways in which culture change is induced by adaptation to environment. This adaptation, an important creative process, is called *cultural ecology*, a concept which is to be distinguished from the sociological concepts "human ecology" or "social ecology." The cross-cultural regularities which arise from similar adaptive processes in similar environments are functional or synchronic in nature.

> But no culture has achieved so perfect an adjustment to its environment that it is static. The differences which appear in successive periods during the development of culture in any locality entail not only increasing complexity, or quantitatively new patterns, but also qualitatively new patterns. Consequently, in the comparison of the history of two or more areas in which the cultural ecological processes are the same, it must be recognized that a late period in one area may be much more like a comparable late or homotaxially similar period in another area than the earlier periods in either area. Cultural development therefore must be conceptualized not only as a matter of increasing complexity but also as one of the emergence of successive *levels of sociocultural integration* . . . . Chapter 4 illustrates the application of this concept at a national-level system (1955, 5).

Three additional concepts central to Steward's cultural ecology were *cultural type, cultural core*, and *form-function*.

> The concept of *cultural type* . . . is based on the two frames of reference previously presented: cultural features derived from synchronic, functional and ecological factors and those represented by a particular diachronic developmental level. Cross cultural regularities are . . . recurrent constellations of basic features—the *cultural core*—which have similar

functional relationships resulting from local ecological adaptations and similar levels of sociocultural integration. ... The concept of culture type is confronted by the apparent difficulty posed by the fact that forms, patterns, or structures differ greatly. Since, however, similar functions maybe served by different forms while similar forms may serve varied functions, the single concept of *form-function* is introduced. (1955, 5–6)

Steward thus understood each cultural type he identified to be a synchronic/ecological *and* diachronic/developmental (and perhaps evolutionary) synthesis. Chapters 6–12 of his book discuss a series of cultural types "presented ... according to their level of sociocultural integration" (1955, 6). These included what he called the "family band" (the lowest level); "patrilineal hunting bands" and "composite hunting bands" with "slightly higher levels of sociocultural integration"; "nonlocalized clans," which "represent a higher level of sociocultural integration than localized lineages" and which "probably developed from such lineages many times in different parts of the world"; and complex civilizations that developed on the basis of irrigation agriculture, writing that "Chapter 11 shows how in each of these areas fundamentally similar cultural ecological adaptations entailed a similar historical sequence" (1955, 7). In Chapter 12, he applies cultural ecology, levels of sociocultural integration, and culture type to "a complex contemporary society, Puerto Rico" (1955, 7).

Steward's approach was not lockstep: an "environmental possibilist," he highlighted "instances where the interrelationship between culture and environment allows considerable latitude or potential variation in sociocultural types. Where latitude is possible, historic factors may determine the nature of society" (1955, 6). Nevertheless, ten years after *Theory of Culture Change* appeared, cultural ecology had been thoroughly picked apart. Some anthropologists criticized Steward for assuming that "cultures," rather than people, might adapt to environments. Others found Steward's positivist analytic goals to be highly problematic. Recent ethnography was also revealing difficulties in the data Steward had used to construct his cultural types.

Still, it is worth lingering a moment over Steward's discussion of the culture core, which can be seen both as a jumping-off point for some later anthropological discussions of cultural change (e.g., dual inheritance theory), and as addressing issues of importance which these later discussions neglect:

> The concept of *cultural core* [is] the constellation of features which are most closely related to subsistence activities and economic arrangements. The core includes such social, political, and religious patterns as are empirically determined to be closely connected with these arrangements. Innumerable other features may have great potential variability because they are less strongly tied to the core. **These latter, or secondary features**, are determined to a greater extent by purely cultural-historical factors—by random innovations or by diffusion—and they give the appearance of outward distinctiveness to cultures with similar cores. **Cultural ecology pays primary attention to those features which empirical analysis shows to be most closely involved in the utilization of environment in culturally prescribed ways.** (1955, 37; boldface added)

That is, natural selection on "cultural variants" seems most applicable to what Steward calls "secondary features" of culture—those less strongly tied to the core, and free to vary "by random innovation or diffusion" (or, perhaps, by natural selection). Features of the culture core, by contrast, are basic to adaptive stability and are not similarly free. Today most anthropologists no longer accept Steward's account of culture cores. However, a satisfactory account of the origin and stabilization of the kinds of cultural features that Steward *attributed* to culture cores remains elusive. In particular, "Darwinian" theories of cultural evolution that emphasize natural selection on cultural variants (e.g., Boyd and Richerson 1985; Richerson and Boyd 2005; Durham 1991) remain unable to account for those key features that Steward saw as indispensable for the culturally mediated adaptation of human organisms to their environments.

Steward's attempt to recast the study of cultural evolution by turning to ecology (rather than, for example, to population biology) directly shaped the next influential anthropological approach to (human) organism-environment interactions: the "new ecological anthropology" of Andrew P. Vayda and Roy Rappaport. Emilio Moran writes that Vayda and Rappaport

> found the concept of the culture core, and the cultural ecological approach, to give undue weight to culture as the primary unit of analysis, and found the presumption that organization for subsistence had causal priority to other aspects of human society and culture to be both untested and premature (Geertz 1963). (Moran 1990, 10)

Moving from *cultural ecology* to *ecological anthropology* thus involved increasing the emphasis on biology relative to culture; as Conrad Kottak explains, "the analytic unit shifted from 'culture' to the ecological population, which was seen as using culture as a means (the primary means) of adaptation to environments" ([1999] 2006, 40). In addition, despite Steward's misgivings about "functional or sociological formulations," his successors embraced an intensified functional ecological analysis. Rappaport defined the ecological population as "an aggregate of organisms having a common set of distinctive means by which they maintain a common set of material relations within the ecosystem in which they participate" (Rappaport 1971, 238; cited in Kottak [1999] 2006, 41). Their theoretical inspiration was cybernetics: systems theory and the role of negative feedback. "Cultural practices were seen as optimizing human adaptation and maintaining undegraded ecosystems" (Kottak [1999] 2006, 40). In this model, two analytic units were basic: (1) the *ecological population*, which might in some cases be said to correspond to a locally named group (such as the Tsembaga Maring of New Guinea) and (2) the *ecosystem*, a set of systemic environmental relationships that regulate themselves by means of negative feedback. Rappaport would later be criticized for his easy identification of ecological populations with locally named groups, for he could offer no explicit criteria distinguishing "ecological populations" from what other anthropologists called "cultures."

In *Pigs for the Ancestors* ([1968] 1984, 4), Rappaport claimed that the Tsembaga *kaiko*, or ritual pig sacrifice, regulates the frequency of warfare among neighboring tribes because it

> operates as a regulating mechanism in a system or set of interlocking systems, in which such variables as the area of available land, necessary lengths of fallow periods, size and composition of both human and pig populations, trophic requirements of pigs and people, energy expended in various activities, and the frequency of misfortunes are included.

Rappaport also insisted that it was important in ecological studies to distinguish between two different models of the environment: the *cognized model*, "the model of the environment conceived by the people who act in it" ([1968] 1984, 238) and the *operational model*, "which the anthropologist constructs through observation and measurement of empirical entities, events, and material relationships" ([1968] 1984, 237). Rappaport maintained that even if these two models overlapped, they were not identical and ought not to be confused.

By the 1980s, however, other anthropologists working in communities that had experienced European or American colonization had begun to adopt views influenced by dependency theory and world systems theory. They argued persuasively that key factors responsible for shaping contemporary ecological practices in the so-called "tribal" societies anthropologists were studying had actually originated *outside those societies themselves*, in colonial metropoles or in the core of the capitalist world system (Frank 1967; Wallerstein 1974; Wolf 1982). Acknowledging the impact of Western imperialism rendered deeply problematic the assumption that "tribal" ecosystems (or "cultures") were timeless, separate, self-contained, self-regulating entities. Accordingly, Rappaport was also criticized for uncritically accepting a "positivist" model of science that ignored history; this realization pushed many ecological anthropologists toward a political economic framework of analysis (Biersack 2006, 7; Goodman and Leatherman 1998).

Still other anthropologists, however, were being attracted by new analytic frameworks coming from biology. In 1963, biologist Niko Tinbergen had published a paper in which he argued that asking "why" any form of animal behavior occurs actually masks four separate questions about (1) the proximate (or immediate) *causal* explanation of the animal's motivation; (2) the *ontogenetic* explanation of the behavior's development across the animal's life span; (3) the *phylogenetic* explanation, tracing the evolutionary history of the species-specific biological systems involved in the particular behavior; and (4) the functional (or ultimate) *adaptive* explanation, showing how performing the behavior influences the individual organism's ability to survive and reproduce (Tinbergen 1963). Tinbergen welcomed the extension of ethological methods to humans (Tinbergen 1963, 430). Biological anthropologist Agustín Fuentes points out, however, that keeping all the "why" questions separate in the study of humans is very difficult, partly owing to a "bias toward the value of the ultimate, or functional, answer ... which is seen as the most important 'level' of analysis in terms of evolutionary understanding (the quest to find human adaptations)" (2009, 29). These difficulties notwithstanding,

> the primacy of interest in Tinbergen's ultimate question combined with a series of mathematical models and perspectives on the role of kin and altruism that arose in the 1960s and 1970s laid the foundation for the most pervasive and influential contribution to the study of the evolution of human behavior since the early 1900s: Wilsonian Sociobiology. (Fuentes 2009, 29).

In my experience, E. O. Wilson's *Sociobiology: The New Synthesis* (1975) landed like a bombshell in four-field anthropology departments across North America. For many anthropologists—particularly, although not exclusively, biological anthropologists—it seemed to be full of bright promise. But many others—especially cultural and linguistic anthropologists (but also some biological anthropologists and archaeologists) were dismayed or angered as Wilson arrogantly proclaimed,

> Sociobiology is defined here as the systematic study of the biological basis of all social behavior. For the present it focuses on animal societies. ... But the discipline is also concentered with the social behavior of early man and the adaptive features of organization in the more primitive contemporary human societies. ... It may not be too much to say that sociology and the other social sciences, as well as the humanities, are the last branches of biology waiting to be included in the Modern Synthesis. (1975, 4)[1]

As Fuentes explains, "[u]biquitous in this 'new synthesis' perspective was the primacy of ultimate explanations, a reliance on relatively linear mathematical models to model natural selection, and reduced concern with the physiological and genetic details of the mechanisms for behavioral adaptations" (2009, 30). Approaching interactions between humans and their environments with such a toolkit, however, could only appear perverse to cultural anthropologists like Marshall Sahlins (1976a), whose understanding of human-environment relations rested on human mobilization of complex, intricately interwoven sets of cultural meanings and practices, resources deeply rooted in history and politics rather than in the genes.[2] Critiques of sociobiology by anthropologists did not abate when sociobiology gave birth to evolutionary psychology, human behavioral ecology, and other variant perspectives (Marks 2009). For anthropologists who take culture seriously, however, Boyd and Richerson's *Culture and the Evolutionary Process* (1985) counts as an indispensable intervention in the debate. Using mathematical models to challenge the mathematical modelers, dual-inheritance theory defended culture in an idiom that sociobiologists found much harder to ignore. Of course, as noted above, anthropologists may still object to dual-inheritance theory on other grounds.

Roy Rappaport did not respond to his critics by turning to sociobiology. On the contrary, the 1984 edition of *Pigs for the Ancestors* contains a 180-page long epilogue in which he addresses a range of complaints, and, in some cases, abandons positions he had formerly defended. Until his death in 1997, Rappaport continued to reject any sharp dichotomization between "nature" and "culture," arguing instead

---

[1] Although disrespect has been expressed on both sides of the divide, the kind of contempt often expressed by sociobiologists and evolutionary psychologists toward their critics has been particularly striking in my experience. Anthropologists and others committed to evolution, but critical of sociobiology, have risked being labeled "anti-science," "anti-evolution," or even "creationist." As a result, many of us have had to adopt the position of "anti-antievolutionists," who resist the critics of evolution, but who are unable to wholeheartedly affirm the hegemonic version of evolutionary theory (see Schultz 2009). The new developments and possibilities discussed in Parts 3 and 4 below may help change this state of affairs.

[2] Sahlins began his career as a cultural evolutionary theorist (e.g. Sahlins and Service 1960), but his views about cultural evolution changed following his experiences in France in the late 1960s (see Sahlins 1976b).

that the human condition involves living "in terms of meanings in a physical world devoid of intrinsic meaning, but subject to causal law" (cited in Biersack 2006, 7); and his later work highlighted the ways in which political economic processes often promoted the "disordering of adaptive structures" of local populations.

Put another way, "Rappaport's intellectual trajectory drew him slowly, tacitly toward political ecology" (Biersack 2006, 8), an approach that attends to the ways human/environment relations are shaped by political and economic processes. As Conrad Kottak explains, "a successor to ecological anthropology is the 'new' ecological, or environmental, anthropology, which blends theory with political awareness and policy concerns" (Kottak [1999] 2006, 40). For Aletta Biersack, the appeal of political ecology and environmental anthropology lies in the way that Marxist analysis, refracted through dependency theory and world systems theory, opens up the possibility of focusing "on human-nature relations in other than adaptationist and reductionist terms," because power is seen as "sociohistorical and structural" (2006, 8). She continues: "The implication for ecology is that the local is subordinated to a global system of power relations and must be understood entirely with respect to that subjection, in terms of what is commonly referred to as capitalist penetration and its effect" (2006, 9).

By the late 1980s, geographers and anthropologists doing political ecology were paying attention to linkages between global and local processes, an approach, Biersack says, that "continues to be productive today" (2006, 12).

In this connection, it is worth considering the legacy of Andrew ("Pete") Vayda, Rappaport's co-creator of the "new ecological anthropology," because his career over the past 50 years illustrates a willingness to grapple with many of the factors that are central to the post-positivist research I review below. One striking feature of Vayda's work has been his ongoing critique of theoretical accounts of the cultural mediation of human relations to their environments, not excluding his own previous views:

> In the 1960s when "cultural ecology" was in vogue, he argued for a "human ecology" instead, and was a leader in the development of systems approaches to human-environment relations. However, in the early 1970s he joined his students in criticizing the teleology and other excesses of systems-based human ecology, arguing instead for an agent-based approach. In recent years, he has taken on widely-held assumptions about the nature—and culture—of explanation in human-environment research, in the course of which he has developed an analytical methodology that is informed by the pragmatic view of scholars like Charles Sanders Peirce, David Lewis, Geoffrey Hawthorn, T. Chamberlin, H. L. A. Hart, and Tony Honoré. ... More generally ... Vayda has been highly critical of holism, essentialism, systems thinking, naïve functionalism, and speculative adaptationism in anthropology and human ecology. ... He has also pointed to the dangers of a priori assumptions and ready-made theories such as those of some cognitive anthropologists and political and spiritual ecologists. (Walters and McCay 2008, 1–2)

A second striking feature of Vayda's scholarship is the extent to which his theorizing has been powerfully informed by his experiences in applied anthropological research, primarily in the forests of Indonesia and New Guinea. Introducing a recent collection of his own essays, Vayda (2009, ix) observes that an original stimulus for some of the essays was my desire to get at the causes of particular phenomena, like intergroup fighting in the mountains of New Guinea and extensive fires in the tropical moist forests of Indonesia. For other essays, the original stimulus was more my being dissatisfied—on logical, empirical, or pragmatic grounds—with research methods widely used or kinds of explanations commonly made in such fields or subfields as political ecology, Darwinian human behavioral ecology, and local knowledge studies. Whatever the stimulus, going beyond my criticisms of the work of others and achieving better explanations and identifying ways of achieving them were among the positive goals I set myself.

Illustrative of this restless field-based critique of theoretical accounts of human-environment connections is an article Vayda co-wrote with Bradley Walters in 1999, entitled "Against Political Ecology." Vayda and Walters were, in fact, *not* urging that a consideration of power relations be eliminated from ecological studies; rather, they were challenging ecological analyses that *privilege the political*, emphasizing instead the importance of *a range of heterogeneous causal factors*, none of which may be excluded *a priori* (Vayda 2009, Chapter 6). Vayda and Walters' critique of narrowly political accounts of causation in ecological studies in anthropology bears a strong family resemblance to Bruno Latour's critique of narrowly "social constructionist" accounts of causation in science studies (e.g., Latour and Woolgar 1986, Postscript). Vayda's "evenemental or event ecology" (Vayda 2009, 13–34)—elsewhere called "progressive contextualization" (e.g., McCay 2008, 5)—bears an equally strong family resemblance both to the "constructivism" of Ludwik Fleck (Smith 2006, Chapter 3) and to actor network theory (e.g., Latour 2005; both Fleck 1979 and Latour 1988 appear in Vayda's 2009 bibliography).

These features of Vayda's legacy are illustrated in Paige West's ethnography *Conservation is our Government Now: The Politics of Ecology in Papua New Guinea* (2006). West follows the fortunes of Gimi people and their neighbors in the eastern highlands of New Guinea between 1994 and 1999, as they are drawn into a "conservation-as-development" project funded by outsiders and designed and implemented by nongovernmental organizations (NGOs) staffed by local and international conservation experts. "[I]t was promised that if Gimi and Pawaia gave their lands for inclusion in the Wildlife Management Area, they would drive cash benefits, access to economic markets for the forest products tied to local biological diversity, and 'development'"; in other words, "conservation was to be the development" (West 2006, 5). By 1994, most Gimi involved in the project had already altered a number of earlier connections to their forests after converting to Seventh Day Adventism, which obliged them to alter their hunting practices to give up pork. But their sense of identity was fluid, and they did not equate these changes with a loss of "traditional" Gimi culture. On the contrary, they expressed to West "the feeling that they had, and have, a choice about which 'traditional' practices they wish to continue and which they wish to abolish" (2006, 66), and by the mid-1990s, many of them wanted "development." However, the "development" they expected to receive in exchange for their cooperation with "conservation" were substantive goods and services (medicine, technology, education for their children), not cash and access to capitalist markets; in the event, neither was forthcoming to the degree anticipated, nor distributed evenly among all members of the community. When the project ended, both Gimi and the conservation practitioners were frustrated and dissatisfied.

In the first chapter, West recounts an event that she witnessed in 1999, a knife fight between two Gimi men, Kelego and Lasini (West 2006, 15ff). Six pages from the end of the final chapter, West returns to the fight, and compares her account of this fight to the account offered by Napoleon Chagnon and Timothy Asch in their classic ethnographic film, *The Ax Fight* (Chagnon and Asch 1975). West draws both on Vayda's critique of Darwinian ecological anthropology and his discussion of progressive contextualization to differentiate her approach from that of Chagnon:

> While Chagnon's goal with his reading of the ax fight is a positivist explanation, he is looking for answers about human nature that can be generalized from the Yanomamo to all people. I spent the past seven years looking for explanations for one fight between Kelego and Lasini and trying to trace out the causal chains that led up to it and the layers of meaning that encompass it (2006, 230.)

West concludes that "The fight was about imbalances, both perceived and real, that have come into being because of the conservation-as-development project" (2006, 231). But tracing the causal chains meant that West had to "try to disentangle the connections between New Guinea and New York, conservation and development, and birds of paradise and commodities" (2006, 4), efforts recounted in the body of her ethnography. West describes historical processes through which Gimi people entered into relations with a variety of outsiders, including colonial administrators, missionaries, linguists, ethnographers, environmentalists, and others. Over time, these relationships transformed both Gimi identity and the cultural practices Gimi people used to mediate their relations to living and nonliving features of their environments. But those environments themselves were expanded and restructured as they were connected to transnational institutions and resources situated within the global capitalist market.

Exposing these entangled processes involved West in both archival research and fieldwork, among both Gimi people and conservation practitioners, both in New Guinea and New York. Multisited research complicated West's understanding of the role of anthropology in analyzing how human groups use culture to mediate connections to their environments. The story she tells "is not a story of 'good guys' and 'bad guys' or even 'the Gimi' and 'the conservationists.' It is a story about the social lives of people associated with a large bit of the forest in Papua New Guinea" (2006, xv). Her final text, she insists, is neither a "translation or legibility-making service" for conservation activists nor "a devastating critique of conservation as a way of knowing and producing knowledge. ... Rather, my goal is to provide an ethnography of the project and perhaps to persuade conservation practitioners, activists, scientists, and others to question the assumptions about nature, culture, and development that underlie many of today's biodiversity conservation efforts" (2006, xviii).

3   New Directions in Anthropological Studies
of Organism-Environment Relations

A rich conversation is in the making between anthropologists and biological theorists who write about the cultural mediation of human connections to their environments, but the exchange could be much more balanced. In this section, I begin by reviewing the way some forms of theoretical biology have begun to shape recent anthropological thinking. But I end by presenting additional anthropological work that, like Paige West's ethnography, merits serious consideration in discussions of the cultural mediation of human-environment connections in theoretical biology, but that so far has not been considered.

Let us start by looking at the recent work of biological anthropologist Agustín Fuentes (2009). He analyzed five contemporary theoretical approaches to the evolution of human behavior that have influenced biological anthropology: Neo-Darwinian (ND) Sociobiology/Human Sociobiology, Human Behavioral Ecology (HBE), Evolutionary Psychology (EP), Gene-Culture Coevolution/Dual Inheritance Theory (DIT), and Memetics. All five, he reminds us, take "Wilson's sociobiology, Hamilton and Trivers' kin selection and reciprocal altruism, and the Dawkinsian genic selfishness as baseline assumptions" (2009, 37). All five claim Darwin as their inspiration, focus on natural selection as the architect of behavior, and tend to de-emphasize other processes of evolution recognized in the modern synthesis, such as gene flow and genetic drift. HBE and EP also emphasize the importance of Ernst Mayr's distinction between ultimate and proximate levels of explanation (2009, 62). But Fuentes is troubled by what all five leave out (Fuentes 2009, 62–63):

> Missing from HBE, EP, DIT, and Memetics is much of the evolutionary anthropological approach pioneered by Sherwood Washburn. In the 50 years since Washburn proposed his "New Physical Anthropology," there has been an explosion in the paleoanthropological data base, resulting in a series of important changes and enhancements of the scenarios for human physical (and social) evolution. Unfortunately, ND-Sociobiology is the only one of these perspectives to regularly exploit both the fossil and archeological records and primate studies as comparative tools. Of the other four, HBE does occasionally incorporate fossil/archeological/primatological datasets (Hawkes et al. 2003) and EP uses assumed Pleistocene selection pressures as its baseline, but neither EP, CIT, or Memetics regularly use fossil or cross-species comparisons in their construction of scenarios and hypotheses for the evolution of human behavior.

To fill in the gaps, Fuentes (2009, 172) incorporates recent work that focuses attention on ontogeny: Jablonka and Lamb's (2005) arguments for "evolution in four dimensions," West-Eberhard's (2003) arguments linking developmental plasticity to evolution, Oyama's (2000) developmental systems approach (see also Oyama et al. 2001), and the niche construction perspective developed by Odling-Smee et al. (2003). In particular, Fuentes (2009, 172–75) is persuaded that niche construction is an evolutionary force that can be tested against the human evolutionary record, and he and two colleagues recently performed such a test, proposing a new explanation for an old puzzle. Between 2.5 and 1 million years ago, the fossil record shows that the genus *Homo* and the genus *Paranthropus* coexisted in eastern and southern Africa, but by one million years ago, *Paranthropus* was extinct, and *Homo* had expanded. Most explanations of this transition attribute it to the superior foraging efficiency of *Homo*, based on increased brain size, tool use, and meat consumption, the sharing of "cultural" information, and (recently) also to niche construction. However, Fuentes and colleagues "propose a model wherein a focus on the role of predation and differential ability to share information and cooperatively modify functional facets of the environment provide an important component of the explanation of the success of the genus *Homo* relative to ... *Paranthropus*" (Fuentes et al. 2010, 436). Using evidence suggesting that both *Paranthropus* and *Homo* were likely vulnerable to the same predators, their model shows how niche construction could have made *Homo* less desirable as prey, shifting predation pressure onto *Paranthropus*, while at the same time providing positive feedback to protective niche-constructing behaviors in *Homo*. Overall, Fuentes recommends abandoning or deemphasizing optimality models, single-trait models, simple proxy measures of fitness, and the focus on DNA; retaining the focus on natural selection (together with niche construction), on the role of symbolic communication and culture, and on past and present environments; and expanding attention to plasticity, multilevel selection on multiple inheritance systems, and the role of behavior as an agent of evolutionary change in humans (2009, 180–186). The outcome, he believes, will be a biocultural approach to the study of human behavior appropriate for the twenty-first century.

Attention to ontogeny has also been central to the work of British anthropologist Tim Ingold. A Cambridge-trained social anthropologist who carried out fieldwork among Sami reindeer herders in Finland, Ingold grew dissatisfied with theoretical proposals treating human life "as merely consequential, the derivative and fragmentary output of patterns, codes, structures or systems variously defined as genetic or cultural, natural or social" (2011, 3). His own work, therefore, "has been driven by an ambition to reverse this emphasis: to replace the end-directed or teleonomic conception of the life-process with a recognition of life's capacity continually to overtake the destinations that are thrown up in its course" (2011, 4). The result is a unique perspective on the relations between humans and their environments that innovatively combines insights from James Gibson, Susan Oyama,[3] Maurice Merleau-Ponty, Martin Heidegger, Alfred North Whitehead, Henri Bergson, Gilles Deleuze, and Félix Guattari (Ingold 1990, 2000, 2007, 2011).

Problematic accounts of the labor process led Ingold to explore the phenomenology of human productive accomplishment, and to conclude that production was not "about transforming the material world, but rather about participating in the world's transformation of itself" (2011, 8). Accordingly, he has taken up the challenge of rehabilitating cosmologies denominated "animist" by Western thinkers: "once we recognize the primacy of movement in the animic cosmos ... we are not required to believe that the wind is a being that blows. ... Rather the wind *is* blowing,

---

[3]Ingold is the only anthropologist who contributed an essay to the DST compendium *Cycles of Contingency* (Oyama et al. 2001).

and the thunder *is* clapping, just as organisms and persons *are* living in the ways peculiar to each" (2011, 73). Ingold has also written insightfully about the evolution of the human foot, and criticizes "the division of labor between hands and feet" that informs most discussions of the evolution of human bipedalism since Darwin (Ingold 2011, chap. 3).

Ingold's ongoing reflections on the relations among anthropology, art, and architecture have attracted wide attention, inside and outside of anthropology. Recently, he has debated anthropologists and archaeologists who study material culture. All of them deplore nature/culture dualism, agreeing that matter has been unjustly neglected by positivist science. Ingold, however, rejects their attempts to generalize about "materiality," insisting that

> so long as our focus in on the *materiality of objects*, it is quite impossible to follow the multiple trails of growth and transformation that converge, for instance, in the stuccoed façade of a building or the page of a manuscript. These trails are merely swept under the carpet of a generalized substrate upon which the forms of all things are said to be imposed or inscribed. I propose that we lift the carpet, to reveal beneath its surface a tangled web of meandrine complexity. (2011, 26)

Rehabilitating the status of matter has also been central for post-processual archaeologists[4] wishing to incorporate into their accounts of the human past the cultural meanings of material artifacts for their makers and users. Because they regularly deal with things that neither speak nor carry written linguistic representations, these archaeologists need methods for studying non-linguistic meaning-making. Post-processual archaeologist Robert Preucel writes that

> material culture, like language, often plays a central role in mediating social identities and relations. However ... material culture does not participate in the same kind of structured system as language. Objects are not words and there is nothing in material culture comparable to syntax or grammar in linguistics. But because material culture has form and substance, it has the power to fix meanings in ways that are not possible in language. (2010, 84)

In the 1970s and 1980s, post-processual archaeologists like Preucel who were disappointed by attempts to adapt Saussurean *sémiologie* for the study of material culture were inspired by the work of linguistic anthropologists like Michael Silverstein, who were questioning Saussure's distinction between *langue* and *parole*. Research in linguistic anthropology showed that linguistic meaning in

---

[4]So-called "processual archaeology" emerged in the 1960s and is closely associated with the work of Lewis Binford (1962). It encompasses a variety of different approaches, but "all share a common processual orientation grounded in cultural evolutionary theory and a systemic view of culture" based on the structuralism of Claude Lévi-Strauss (Preucel 2010, 94). "Post-processual archaeology" encompasses a variety of different approaches sharing "a common dissatisfaction with the scientistic approach of much of processual archaeology, particularly its focus on positivism and general laws of human behavior. In its place they adopt hermeneutic methods and emphasize the social salience of ideology and power," commenting, "as an empirical social science which privileges material culture, archaeology retains a strong modernist core and resists full colonization by poststructuralism and postmodernism" (Preucel 2010, 123).

contexts of use depended heavily on speech, or *parole*, rather than on idealized symbolic meanings supposedly encoded in *langue*. But this meant that linguistic anthropologists needed a method of analysis that would allow them to study dimensions of meaning communication that were not purely symbolic. They found what they were looking for in the semiotics of Charles Sanders Peirce; some cultural anthropologists and post-processual archaeologists soon joined them. What resulted was the emergence of a pragmatic anthropology critical of the limitations of symbolic, structural and cognitive anthropology, but also resistant to the poststructural claims about the radical ambiguity of meaning (Preucel 2010, Chapter 4).

Scholars in many fields are familiar with Peirce's tripartite division of signs into icons, indices, and symbols. Michael Silverstein (1976) proposed that each of these sign functions constitutes a separate mode of meaningfulness, and argued that indices are indispensable for the study of language in use. Silverstein and others later demonstrated a variety of ways in which indexicality is mobilized ideologically by speakers to modify linguistic structures in contexts of use (Silverstein 1985). When Preucel and other archaeologists reviewed Peirce's writings about signs, they discovered that by the time of his death in 1906, Peirce had elaborated a typology of at least 66 signs that linked icons, indices, and symbols both to their interpretants and to other signs (Preucel 2010, 56–60). In a historical archaeology project at Brook Farm, Massachusetts, Preucel uses Peirce's typology to explicate a range of different kinds of meanings mediated for their original Transcendentalist residents by the buildings they used and built. He also shows how these meaningful architectural mediations were undermined when later residents, committed to Fourierism, and with different class origins, promoted a different kind of communal architecture at odds with Transcendental cultural practices. Preucel concludes that the varied buildings used and constructed by Transcendentalists were "a material expression of the Transcendentalist celebration of the individual in society" that also exercised "house agency" as they "actively engendered certain habits of thought and social practices at the core of Transcendentalism" (2010, 209).

Semiotic archaeology is not the only variety of post-processual archaeology that investigates relations between human organisms and their environments, but it is a provocative "dissident" version. Another dissident version is the "social archaeology" tradition associated with Ian Hodder. Over the past 30 years Hodder's career has taken him from processual to post-processual archaeology[5] and he is surely the most influential post-processual archaeologist at work today. Hodder is probably best known among anthropologists who are not archaeologists for his work in ethnoarchaeology: in the 1970s, he carried out ethnographic fieldwork in several East African societies to test the correlation between distributions of material artifacts and the social identities of their makers and users. Hodder's conclusion that such connections were unreliable set him apart from Lewis Binford, who also carried out ethnoarchaological research, but drew the opposite conclusion.

---

[5] A term he coined (Preucel 2010, 126).

Over the years, Hodder has published a series of texts in which he has relentlessly explored the consequences for archaeology that follow from challenges to its former identity as a disinterested scientific enterprise. In recent years, these challenges have come not only from indigenous communities who connect archaeology with colonial domination and expropriation, and from government laws mandating repatriation of human remains and artifacts, but also in the form of epistemological challenges from science studies scholars like Bruno Latour. While acknowledging the close ties that still bind many processual archaeologists to the tenets of positivist philosophy of science, Hodder has chosen to embrace the challenges of reflexive scrutiny and the critique of positivist science. He has addressed in detail the issues surrounding hermeneutical analysis in science, arguing that even though the whole of an archaeological site is understood in relation to its parts, "this circle of part-whole relationships is not vicious. . . . Rather, the objects of study can cause us to change our ideas about the whole or about the relationship between the parts. This circle can best be described as a spiral" (1999, 33). He has incorporated ideas from science studies to open up archaeological concepts like the *chaîne opératoire*, which specifies the sequence of practices that produces particular material artifacts (1999, 76). Hodder honors the skills archaeologists have developed to trace long-term and large-scale cultural processes, but he insists they must also develop narrative techniques for interpreting, whenever possible, the human activities they are able to reconstruct at a human scale: "both are needed in an archaeology which accepts diversity, uncertainty and relationality in human behavior" (1999, 147).

Because he acknowledges, but goes beyond, the resolutely local, phenomeno-logical focus of Ingold, Hodder cannot avoid coming to terms with heterogeneous global flows of wealth, commodities, people, images, and ideologies that have been unleashed since the end of the Cold War. Indeed he must do so, for his ongoing archaeological project at Çatalhöyök, in Turkey, is sustained by these flows: it is financed by private capital, employs local and international workers, requires the ongoing support of local and national governments, attracts tourists from Turkey and elsewhere, and for some years has had a presence on the internet (http://www. catalhoyuk.com/). Hodder has paid close attention to the work of Arjun Appadurai, an anthropologist whose book *Modernity at Large* (1996) has profoundly shaped cultural anthropologists' understanding of these global flows (Appadurai et al. 2001). Like Ingold, Appadurai turns to Deleuze and Guattari's *A Thousand Plateaus* (1988) for language capable of articulating "the special problems that beset the production of locality in a world that has become deterritorialized" (1996, 188). But the global lines of flight Appadurai describes generate heterogeneous, hybrid forms of movement spun out of rootlessness, alienation, and transgenerational instability of knowledge, with both points of departure and points of arrival in cultural flux (1996, 29, 43–44).

Nevertheless, Appadurai also argues that an upside to globalization can be perceived when new global technologies and connections are mobilized to solve old problems (1996, 43). This phenomenon may be glimpsed at the Çatalhöyök

Research Project, where Hodder and his collaborators have worked for some years to develop and institutionalize reflexive archaeological practices that now are mediated by a sophisticated computer database providing access, in different ways, to field staff, laboratory specialists, and internet viewers. Perhaps most innovative of all is the project's employment of cultural anthropologists specializing in science studies, who carry out participant observation on the entire research process and feed back their insights into the ongoing project (Hodder 1999, chap. 10).

I conclude with one final example illustrating my conviction that archaeology may be the most lively source of innovative thinking in contemporary anthropology. In some ways, Nicole Boivin's recent volume *Material Cultures, Material Minds: The Impact of Things on Human Thought, Society, and Evolution* (2008) brings my observations in this section full circle. Trained at Cambridge University in the Hodderian social archaeology tradition, Boivin insists that the physicality of matter gives things agency that is independent of human organisms. While she was studying domestic space in rural Rajasthan, India, she noticed that

> much of the way that houses assumed a social and symbolic role relied on the use of soil to create them. ... Mud houses are infinitely malleable, and are constantly plastered and replastered in ways that enable them to acquire a new appearance, texture and feel. ... I thus began to think about the first mud houses and how they may accordingly have played a role in generating new symbolic and social possibilities within prehistoric society (2008, 133–34).

Boivin eventually concluded that "soil was an active agent in the process of Neolithicization in the eastern Mediterranean, among many other active agents, both human and non-human" (2008, 138), and she found Tim Ingold's arguments helpful for imagining how domesticated species and artifacts might emerge "as a result of the 'mutual involvement of people and materials in an environment' in which outcomes cannot always be anticipated" (2008, 156). She acknowledged that "locating agency is a complex exercise that probably demands a new way of thinking about it, as well as about humans and things" (2008, 168), and she found such a new way of thinking in actor network theory (2008, 176). Combining insights from Tim Ingold and Bruno Latour, she then determined that "the realms of technology and environment become difficult to differentiate" (2008, 178). This realization led to an extended exploration of niche construction, development, and cognitive plasticity, in which Odling-Smee et al. (2003) and Oyama et al. (2001) are prominent sources (2008, 197, 220). Boivin closes her discussion by urging cognitive scientists to pay attention to archaeologists and anthropologists, and for niche construction theorists to talk to social anthropologists: "Material culture, which by very definition straddles [the social sciences and humanities,] demands an integrated approach that brings these very different models together" (2008, 229). In this way, citing many of the same sources who inspired Fuentes, Boivin likewise echoes Fuentes's call for an integrative anthropological approach that is "holistic, messy, but potentially highly profitable" (Fuentes 2009, 249).

4   Integrating Post-positivist Anthropology into Theoretical Biology: A Proposal

In this section, I propose a theoretically informed way of articulating neglected work in post-positivist anthropology into analyses by theoretical biologists of the cultural mediation of human-environment connections. Paige West's work in Papua New Guinea, discussed in Sect. 3, is a good place to begin. West (2006, xvii) writes that she has begun to see her role and the purpose of her work as related to a "new ethnography of development" that "takes seriously the governmentality of projects—the fact that social lives, environments, and subjects come to make and be made by the productive power of the structures created by projects (Foucault 1977)—and the social interactions during all sorts of projects (be they conservation, development, or resource extraction) which create new communities (Golub [2006])." Because projects like this are remaking people's environments all over the world, ignoring them in accounts that attempt to describe the cultural mediation of a human population to its environment cannot continue. The highland village of Maimufa where West carried out much of her fieldwork was a hybrid community, consisting not only of Gimi people, but also of numerous non-Gimi from Australia, the United States, and elsewhere in Papua New Guinea, jointly engaging with Gimi people and their neighbors in the conservation-as-development project. West carried out participant-observation among the conservation scientists as well as the Gimi residents because the causal interventions of the scientists could not be ignored:

> The value of the eagle is not in and of the eagle, though its commodification might make it seem so—it is a value produced by a set of social relations of production in science and in the imagination of scientists. And what of the labor and value that went into the eagle that is forgotten as it becomes a commodity? That labor is the labor of scientific practice (Latour 1987:7), and the nature of the bird is its relation to all the processes of the forest that it influences and that influence it. (2006, 212)

Or, to put it another way, the eagle is an actor network. Latour (2011, 797–798) has recently defined an actor network as follows:

> In its simplest but also in its deepest sense, the notion of network is of use whenever action is to be redistributed. ... Take any object: At first, it looks contained within itself with well-delineated edges and limits; then something happens, a strike, an accident, a catastrophe, and suddenly you discover swarms of entities that seem to have been there all along but were not visible before and that appear in retrospect necessary for its sustenance. You thought the Columbia shuttle was an object ready to fly in the sky, and then suddenly, after the dramatic 2003 explosion, you realize that it needed NASA and its complex organizational body to fly safely in the sky. ... The action of flying a technical object has been redistributed throughout a highly composite network where bureaucratic routines are just as important as equations and material resistance. ... What was invisible becomes visible, what had seemed self-contained is now widely redistributed. ... the search for the production of object and of objectivity is totally transformed now that they are portrayed simultaneously in the world and inside their networks of production.

Let us now return to Nicole Boivin, who pulled together insights from Tim Ingold, Bruno Latour, developmental systems theory, and niche construction. As an archaeologist, the affinities between actor-network thinking and niche construction helped her understand that the physicality of matter gives things agency independent of human organisms. DST theorist Susan Oyama is also sensitive to actor-network thinking in her own thinking about developmental systems, noting that the "swarms of entities" to which Latour refers above (called *actants* in actor network theory) bear a family resemblance to the "interactants" Oyama identifies as components of developmental systems (2000, 123). Actor-network theory appeals to her for an additional reason as well:

> Latour (1987:71–72) has described the scientist as the spokesperson for that which is studied. One of the many reasons I have found it worthwhile to think and write in developmental systems terms is that it allows me to speak for the background—the mute, manipulated materials, the featureless surround. Sometimes the peripheral is the political. (2000, 126)

Developmental systems theory and niche construction seem to require joint consideration (Oyama et al. 2001), because niche construction draws attention to the ways in which organisms make themselves, in part, by making their own environments. With Boivin, I agree that niche construction and actor-network theory also require joint consideration, and suggest that this may be more easily facilitated once it is recognized that both views rely on the same mechanism. According to John Odling-Smee and his colleagues, a major motivation for their development of a theory of niche construction was the desire to link ecological studies that focused on abiotic processes with ecological studies that focused on biotic processes. Bruno Latour has written: "As soon as you start to have doubts about the ability of social ties to durably expand, a plausible role for objects might be on offer" (2005, 75). Odling-Smee and his colleagues apparently came to a similar conclusion, which led them to propose the concept of an "artifact,"

> a third kind of object in ecosystems that is neither biotic nor conventionally abiotic, but intermediate between the two. Artifacts are not alive, yet they can only be built by living organisms. Also, once built, they are likely to respond to niche-constructing organisms in a different way from either biota or raw abiota. (Odling-Smee et al. 2003, 190)

The concept of artifact allowed them to propose a second concept, the environmentally mediated genotypic association (EMGA), in which constructed artifacts mediate between one population of organisms and another by modifying the selection pressures experienced by the second population:

> If, in a single population, genetic variation is expressed in a niche-constructing phenotype that affects natural selection acting on other genes in the same population, then the population will merely codirect its own evolution. However, if the niche construction modifies natural selection acting on genes in a second population, then the first population will now codirect the evolution. Conceivably, the induced change in the second population could feed back to the first population in the form of another modified natural selection pressure. The two populations would therefore coevolve through niche construction. (2003, 23)

To me, an EMGA looks suspiciously like a stripped-down proto-actor-network for two reasons: (1) because it is a heterogeneous assemblage linking together living and non-living actants within an ecological network, and (2) because each actant contributes its own causal influence to the network's activity—i.e., serves as a mediator—rather than serving as an intermediary that merely transports causation without affecting it (Latour 2005, 39). Biota, abiota, and artifacts would all seem to be mediators, rather than intermediaries, since they all have the capacity both to respond to further niche construction and to modify natural selection pressures (Odling-Smee et al. 2003, 191; see also Barker and Odling-Smee, this volume).

Latour also contrasts the way that an assemblage of heterogeneous mediators can be stabilized (or black-boxed) and turned into a unified whole that acts as one and is capable of transporting agency without affecting it. But black-boxed intermediaries can degenerate into networks of mediators, as, for example, when complex pieces of technology break down. Odling-Smee et al.'s EMGAs appear to have these properties. First, "niche-constructing organisms work in open systems," which means that they can "potentially drive some selected components of their environments in both thermodynamic directions, by either locally increasing or locally decreasing entropy levels." Second, "like organisms, artifacts demonstrate negative rather than positive entropy because they are usually quite highly organized; yet, unlike organisms, they have no ability to defend their own organization nor to prevent their own dissipation. Artifacts are therefore likely to demand repetitive niche construction from organisms to maintain them" (2003, 190).

If these parallels are persuasive, actor-network theory might provide a bridge that allows work in post-positivist anthropology to be articulated with developmental systems theory and niche construction in theoretical biology—thereby allowing "history" to be incorporated into discussions of "evolution." Actor-network thinking already informs ethnographic studies in science, technology, and medicine. But it is also implicit in the ethnography of development. Paige West draws readers' attention to the "abiotic" artifacts that sustained life in her field settings. The cultural mediations she describes involved not just "humans" in the lump, but a specific heterogeneous community of humans composed of Gimi people and outsiders from Australia, the United States, and elsewhere in Papua New Guinea; and they are connected not just to "the environment" in the lump, but to specific mountains and forests, to birds of paradise and trees with harpy eagle nests, to game animals, and to swiddens. These heterogeneous living actants intertwined with heterogeneous nonliving actants: the tools of the hunters and farmers, such as bush knives; imported tinned fish that replace the pork they no longer eat; dwellings for residents, five church buildings, a health post perennially out of medicine, and a school. Particularly salient are the village airstrip and planes run by the Seventh Day Adventist Church that provide the community's sole link to the outside world in the absence of roads: "The point cannot be made too strongly—everything that comes to Maimafu comes on an airplane ... The village airstrip is the site of new things, ideas, people, money" (2006, 76). The airstrip is also the site where important goods, like locally grown cash-crop coffee, go out: "all residents of Maimafu have to pay freight charges ... to the missionary planes that pick the coffee up and take it to Goroka [the provincial capital]. ... The airfreight charges paid to the mission planes fluctuate according to the price of fuel, thus tying Maimafu and other rural places that grow 'airstrip coffee' to the global political economy of oil" (2006, 106).

In her description of the knife fight, West also observed that Kelego was wearing a bath towel around his neck that had been given to him by a visiting biologist. "It is not the material nature of the towel that is most important to him," she writes. "Rather, the importance ... is the meaning of the exchange with the biologist [showing] that he has a tie to conservation and to someone who is somewhere else" (2006, 15). And yet, as Latour reminds us, the towel as a material actant plays an indispensable role as one of those objects that enables social ties to durably expand.

Niche-construction, explicitly informed by actor-network theory, would therefore consider the role of "the social" in the production of space, but "the social" would be reconceptualized in terms of "collectives," in which humans are attached to nonhumans, living and nonliving, physical and nonphysical (Latour 2005). Acknowledging this would mean, among other things, that organisms, cultures, and environments would need to be approached as emergent hybrid products of "natureculture" (Haraway 2008, 6–7; Latour 1993, 7). For example, it would mean acknowledging the naturalcultural heritage of Gimi country itself: as West argues, "The biodiversity that exists in and around Maimafu is the by-product of human habitation and use. ... The people of Maimafu, through the subsistence patterns that the NGO wishes to curtail, produced the landscape in which they live. So there is, therefore, no 'pristine condition' to preserve" (2006, 178).

A natural-cultural, actor-network understanding of niche construction might help resolve the problems faced by Steward and other analysts, inside and outside of anthropology, who have struggled to fit culture and history into discussions of human (cultural) adaptation and (cultural) evolution. For example, there was nothing predestined about Gimi country becoming the location of a biodiversity conservation project; it was a serendipitous development, connected to the fact that the husband of an ethnographer working among Gimi people in the 1970s took an interest in birds of paradise (2006, 130–131). But that contingent event led to a conservation-as-development project that mobilized features of the naturalcultural constructed niches of Gimi people and of outside conservation practitioners; both were "folded into each other," leading to the emergence of a powerful hybrid naturalcultural construct, the Crater Mountain Wildlife Management Area, (2006, 32). The results of that process, for good and for ill, could not easily foreseen or controlled, but may be explored and explained in part by post-positivist ethnographers and their allies.

Cultivating "naturalcultural" thinking by elaborating niche construction with insights from actor network theory could help biological theorists grapple with a range of issues tied to the very basic connections that organisms, particularly human organisms, forge with their environments. As Ian Hodder has recently observed,

> the brute matter of things has effects on us that go beyond social meaning. We cannot reduce things solely to the relational, to a semiotics of things. To do so undermines the power of things to entrap, and particularly to trap the more vulnerable whether these be the victims of the AIDS virus, the work gang bound by chains, the women bound by child rearing, the populations bound by global agricultural systems. ... There is much to be done in terms of understanding the different paths we have taken as humans, caught up in our varied ways with things. But the big picture is clear. Since a dependence on made things became

an evolutionary pathway, there has been one long movement, initially slow, but speeding up exponentially as the strands of human-thing entanglement lengthened and intensified. (2012, 220)

If attachments to things are part of our evolutionary pathway, expecting to escape from them, in theory or in life, is futile. Rather, the task, as Latour tells us, "is no longer a matter of abruptly passing from slavery to freedom by shattering idols, but of distinguishing those attachments that save from those that kill" (2010, 61).

Acknowledgments I would like to thank Gillian Barker, Eric Desjardins, and Trevor Pearce for inviting me to present an earlier version of this paper at the 2010 ISHPSSB Off-Year Workshop, *Integrating Complexity: Environment and History*, at the University of Western Ontario, Canada, October 7–10, 2010. I benefited greatly from their feedback and that of other conference participants. Robert Lavenda and Daniel Lavenda provided helpful observations as I revised the paper for publication. Final responsibility for the views expressed, however, rests solely with me.

References

Abu-Lughod, Lila. 1991. Writing against culture. In *Recapturing anthropology*, ed. Richard Fox, 137–162. Santa Fe: SAR Press.

Appadurai, Arjun. 1996. *Modernity at large: Cultural dimensions of globalization*. Minneapolis: University of Minnesota Press.

Appadurai, Arjun, Ashish Chadha, Ian Hodder, Trinity Jackman, and Chris Witmore. 2001. The globalization of archaeology and heritage: A discussion with Arjun Appadurai. *Journal of Social Archaeology* 1: 35–49.

Biersack, Aletta. 2006. Reimagining political ecology: Culture/power/history/nature. In *Reimagining political ecology*, ed. Aletta Biersack and James B. Greenberg, 3–40. Durham: Duke University Press.

Binford, Lewis R. 1962. Archaeology as anthropology. *American Antiquity* 11: 217–225.

Boivin, Nicole. 2008. *Material cultures, material minds: The impact of things on human thought, society, and evolution*. Cambridge: Cambridge University Press.

Boyd, Robert, and Peter J. Richerson. 1985. *Culture and the evolutionary process*. Chicago: University of Chicago Press.

Chagnon, Napoleon, and Timothy Asch. 1975. *The Ax Fight*. Watertown: Documentary Educational Resources (distributor).

Childe, V. Gordon. 1951. *Man makes himself*. New York: New American Library.

Deleuze, Gilles, and Félix Guattari. 1988. *A thousand plateaus: Capitalism and schizophrenia*. Minneapolis: University of Minnesota Press.

Durham, William H. 1991. *Coevolution: Genes, culture, and human diversity*. Stanford: Stanford University Press.

Fleck, Ludwik. 1979. *Genesis and development of a scientific fact*. Chicago: University of Chicago Press.

Foucault, Michel. 1977. *Discipline and punish: The birth of the prison*. New York: Vintage.

Frank, Andre Gunder. 1967. *Capitalism and underdevelopment in Latin America: Historical studies of Chile and Brazil*. New York: Monthly Review Press.

Fuentes, Agustín. 2009. *Evolution of human behavior*. New York: Oxford University Press.

Fuentes, Agustín, Matthew A. Wyczalkowski, and Katherine C. MacKinnon. 2010. Niche construction through cooperation: A nonlinear dynamics contribution to modeling facets of the evolutionary history in the genus *Homo*. *Current Anthropology* 51: 435–444.

Geertz, Clifford. 1963. *Agricultural involution*. Berkeley: University of California Press.

Golub, Alex. 2006. Making the Ipili feasible: Imagining global and local actors at the Porgera gold mine, Enga Province, Papua New Guinea. Ph.D. dissertation, University of Chicago.

Goodman, Alan H., and Thomas L. Leatherman (eds.). 1998. *Building a new biocultural synthesis: Political-economic perspectives on human biology*. Ann Arbor: University of Michigan Press.

Haraway, Donna J. 2008. *When species meet*. Minneapolis: University of Minnesota Press.

Hawkes, Kristen, J.F. O'Connell, and Nicholas G. Blurton Jones. 2003. Human life histories: Primate trade-offs, grandmothering socioecology, and the fossil record. In *Primate life histories and socioecology*, ed. Peter M. Kappeler and Michael E. Pereira, 204–227. Chicago: University of Chicago Press.

Hodder, Ian. 1999. *The archaeological process: An introduction*. Oxford: Blackwell.

Hodder, Ian. 2012. *Entangled*. Oxford: Wiley-Blackwell.

Ingold, Tim. 1990. An anthropologist looks at biology. *Man (N.S.)* 25: 208–209.

Ingold, Tim. 2000. *The perception of the environment: Essays on livelihood, dwelling and skill*. London: Routledge.

Ingold, Tim. 2007. *Lines: A brief history*. London: Routledge.

Ingold, Tim. 2011. *Being alive: Essays on movement, knowledge and description*. London: Routledge.

Jablonka, Eva, and Marion J. Lamb. 2005. *Evolution in four dimensions: Genetic, epigenetic, behavioral, and symbolic variation in the history of life*. Cambridge, MA: MIT Press.

Kottak, Conrad P. [1999] 2006. The new ecological anthropology. In *The environment in anthropology*, ed. Nora Haenn and Richard Wilk, 40–52. New York: New York University Press.

Latour, Bruno. 1987. *Science in action*. Cambridge, MA: Harvard University Press.

Latour, Bruno. 1988. The politics of explanation: An alternative. In *Knowledge and reflexivity: New frontiers in the sociology of knowledge*, ed. Steve Woolgar, 155–176. London: Sage.

Latour, Bruno. 1993. *We have never been modern*. Cambridge, MA: Harvard University Press.

Latour, Bruno. 2005. *Reassembling the social: An introduction to actor-network-theory*. Oxford: Oxford University Press.

Latour, Bruno. 2010. *On the modern cult of the factish gods*. Durham: Duke University Press.

Latour, Bruno. 2011. Networks, societies, spheres: Reflections of an actor-network theorist. *International Journal of Communication* 5: 796–810.

Latour, Bruno, and Steve Woolgar. 1986. *Laboratory life: The construction of scientific facts*. Princeton: Princeton University Press.

Marks, Jonathan. 2009. *Why I am not a scientist: Anthropology and modern knowledge*. Berkeley: University of California Press.

McCay, Bonnie J. 2008. An intellectual history of ecological anthropology. In *Against the grain: The Vayda tradition in human ecology and ecological anthropology*, ed. Bradley B. Walters, Bonnie J. McCay, Paige West, and Susan Lees, 11–26. Lanham: AltaMira.

Moran, Emilio F. 1990. Ecosystem ecology in biology and anthropology: A critical assessment. In *The ecosystem approach in anthropology: From concept to practice*, ed. Emilio F. Moran, 3–40. Ann Arbor: University of Michigan Press.

Morgan, Lewis Henry. 1963. *Ancient society*. Cleveland: Meridian Books.

Odling-Smee, F. John, Kevin N. Laland, and Marcus W. Feldman. 2003. *Niche construction: The neglected process in evolution*. Princeton: Princeton University Press.

Oyama, Susan. 2000. *Evolution's eye: A systems view of the biology-culture divide*. Durham: Duke University Press.

Oyama, Susan, Paul E. Griffiths, and Russell D. Gray (eds.). 2001. *Cycles of contingency*. Cambridge, MA: MIT Press.

Preucel, Robert W. 2010. *Archaeological semiotics*. Malden: Blackwell.

Rappaport, Roy A. [1968] 1984. *Pigs for the ancestors: Ritual in the ecology of a New Guinea people*, enlarged ed. New Haven: Yale University Press.

Rappaport, Roy A. 1971. Nature, culture, and ecological anthropology. In *Man, culture, and society*, ed. Harry L. Shapiro, 237–268. New York: Oxford University Press.

Richerson, Peter, and Robert Boyd. 2005. *Not by genes alone*. Chicago: University of Chicago Press.

Sahlins, Marshall D. 1976a. *The use and abuse of biology: An anthropological critique of sociobiology*. Ann Arbor: University of Michigan Press.

Sahlins, Marshall D. 1976b. *Culture and practical reason*. Chicago: University of Chicago Press.

Sahlins, Marshall D., and Elman R. Service (eds.). 1960. *Evolution and culture*. Ann Arbor: University of Michigan Press.

Schultz, Emily. 2009. Resolving the anti-antievolutionism dilemma: A brief for relational evolutionary thinking in anthropology. *American Anthropologist* 111: 224–237.

Silverstein, Michael. 1976. Shifters, linguistic categories, and cultural description. In *Meaning in anthropology*, ed. Keith H. Basso and Henry A. Selby, 11–55. Albuquerque: University of New Mexico Press.

Silverstein, Michael. 1985. The functional stratification of language and ontogenesis. In *Culture, communication, and cognition: Vygotskian perspectives*, ed. James V. Wertsch, 205–235. Cambridge: Cambridge University Press.

Smith, Barbara Herrnstein. 2006. *Scandalous knowledge: Science, truth and the human*. Durham: Duke University Press.

Steward, Julian H. 1955. *Theory of culture change*. Urbana: University of Illinois Press.

Tinbergen, Niko. 1963. On the aims and methods of ethology. *Zeitscrift für Tierpsychologie* 20: 410–433.

Trouillot, Michel-Rolph. 2002. Anthropology and the savage slot: The poetics and politics of otherness. In *Recapturing anthropology*, ed. Richard Fox, 17–44. Santa Fe: SAR Press.

Tylor, Edward Burnett. 1958. *Primitive culture*. New York: Harper and Row.

Vayda, Andrew P. 2009. *Explaining human actions and environmental changes*. Lanham: AltaMira.

Wallerstein, Immanuel. 1974. *The modern-world system: Capitalist agriculture and the origins of the European world-economy in the sixteenth century*. New York: Academic.

Walters, Bradley B., and Bonnie J. McCay. 2008. Introduction. In *Against the grain: The Vayda tradition in human ecology and ecological anthropology*, ed. Bradley B. Walters, Bonnie J. McCay, Paige West, and Susan Lees, 1–10. Lanham: AltaMira.

West, Paige. 2006. *Conservation is our government now: The politics of ecology in Papua New Guinea*. Durham: Duke University Press.

West-Eberhard, Mary Jane. 2003. *Developmental plasticity and evolution*. Oxford: Oxford University Press.

White, Leslie. 1949. *The science of culture*. New York: Grove.

Wilson, Edward O. 1975. *Sociobiology: The new synthesis*. Cambridge, MA: Harvard University Press.

Wolf, Eric R. 1982. *Europe and the people without history*. Berkeley: University of California Press.

Part II
Contested Models

Adaptation, Adaptation to, and Interactive Causes

Bruce Glymour

Abstract This paper develops alternative ways of understanding 'adaptation to' specific environmental conditions, with particular attention to the explanatory power offered by differing conceptions, the concomitant epistemic demands they make of explanations, and the models such explanations employ. It is shown that explanations of adaptation to particular environmental conditions can satisfy important intuitions only if the environmental conditions to which phenotypes are adapted are interactive causes of fitness. However, taking this constraint to be both necessary and sufficient for 'adaptation to' imposes epistemic burdens on our explanatory practice, and risks violating yet other intuitions. The paper briefly explores the consequences of the constraint for the idea that selection requires shared environments, the idea that selection requires a homogeneous environment, the idea that phenotypes may be extended, and the idea that niches may be constructed.

1 Introduction

Adaptation is a relational concept: a trait cannot be an adaptation without being an adaptation to some environment. Hence, to identify a trait as an adaptation is to imply the existence of some relation between environment and trait. The theory of evolution by natural selection identifies that relation: a trait is an adaptation to an environment only if the environment selected for the trait as against alternatives. This leaves it an open question whether such selection relates any given adaptation to specific features of an environment, or instead relates the adaptation to the environment as an undifferentiated whole. The issue is of some moment, if only because biological practice invites both readings of 'adaptation to.'

B. Glymour (✉)
Department of Philosophy, Kansas State University, Manhattan, KS 66506, USA
e-mail: glymour@ksu.edu

G. Barker et al. (eds.), *Entangled Life*, History, Philosophy and Theory
of the Life Sciences 4, DOI 10.1007/978-94-007-7067-6_6,
© Springer Science+Business Media Dordrecht 2014

On the one hand, with some frequency biologists venture hypotheses about the particular environmental conditions to which traits are adapted, as e.g. the claim that sex is an adaptation to parasitism (Levin 1975; Maynard Smith 1976; Jaenike 1978). If adaptations are not, at least sometimes, adaptations to particular environmental conditions, then such hypotheses and tests of them make no sense. On the other hand, one does not typically specify environmental conditions when estimating fitnesses—though norms of reaction for fitness, i.e. fitness functions, can be and sometimes are estimated from data, there are many perfectly standard population genetic models employing fitnesses that implicitly condition on the whole of the environment, whatever it may be. Or again, some putative selection processes, e.g. pure r-selection, seem to depend on no specific environmental feature (see e.g. Lennox and Wilson 1994). To the extent that r-selection can drive adaptations, e.g. in life history strategies, those adaptations are arguably evolved responses to the environment as a whole, rather than to any specific environmental condition.

Even if we accept the idea that adaptation is adaptation to one or another set of particular environmental conditions, there remains a further question about just what causal or nomic relations must hold between phenotype and environmental condition if it is to be sensible to speak of the phenotype as an adaptation to that condition. There are a number of alternative possible requirements that might be imposed, and the choice among them will have consequences for both our epistemic and explanatory practices. Hence, it is of some importance to ask in what sense, and to what extent, a particular environmental condition must cause (generate, explain?) selection on a phenotype if the phenotype is count as an adaptation to that environmental condition. In this paper I consider some fairly intuitive constraints on the explanatory role of appeals to 'adaptation to,' and explore the extent to which those constraints require 'adaptation to' to imply interactions between traits and environmental conditions.

I begin by adumbrating two arguments, given in detail elsewhere (Glymour 2011), and then draw out some implications of the respective conclusions. The first argument aims to show that it makes sense to speak of an adaptation *to* a particular environmental condition only if that condition interactively causes survival or reproductive success. The second argument aims to show that, in consequence, it is possible to identify the environmental conditions to which an adaptation is adapted only if one measures and models the causal influence of the environmental condition on survival or reproductive success. If the arguments are correct, they have a number of implications. Among them are constraints on an understanding of 'environment' suitable for representing relevant organism-environment interactions; some unavoidable choices about which environmental conditions are to be taken as essentially explanatorily relevant; and finally some limitations on standard methods for measuring the strength of selection. I begin with some preliminaries and then rehearse the central arguments. In the subsequent sections I explore the above mentioned implications.

2   Preliminaries

I will, for ease, confine my attention to phenotypic adaptations carried by individual organisms. There is nothing in what is to follow that prevents extensions, with some modification, to either the genic or the population/species level. But extensions of either sort do introduce complexities that require more space to deal with than is here available. In consequence, I will employ models of natural selection in which selection acts on individuals. Nothing at all hinges on this second choice of modeling level—the same results for phenotypic adaptations can be got by modeling selection at other levels, but the treatment would thereby be made unnecessarily complex.

The arguments to follow turn on the range of possible causal structures governing survival and reproduction in biological populations. The essential causal relations are between phenotypic variables, environmental variables and individual survival and reproductive success. I will use W, which I will call fitness, as the relevant effect variable throughout. In the examples to follow, W is calculated as actual or expected reproductive success, but no metaphysical commitment is intended thereby. The reader may take those calculations to be estimates of fitness in whatever sense she prefers to understand it, so long as fitness so understood depends on the joint probability density over survival and/or reproductive success.

I assume that phenotypic variables are unproblematic. No such assumption is possible with respect to environmental variables. More will have to be said later about these, but for the moment the following will be enough. There are two ways to measure an environment, either by its net effect on fitness (in whatever sense) or by the presence/absence or magnitude of some particular property. I will call variables of the first kind *measures of environmental quality* and say that they offer a qualitative representation of the environment (though the variables may well be real-valued, what these values represent is something about the quality of the environment from the perspective of the organism). I will call variables of the second kind *environmental variables*, and say that they offer an explicit rather than qualitative representation of the environment.

I adopt the now standard language of graphical causal modeling, according to which causation is an asymmetric dependence relation between variables (Pearl 2000; Spirtes et al. 2000). A variable P is said to be a direct cause of a variable W relative to some set of variables ($\mathbf{V}$, P, $W \in \mathbf{V}$) and background conditions $\mathbf{B}$ when there is some pair of interventions on P, holding all other variables except W in $\mathbf{V}$ constant, such that the probability distribution or density over W differs across the interventions. Such direct causal relations will be represented in graphs as arrows directed from the cause to the effect. Interactive or context-dependent causation is a special case of causation. P and E are interactive causes of W, relative to $\mathbf{V}$ and $\mathbf{B}$, if and only if P and E are both direct causes of W and for some moment of the distribution or density over W, there is some pair of interventions on P (or E) and some pair of values for E (or P) such that the difference in the value of the moment of the distribution or density over W between the two interventions on P (or E) given the first value of E (or P) is unequal to the difference in the value of the moment between the two interventions on P (or E) given the second value of E (or P). Informally, the effect on W of a change in one cause depends on the value of the second cause. For example, pressure and volume are interactive causes of temperature because the effect on temperature of a unit change in pressure depends on the volume.

The Japanese camellia (*Camellia japonica*) and its predator the camellia weevil (*Curculio camellia*) provide a biological example. In order to oviposit on the camellia seeds, the weevil bores a hole in the camellia fruit. This selects for a thicker pericarp. However, the strength of this selection depends on the length of the weevil's rostrum, and both traits vary within and among local populations. The equation relating the probability of boring success (PBS), and hence the probability that a seed is predated, to these traits is given by $\mathrm{PBS} = 1/\left[1 + e^{(0.819p+0.471t-4.18)}\right]$, where p is the pericarp thickness and t is the rostrum length (cf. Toju and Sota 2006). The contribution of pericarp thickness to the fitness of a given camellia plant thus depends on a locally varying environmental condition—the rostrum length characteristic of the local weevil population.

When interactive causal dependencies are mathematically modeled, i.e. when W is written as some function of its causes, interactive causes will appear together in at least one term on the right hand side of the equation (often, though as above not always, this term is a product of the causal variables). In such cases the contributions of the two (or more) causes are not separable; if P and E are *not* interactive causes of W, then it will be possible to write W as a function of P and E (and perhaps other variables) in such fashion that the terms containing E do not contain P, and vice-versa. In this case the contributions of the two causes are separable. It follows that interactive causal connections are symmetric in the following sense: if P is an interactive cause of W with E, then E is an interactive cause of W with P. It will sometimes be useful to attend to only one cause of such a pair. When necessary I will therefore write that P (or E) is the interactive cause and E (or P) the context; the difference is entirely pragmatic.

Technical preliminaries done, a philosophical preliminary is in order. In what follows I advance a (partial) conception of what it is for a phenotypic trait to be an adaptation to some but not other environmental conditions. As such, I'm engaging in a species of conceptual analysis. But I wish to be as clear as possible about just what species of conceptual analysis I intend. The explanatory power of language depends in part on how we use language to represent the world. Scientific terms, in particular, inherit their explanatory power from the fact that in using them we denote, more or less systematically, real physical, causal, nomic, or statistical features of the world. There are any number of features of causal and statistical structure that might be counted as explanatorily relevant to evolutionary outcomes, depending on which features of which outcomes one takes to be in need of explanation and on what intuitions one has about the kind of information a satisfactory explanation ought to offer. Thus, in my view, there is no fact about which phenotypes are and are not adaptations to particular environmental conditions, prior to a choice about what we will mean by 'adaptation to.' And that choice is in large measure open—there are many explanatorily relevant features of causal or statistical structure that we could choose to denote by 'adaptation to.' In consequence, I do not aim in what follows to specify any fact about what we do mean by 'adaptation to,' and still less to specify either what we ought to mean by that locution or what adaptation to *is*, metaphysically speaking.

The aim is rather this. Depending on which phenomena we choose to denote by 'adaptation to,' different kinds of information will be required to explain why any given phenotype is an adaptation to any particular environmental condition; similarly, an appeal to the fact that a phenotype is an adaptation to some particular environmental condition will itself carry some explanatory power, but what that power is will depend on which features of causal or statistical structure our usage of the term 'adaptation to' systematically respects. Further, the epistemic demands imposed by explanations of the adapted nature of a phenotype in turn depend on these facts about usage. And, roughly, the more intricate the physical distinctions we choose to denote with the term 'adaptation to,' the more explanatory power this usage has, but also the greater the epistemic demands on correct usage. The aim of the conceptual analysis to follow, then, is to clarify the range of choices available.

Specifically, I will argue that the choice to use 'adaptation to' in ways that track a natural but quite minimal structural distinction deprives the locution of any explanatory power beyond that already inherent in our usage of 'adaptation,' while a choice to track other features of causal and statistical structure confers a particular explanatory power on the locution, but at an epistemic cost. I will further point to yet more intricate structural features of interactions between organisms and their environments that one might wish to respect, in that doing so would endow the 'adaptation to' locution with yet more explanatory power. But I will take no stand on whether such further constraints on 'adaptation to' offer a particularly efficient regimentation of our language. It will be enough, here, to point to the choices that are open to us.

3 "Adaptation to" and Interactive Causation

I assume that to say of a phenotype that it is an adaptation to some particular environmental circumstance carries more explanatory power than to say of the phenotype simply that it is an adaptation. And I further assume that this extra explanatory power depends on the contrast between the environmental conditions to which the adaptation is an adaptation and those conditions to which it is not an adaptation. That is, the idea of 'adaptation to' is explanatorily useful only if for some adaptations there are features of the environment to which the adaptation is an adaptation, and others to which it is not an adaptation. For example, if hypsodonty (having high-crowned teeth) is an adaptation to the siliceous phytoliths (hard mineral particles contained in plant tissues) of grass, it had best not also be an adaptation to every other feature of grassland environments. For were it, then that hypsodonty is an adaptation to the siliceous phytoliths of grass would imply no more than that (a) hypsodonty is an adaptation and (b) it evolved in populations inhabiting grassland environments; but exactly this would similarly be implied by (and imply) the fact that hypsodonty is an adaptation to the dusty conditions of grassland habitats, and similarly to any and every other feature of such environments.

If adaptation is to be to some but not all features of an environment, we must have some principled way of distinguishing those features of the environment to which species adapt from other features of the environment. Henceforth I will call these conditions, the conditions to which an adaptation is an adaptation, the *adapting conditions*. For preference, any principle we employ to individuate adapting conditions from other features of an environment should respect certain constraints suggested by the explanations we give of adaptations and the kind of explanatory and inferential use we make of them. Among these are three intuitions that are both deep and fairly central to explanatory practices in biology.[1] First, environmental conditions are supposed to explain (in part) the fixation of those phenotypes that are adaptations to those conditions—the adapting conditions should play some central role in a full-bodied explanation of phenotypic adaptations to them.[2] I will call this the intuition ACEA (adapting conditions explain adaptations).

Second, such explanations are (at least potentially) doubly contrastive: they explain the fixation of adaptive trait values *as opposed to* alternative trait values, and they do so by appeal to one *rather than another* set of environmental circumstances. Just so, if hypsodonty is an adaptation to siliceous phytoliths in grass, we explain why horses and cows evolved high- rather than low-crowned teeth by appeal to the fact that grass has siliceous phytoliths rather than to the fact that herbivores in grasslands consume large quantities of dust when grazing. More narrowly, an appeal to environmental conditions $E = \text{e}$ to explain the fixation of phenotype $P = \text{p}$ is warranted only if there are alternative circumstances $E = \text{e}'$ and phenotype $P = \text{p}'$, such that had $E = \text{e}'$ obtained, the phenotype p' would (probably) have been maintained at a non-zero frequency in the population. I will call this intuition 'DC' (explanatory appeals to adaptations to particular environmental conditions are doubly contrastive).[3] The pair of contrasts will play a crucial role in the arguments to follow, and so for ease of reference I will call the alternative phenotypes permitting the first contrast *contrasting phenotypes*, and the alternative environmental conditions underwriting the second contrast *contrasting environmental conditions*.

---

[1] Space prevents a careful development of these intuitions from primary sources. But readers who do not find them obvious might usefully consider discussions of particular adaptations, such as Heywood (2010), McFadden (1992), Hunt (1994), and Wheeler (1991).

[2] I omit consideration of traits, genetic or phenotypic, which are in some important sense adaptive, but such that selection cannot drive the trait frequency to that expected from the mutation rate characteristic of the relevant genetic loci. Examples here include the sickle-cell allele. The issues here are important, but beyond the scope of this essay.

[3] Those puzzled by this intuition may consult van Fraassen (1980) for discussion of the first contrast (between alternative outcomes), and Glymour (1998, 2007) for discussions of the second contrast (between alternative causes or processes).

Thirdly, I suppose that the relevant explanations are selective—i.e., that the environmental conditions appear in such explanations as part of the description of a selection process rather than a drift process. I will call this intuition AEASE (adaptive explanations are selection explanations).

Given these explanatory intuitions, how are we to differentiate between environmental conditions that are and those that are not adapting conditions? One obvious individuating principle suggests itself. Some features of the environment cause survival and reproductive success, while others do not. If E_1 and E_2 are variables that measure the presence or magnitude of some environmental feature, where E_2 causes fitness but E_1 does not, then arguably no evolved phenotypic trait can be an evolutionary response to the presence or magnitude of E_1 in the environment, and hence the phenotypic trait cannot be an adaptation to the value of E_1 characteristic of the environment.[4] We might then identify the features of the environment to which an adaptation is an adaptation with those features of the environment that cause fitness; call this the *Causal Condition*.

Unfortunately, the Causal Condition will, in certain circumstances, identify particular environmental features as adapting conditions despite the fact that an explanatory appeal to those features would run afoul of the three intuitions mentioned above. Say that the distribution of an environmental variable is *homogeneous* if for any value of that variable, the proportion of one phenotype characterized by that value is equal to the proportion of any other phenotype characterized by that value. Thus if P is a phenotypic trait variable with values 1 and 2, while E is a discrete valued environmental variable, the distribution of E in a population is homogeneous when, for any value e of E, if 1/nth of the $P = 1$ phenotype is in $E = e$, then 1/nth of the $P = 2$ phenotype is in $E = e$. So, for example, if we quantify local populations of the camellia weevil as having short ($E = s$) or long ($E = l$) rostra and camellia plants as having thin ($P = 1$) or thick ($P = 2$) pericarps, the distribution of E for the metapopulation of plants is homogenous when the proportion of thin pericarp plants beset by short rostrum weevils is the same as the proportion of thick pericarp plants beset by short rostrum weevils, and the proportion of thin pericarp plants beset by long rostrum weevils is the same as the proportion of thick pericarp plants beset by long rostrum weevils. Say that an environment is homogeneous if all environmental causes of W are homogeneously distributed. If E is a cause of W, but it is not the case that this connection is interactive with P as a context, then one of two things will be true. If the actual and contrasting environments are homogeneous, an explanatory appeal to E as an adapting condition for whichever value of P is

---

[4]Recollect that on the conception of causation here employed, causal relations hold between variables, and to say that E causes W is to say that by changing E one can change (the probability density over) W; hence there will be values of E that increase the value of W, and other values of E that decrease the value of W. Loosely, the causes of an outcome include both producers and preventers of that outcome.

fixed will violate DC (i.e. will not be doubly contrastive). Conversely, if the relevant contrasting environment is not homogeneous such an explanatory appeal will violate AEASE (i.e. will not be a selection explanation).

To see this, consider a population of ants invading a valley. Some ants suffer when soil moisture content is too high, while others are relatively resistant; call this trait P, with values 1 and 2 respectively for the resistant and non-resistant types. Ants of both types prey on seeds, and the non-resistant type is slightly more efficient at finding and processing seeds. Further, both types are equally sensitive to the lowest soil temperature in winter. Denote the winter-minimum soil temperature by E_1, with binned values from 0 (below freezing) to 5 (above 25 °C), and the local soil moisture content by E_2, again with values 0 (very dry) to 5 (very wet). Colony fecundity (i.e. the number of daughter queens sent out in a given year) is given by the equation $W = 20 + 3P + 2E_1 - 2E_2P$. Initially, the valley is unoccupied, with far more potential colony sites than offspring colonies, so all daughter colonies survive. Once the valley fills, fecundity is still determined by the equation for W, and old colonies are replaced by offspring colonies at random from among all offspring colonies, with a probability that is independent of the types of both old and offspring colonies, so as to hold the population size constant at K, whatever it may be, for the valley.

Suppose ants of both types initially invade the value under fairly good conditions—E_1 values are at 4 for every ant colony, and E_2 values are at 1 for every ant colony (call this Environment 1). But as time passes, these values fluctuate. Consider first uniform changes in E_1, with E_2 constant. Intuitively, uniform changes in E_1 can change the rate of evolution, because a change in E_1 will influence the reproductive success of both types. But since this influence will affect both types equally, it can change the magnitude of selection coefficients, but it cannot change which of the two types is fitter. Numbers may help.

Using fecundity as our measure of fitness, the fitness of any given colony is given by the equation $W = 20 + 3P + 2E_1 - 2E_2P$. Thus, for our ants in the initial environment, the resistant strain will have a fitness of $20 + 3(1) + 2(4) - 2(1 \cdot 1) = 29$ while the non-resistant strain has a fitness of $20 + 3(2) + 2(4) - 2(2 \cdot 1) = 30$. If we relativize to the fittest type, then initially the non-resistant strain has relative fitness 1 while the resistant strain has relative fitness $29/30 = 0.967$. If the environment is invariant, i.e. remains fixed at Environment 1, the non-resistant type will, slowly, go to fixation. Imagine now that winter minima decrease, so that $E_1 = 1$ for every colony in every generation (call this Environment 2). Then the absolute fitnesses become $20 + 3(1) + 2(1) - 2(1 \cdot 1) = 23$ and $20 + 3(2) + 2(1) + 2(1 \cdot 2) = 24$ for the resistant and non-resistant types respectively. This leads to a decrease in the relative fitness for the resistant type, to 0.958: selection is slightly stronger now, and the pace of evolution has quickened. But notice that we have not changed which of the two types is fitter. And in fact, there is no change in the value of E_1 that could produce such a reversal of fitness, for the very reason that E_1 causes W independently of, i.e. without interacting with, P.

Say that two environments differ uniformly (or that a change from one to the other is uniform) if, for each cause of fitness E and for all individuals i, j in the population, the difference between the E values for i in the two environments equals the difference between the E values for j in the two environments (in consequence, uniform changes on a homogeneous environment result in another homogeneous environment). Appeals to the causal role of the variable E_I in producing fitness in our ant population cannot contrastively explain why one type is fitter than another (and hence why that type evolves), so long as the contrasting environment (i.e. alternative distribution of E_I over types) differs only uniformly from the actual distribution of E_I. Contrasting homogenous environments differ uniformly, and so relative to contrasting homogeneous environments, appeals to E_I cannot explain why one rather than another phenotype evolves. This is a general feature of non-interactive causes: non-interactive environmental causes of fitness can explain why one rather than another contrasting phenotype evolves only by appeal to an actual or contrasting non-homogeneous environment, i.e. a situation in which one type differentially inhabits the better local habitats.

To see how an explanatory appeal to a non-uniform change in the environment works, suppose our study population moves from Environment 1 ($E_1 = 4$, $E_2 = 1$) to the following non-homogeneous environment (Environment 3): all $P = 1$ individuals are in $E_1 = 5$, $E_2 = 1$ habitats (so their realized fitnesses will be $20 + 3(1) + 2(5) - 2(1 \cdot 1) = 31$), while all $P = 2$ individuals are in $E_1 = 0$, $E_2 = 1$ habitats (so their realized fitnesses will be $20 + 3(2) + 2(0) - 2(1 \cdot 2) = 22$). Now the resistant type has the higher fitness, and (supposing this distribution of types to local habitats is constant), resistance evolves. This result, the evolution of $P = 1$ (resistance) rather than $P = 2$ (non-resistance) can be explained contrastively by appeal to the new rather than old distribution of E_1. But if any such explanation of the fixation of $P = 1$ also treats $P = 1$ as an adaptation to the environment, as characterized by the distribution of E_1, the explanation will be fallacious, and will violate AEASE.

Intuitively, the problem is that resistance has nothing to do with the success of the resistant type; that success rather derives from the fact that the resistant types more commonly experience higher winter minimum temperatures. More precisely, either the new, non-homogenous distribution of E_1 arises by chance or as a result of some other, behavioral, phenotype perfectly correlated with P. If the former, we have a case not of selection, but of drift (see Brandon 1990), in violation of AEASE. If the latter, then there is selection, but in favor of the behavioral phenotype that leads $P = 1$ individuals to favorable habitats and $P = 2$ phenotypes to unfavorable habitats; $P = 1$ has been sorted rather than selected. Again, AEASE has been violated, and in neither case is $P = 1$ an adaptation to the environment.[5]

Thus the Causal Condition fails because the structural, causal features it employs to sort adapting from non-adapting conditions are insufficiently explanatorily powerful. In particular, the Causal Condition can be satisfied by homogeneous

---

[5]Note that it matters here not at all whether the behavioral trait in question produces the non-homogenous environment by habitat selection, or by niche construction—in either case, it is not P, but the phenotypic cause of E_1 that is the immediate focus of selection, and hence the immediate locus of adaptation.

environmental conditions that cannot contrastively explain the evolution of one rather than another phenotype, except by appeal to contrasting non-homogeneous environments. Doubly contrastive selection explanations are impossible in such cases because the same phenotype would evolve under any uniform change in the environmental conditions. The same failure does not beset an alternative condition, which I will call the Interactive Causal Condition. According to this condition a phenotype $P = \mathrm{p}$ can be an adaptation to an environmental condition $E = \mathrm{e}$ only if E is an interactive cause of fitness with P.

To see how the Interactive Causal Condition avoids the problems besetting the Causal Condition, consider again our ants. Suppose, as before, our ants begin in Environment 1, with relative fitnesses of 0.967 for the resistant type and 1 for the non-resistant type. Now suppose the environment changes uniformly so that $E_2 = 4$, while $E_1 = 4$ remains constant (call this Environment 4). The absolute fitness of the resistant type then becomes $20 + 3(1) + 2(4) - 2(4 \cdot 1) = 23$ while that of the non-resistant type is $20 + 3(2) + 2(4) - 2(4 \cdot 2) = 18$. The relative fitnesses are now 1 for the resistant type and 0.783 for the non-resistant type: selection strongly favors resistance, and hence an appeal to the new (homogeneous) environment ($E_2 = 4$) explains the evolution of resistance. While uniform changes in E_1, can influence the rate at which evolution occurs but not its eventual outcome, uniform changes in E_2 can influence evolutionary outcomes, because they can change which type is fitter. Thus, in this scenario, we may say that resistance, $P = 1$, is an *adaptation to* the (homogeneous) distribution of E_2 in this sense: had that distribution been different (though still homogeneous), $P = 2$ would not have evolved to fixation. Because the contrasting environment (e.g. Environment 1) is homogenous, the different evolutionary outcomes our population would experience in the two scenarios (Environment 4 versus Environment 1) are a consequence of selection on P, rather than an artifact of chance or selection on some correlated trait.

4   Explanation, Inference, and Representation

Though more demanding than the Causal Condition, the Interactive Causal Condition remains a fairly minimal constraint on usage, and thus underwrites only limited explanatory power. Insofar as we use 'adaptation to' in ways that respect the Interactive Causal Condition, by 'the phenotype $P = 1$ is an adaptation to environmental condition $E_2 = 4$' we imply that E_2 is an interactive cause, with P, of fitness in the relevant population. It follows from this that a certain kind of counterfactual is true, namely that there is some (possible) homogenous environment E in which $P = 1$ evolves to fixation, and some other environment E' in which it does not, where E and E' differ only in a uniform change in the distribution of E_2. This makes possible doubly contrastive explanations of adaptive phenotypes, for example: P is fixed at 1 rather than at 2 because $E_1 = 4$ rather than 1. But this power imposes an epistemic cost. If one is to diagnose from observational data the fact that E_2 is an interactive cause, with P, of W, two things must be true. First, the observations and models fitted to them must include measures of E_2, and second, E_2 must vary over individuals in the sample. I consider the two points in turn.

Population biologists employ two different ways of representing environments in mathematical models. The first, in effect, conditions on features of the environment which are thereby presupposed to be common to all members of the modeled population; the second explicitly introduces variables whose values denote the presence, absence, or magnitude of specific features of the environment. For example, wildlife biologists often employ summary measures of environmental quality, related to the expected rate of reproduction for a focal species occupying the environment (see Johnson 2007 for a review). Similarly, logistic growth models employ the parameters r and K, both of which are, in this sense, measures of habitat quality. In the same way, in simple population genetic models employing a single fitness or selection coefficient for each genotypic class, the fitnesses or selection coefficients are in effect a measure of habitat quality, from the perspective of each genotype. More complicated models, e.g. those employing contextualized fitnesses (sensu Kerr and Godfrey-Smith 2002) or niches (Levene 1953), specify fitnesses for genotypes in more narrowly circumscribed environments. However, such measures of environmental quality, whether or not they are niche or genotype specific, do not specify or measure those features of the environment that are causally responsible for the differences in fitness, intrinsic rate of increase, or carrying capacity. When such models are fitted to data they can, sometimes, be used to diagnose the presence of an interactive environmental cause of fitness (this for example is one thing Brandon's phytometer studies do; see Brandon and Antonovics 1996). But neither the models nor the component measures of environmental quality can by themselves be used to identify which features of the environment are in fact interactive causes of fitness. Thus, while such models allow one to identify a phenotype as an adaptation, they will not permit one to identify the adapting conditions to which the phenotype is, in fact, an adaptation.

This limitation arises in the following way. Suppose we gather data which include measures on individuals of components of fitness (fecundity, survival, or what have you), measures of individual phenotype (size, height, coloration, or what have you) and location (position on a transect or grid, say), but not specific values for specific environmental conditions obtaining at that location. We can, for each location, calculate type-specific mean values for our measures of fitness and note differences in them. And we can look, in particular, for pairs of locations in which the ordinal relation between type fitnesses is reversed (i.e. one type is fitter in one location while another is fitter in a different location). One way to account for such reversals is to appeal to some changing environmental condition which is, with phenotype, an interactive cause of fitness. But, necessarily, we will have no evidence about just what this condition is, since very many environmental conditions will differ between locations and we will have measured none of them on individual organisms. And even the inference that such an interactive environmental cause varies in value over locations is suspect, for there is another way to explain such reversals in the ordinal relations between type fitnesses. If there are two or more non-interactive causes of fitness, and they have non-homogeneous distributions over the types in each location, so that one type experiences better conditions in one location, the other better conditions in the second location, the fitness relations may be reversed without the presence of an interactive cause. To rule out this sort of case it is generally necessary to actually measure the relevant environmental conditions, either in the wild or in experimental contexts in which the conditions of interest are controlled (or manipulated).

It is now common practice to introduce explicit measures of particular environmental conditions into one's models in evolutionary ecology; it is becoming common in population demography and population management as well (see Caswell 2001; Guissan and Thuiller 2005). Given joint measures on environmental conditions and components of fitness, it is possible to test from observational (rather than experimental) data hypotheses about the causal influence of specific environmental variables on fitness, and if such a causal connection is found, to further test the hypothesis that the dependence is interactive with one or another phenotypic feature. The first epistemic price of such tests is the requirement that environmental conditions actually be measured and represented explicitly in one's model. Thus, the judgment that a phenotype is an adaptation to some specific set of environmental conditions requires that those conditions be explicitly represented as the value of a measurable (or at any rate estimable) variable, and can be warranted only by data that include measures of those variables.

What is more, that a given environmental variable is a cause of fitness can be determined from observational data only if those conditions vary over sample membership. While it is true that correlation is not causation, it is also true that the one statistical signature characteristic of the absence of a causal dependency is statistical independency (e.g. the absence of a correlation). Associations between variables can be detected in a sample only if both variables vary over the sample membership—without variance there can be no covariance. Hence, the hypothesis that an environmental variable causally influences fitness cannot be tested against data in which the environmental variable is constant. This is the second epistemic price of tests for interactive causal dependencies between environment, phenotype, and fitness. It is not trivial, for there may be minimal variation over individuals or sub-populations of a species with respect to large scale environmental features; this in turn may require that data be gathered over long time periods so that the requisite variation will appear as temporal, inter-generational variation rather than intra-generational variance in environmental conditions. There is a conceptual price here as well. It is sometimes thought that selection requires a common environment.[6] Whether or not this is so, the kind of selection that drives adaptation to particular environmental conditions will be undiscoverable if all organisms in the population are subject to identical environmental conditions. We will recur to this point below.

---

[6]This is an implicit consequence of any view that pairs dispositional fitnesses with the standard view that selection requires heritable differences in fitness. It is sometimes made explicit, as e.g. in (Brandon 1990).

5   Limits and Extensions

I said above that the Interactive Causal Condition imbues appeals to 'adaptation to' with a limited explanatory power, but did not elaborate on the nature of the limits. I do so now. If we confine our usage of 'adaptation to' so as to respect the Interactive Causal Condition, it will follow from the claim that $P = 1$ is an adaptation to $E_2 = 4$ that there is some (possible) homogenous environment E in which $P = 1$ evolves to fixation, and some other environment E' in which it does not, where E and E' differ only in a uniform change in the distribution of E_2, and such counterfactuals make possible the doubly contrastive explanations of the form 'the phenotype is fixed at this value rather than that because the environment was characterized by this condition rather than that.' But in fact very few populations inhabit homogeneous environments, and for any phenotype of interest it is likely that there are some adapting conditions that do not have a homogeneous distribution, over time and space, in the adapting population.[7] This raises some puzzles about which potential contrasting environments are explanatorily relevant. To explore the implications of such non-homogeneous distributions, we need to expand our conception of the environment in a number of ways.

First, we need two distinct conceptions of the environment occupied by an individual organism. Let E be a vector $< E_1, E_2, \ldots E_n >$ of environmental variables. I will say that a set of values for each of the E_i, as measured on an individual organism j (thus, $E(j) = e = <E_1(j) = e_1, E_2(j) = e_2, \ldots E_n(j) = e_n>$), comprises the *narrow individual environment* occupied by individual j. Let $\mathbf{P}_j(E)$ be a probability density over E for j, characterizing for each possible narrow individual environment the chance that individual j comes to occupy that environment. I will call $\mathbf{P}_j(E)$ a *wide individual environment*. Second, we need two distinct conceptions of the environment occupied by a population. Let $\mathcal{F}_p(E)$ be a frequency distribution of individuals in population p over narrow individual environments; I will call $\mathcal{F}_p(E)$ the narrow population environment. Finally, let $\mathbf{D}_p(E)$ be a probability density over all possible frequency distributions $\mathcal{F}_p(E)$. I will call $\mathbf{D}_p(E)$ a *wide population environment*. It will be helpful in what follows to relativize population environments to phenotypically specified classes in a population. To that end I will write $\mathcal{F}_c(E)$ to represent the frequency distribution over narrow individual environments of individuals in the cth class of the population p, and $\mathbf{D}_c(E)$ to represent the probability density over such frequency distributions.

Let us reconstruct the explanatory import of the 'adaptation to' locution, given the Interactive Causal Condition, but now employing the above conceptions of 'environment.' Suppose that every individual in the population occupies an identical wide individual environment, with $E(i)$ probabilistically independent of $E(j)$ for any two individuals i and j in the population. It follows that for any pair of similarly sized classes c and c' in the population, $\mathbf{D}_c(E) = \mathbf{D}_{c'}(E)$, even though

---

[7]Marshall Abrams has in conversation pressed various critical points regarding actual non-homogeneous distributions of adapting conditions. Though what I say will doubtless leave him unsatisfied, his worries influenced some of what follows and I thank Marshall for pressing them.

individuals may differ in their narrow environments, and hence even though the frequency distributions over the classes c and c′ might differ (i.e. $\mathbf{F}_c(E) \neq \mathbf{F}_{c'}(E)$). When $\mathbf{D}_c(E) = \mathbf{D}_{c'}(E)$, I will say that the wide population environment $\mathbf{D}_p(E)$ is homogeneous. Say that $\mathbf{D}_p(E)$ and $\mathbf{D}'_p(E)$ differ uniformly if and only if both are homogeneous. For cases in which the actual wide population environment is homogeneous, the actual wide population environment and some subset (not necessarily proper) of the remaining homogeneous environments will comprise a set of reasonable contrasting environments. What is required is that the actual and contrasting environment(s) differ in the expected (homogeneous) distribution of one or more interactive environmental causes. Given such homogeneous wide population environments, the explanation of the fixation of P at 1 rather than 2 can appeal to the fact that the adapting conditions have expected values determined by $\mathbf{D}_p(E)$ rather than different expected values, determined by a different homogeneous wide population environment $\mathbf{D}'_p(E)$, where, had $\mathbf{D}'_p(E)$ been the actual wide population environment, we would not have expected P to fix at 1.

The forgoing recapitulation presupposes that the adapting population occupies a homogeneous wide population environment, and that supposition will, for various reasons, often not be satisfied. Such cases raise a number of puzzles; before turning to them, it is worth pausing to expand on the conceptual puzzle noted at the end of Sect. 4. The idea that selection requires a common environment is certainly implicit in standard readings of relatively simple population genetic models (the charge of illicit averaging, for example, depends on it). Explanations of adaptations to particular adapting conditions, as above, also depend on a shared environment, when the environment is understood as a wide population environment—i.e. a probability density over the proportion of each geno- or phenotype characterized by specific values for environmental causes of fitness. But such explanations are available even when the actual proportion of types in particular conditions varies quite radically from type to type. Hence, the relevant classes may differ in their narrow population environments, and individual organisms may differ in both their wide and narrow individual environments (and in fact, as noted at the end of Sect. 4, the last such difference is essential for the possibility of detecting the kind of selection that drives adaptation to particular conditions). Hence, to make sense of adaptation to specific environmental conditions, we must relinquish the idea that selection requires shared environmental conditions. This will be relevant below.

Let us now recur to the case in which the actual wide population environment is not homogeneous. The Interactive Causal Condition insists that for a phenotype to be an adaptation to some environmental condition, the phenotype must be encoded as some value of a trait variable P_a and the environmental condition as the value of an environmental variable E_a where P_a and E_a are connected to fitness W by a causal structure of the kind depicted in Fig. 1.

Here f is the functional dependence of W on E_a, which is controlled by P_a, and P_a may or may not also directly influence W (represented by the dotted arrow). One important way in which homogeneity can fail involves a probabilistic association between environmental conditions and phenotype, represented by a double-ended edge in Fig. 2.

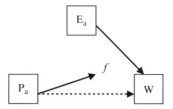

Fig. 1 A simple interactive dependency. A simple causal structure satisfying the interactive causal condition; the *dashed edge* represents a possible but non-essential secondary, direct path by which P_a may contribute to W independently of E_a

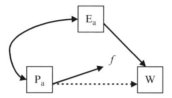

Fig. 2 An interactive dependency with an unexplained association between causes. E_a and P_a interactively cause W, and are themselves associated in virtue of an unspecified causal connection between them and/or some unmeasured common cause

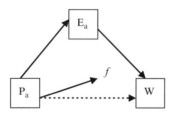

Fig. 3 Interactive phenotypic cause of an environmental condition. P_a both causes E_a and interactively controls the influence of E_a on W

This sort of association may arise in three ways: E_a may cause P_a, P_a may cause E_a, or they may share some common prior cause. Phenotypic plasticity offers an example of the first kind of case, and I discuss it in my (2011). The third kind of case involves more complexities than I have space to deal with, but any resolution of such cases requires as a background some view of the second kind of case, which I will therefore briefly address here. Graphically, the second kind of case can be represented by Fig. 3.

This structure arises whenever the phenotypic composition of a population causally influences (or in the case of frequency dependent selection, constitutes) an adapting condition. Such structures also represent one kind of 'extended phenotype' (Dawkins 1999) and one kind of 'constructed niche' (Odling-Smee et al. 2003), in

Fig. 4 Causal structure with
a landscape variable. The
landscape variable L causes
E_2, which is both an effect of
P and an interactive cause
with P of W

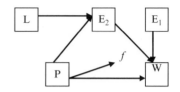

that they occur whenever the phenotype of an individual causally influences the narrow individual environment occupied by the individual.[8]

Consider again our ant population, but now attending to variations in the valley's landscape. At least initially, the landscape includes many more potential colony sites, or locations, than there are progeny seeking such sites. For ease, suppose that all locations in the landscape are characterized by a similar lowest annual soil temperature, encoded by the value 2 for the variable E_1. We will, however, allow E_2 to vary uniformly, so that 1/6 of the locations have a soil moisture content falling in the range denoted by $E_2 = 0$, 1/6 have an E_2 value of 1, and so on. We will encode this shared landscape-level environmental property (the frequency distribution of E_2 values over locations) with the variable L; every individual in the population shares the same value for L because all inhabit the same landscape. The reproductive success (new colonies established) is governed, as in the examples above, by the equation $W = 20 + 3P + 2E_1 - 2E_2P$. But now we will not only let P directly cause fitness (as per the second term in the equation for W), and govern the contribution E_2 makes to fitness (as per the fourth term in the equation for W), but also further suppose that P causes E_2: new queens with the $P = 2$ phenotype simply choose the first location they happen across, while those with the $P = 1$ phenotype are slightly more likely to choose drier locations. In particular, let $\Pr(E_2(i)=n)=1/6+b$, where the preference for drier habitats is quantified by b, with $b=(P(i)\text{-}2)(n\text{-}2.5)/15$. Hence, the probability that a $P = 2$ queen occupies a location with $E_2 = n$ is 1/6, for all values of n. But the probability that a $P = 1$ queen occupies a location with $E_2 = n$ varies: when n = 0, this probability is at its highest, $1/6 + 1/6 = 10/30$; for n = 1 the probability is $1/6 + 1/10 = 8/30$; for n = 2 the probability is $1/6 + 1/30 = 6/30$; for n = 3 the probability is $1/6 - 1/30 = 4/30$; for n = 4 the probability is $1/6 - 1/10 = 2/30$; and for n = 5 the probability is $1/6 - 1/6 = 0$. Thus, resistant ($P = 1$) ants enjoy a kind of double advantage—they are less likely to find themselves in especially wet soil conditions, and better able to deal with those conditions when they do occupy them. The causal structure for the system, including the landscape variable L, is given in Fig. 4.

---

[8] Both ideas remain largely metaphorical, and hence conceptually quite rich. Consequently, it is not the case that Fig. 3 represents the causal structure operative in any realization of either metaphor. It is rather that any system for which the structure in Fig. 3 holds is a system in which the E_a value counts as an extended phenotype, in one sense of that term, and as a constructed niche, in one sense of that term.

Fig. 5 Causal structure with a landscape variable, omitting mediating environmental variable. The causal structure from Fig. 4, with E_2 omitted; L is now an interactive cause with P of W

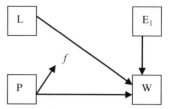

The respective fitnesses of the two types, calculated as expected per-capita reproductive success, are then 24.33 and 20 for the $P = 1$ and $P = 2$ phenotypes. Barring drift, the population fixes at $P = 1$, and clearly it does so as a result of selection on P, and for $P = 1$ in particular. And clearly, P is not an adaptation to $E_1 = 2$, for $P = 1$ would fix under any uniform change in the distribution of E_1. E_1 influences only the rate at which evolution occurs, not its eventual outcome. We might, employing the Interactive Causal Condition as not only necessary but sufficient, claim that P is an adaptation to E_2, since P and E_2 are interactive causes of W. And there is at least the following to say for that option: had the distribution of E_2 been different in specifiable ways, $P = 1$ would not have evolved (barring drift); what is more, some of those alternative distributions of E_2 are homogeneous. If, for example, $\mathbf{D}_{p=2}$ were biased in the same way $\mathbf{D}_{p=1}$ is, then $P = 2$ would evolve; similarly, were the wide population environments for both the $P = 1$ and the $P = 2$ phenotypes Environment 1, then again $P = 2$ would fix.

On the other hand, the *actual* distribution of E_2 over phenotypic classes, i.e. the narrow population environment, will not be homogeneous in any generation (again, barring drift in the form of sampling error). Moreover, these biases are not accidental, which suggests first that the net effect of the bias should not be chalked up to drift, and second that we might be attending to the wrong causal variables. We might then prefer to say that $P = 1$ is not an adaptation to E_2, but rather to whatever causes the biased distribution of E_2. But—and this is perhaps even worse—P is itself just such a cause, since the P value of any individual in our population is a cause of the E_2 value for that individual. And it seems, at least, disconcerting to say of a phenotype that it is an adaptation to an environmental condition that it itself causes. Here again the notions of an extended phenotype and constructed niche seem relevant.

One might, instead, hold that $P = 1$ is an adaptation not to E_2, but to L, since L is the other cause of the biased distribution of E_2 over phenotypes. What is more, this is consistent with the Interactive Causal Condition. For if we simply ignore the variable E_2, the causal structure in the system reduces to that in Fig. 5, and with respect to that structure, L is an interactive cause, with P, of W. On the other hand, if we say only that $P = 1$ is an adaptation to L, our description of the situation omits the explanatorily relevant fact that P not only influences the value that E_2 takes, it influences as well the degree to which that value, whatever it is, in turn influences W.

Other options are possible. One might hold that $P = 1$ is an adaptation to each of L and E_2 individually—this follows if we insist that the Interactive Causal Condition is both necessary and sufficient for adaptation to a particular environmental condition. Somewhat differently, we could hold that $P = 1$ is an adaptation to the conjunction of L and E_2. Either option entails that the selection pressure driving adaptation to a particular environmental condition can occur even when the actual distribution of the relevant environmental condition is not homogeneous, i.e. even when environments are not shared. Indeed, either view implies that selection does not require even that phenotypes share wide population environments.

Very differently, one might insist that adaptation to a particular environmental condition requires not only that the Interactive Causal Condition be satisfied, but requires as well that the putative adapting condition have a homogeneous distribution. Motivation for this view can be found in the idea that selection requires a shared environment, for that idea is preserved on this conception of 'adaptation to' by the requirement that adapting conditions have an actual homogenous distribution over phenotypes. On the other hand, any such requirement will impose exacting epistemic demands on explanatory appeals to 'adaptation to,' and will severely restrict the number of adaptations that are, in fact, adaptations to particular adapting conditions. Nonetheless, the requirement would nicely circumvent the above quandaries.

My own inclination is to regard the Interactive Causal Condition as a necessary and sufficient constraint on 'adaptation to,' i.e. to say that a phenotype is an adaptation to a particular environmental condition just in case the phenotype is an adaptation and an interactive cause with the condition of fitness. But I don't hope to defend my preferences here. In my view there is no independent fact about what P is or is not an adaptation to, with respect to which our analysis can get things right or wrong. Rather, there is a set of causally and statistically distinguishable ways in which phenotypes and environmental conditions can interact to produce fitnesses, and we can choose to use 'adaptation to' in ways that respect more or fewer of those distinctions, with concomitant implications for the explanatory power and epistemic demands of the 'adaptation to' locution. Our choices should, however, be informed, and I hope that the forgoing has limned some of the consequences of some of the available choices. I do wish, however, to close in the next section by attending to one further epistemic consequence of the Interactive Causal Condition.

6   Measuring Selection

There are several broadly distinguishable ways of measuring the strength of selection. One set of measures quantifies the strength of selection by the evolutionary change, *i.e.*, the change in type frequencies, engendered by a selection process.[9] Among such measures are selection differentials and the response to selection, for example. Unfortunately, these statistics are not well suited as measures of

---

[9]Though heritable variation in fitness is commonly taken to be either a necessary or a necessary and sufficient condition for selection (e.g. Lewontin 1970), on some views selection just *is* differential reproductive success (see e.g. Eldredge 1986 or Grant 1991) or differential fitness of types (e.g. Schluter 1988).

the strength of selection driving adaptation to particular environmental conditions, because when treated as measures of selection they confound the effect of the adapting conditions with the effect of other environmental features that, while influencing survival and reproductive success, are not adapting conditions.

To illustrate, consider again our ants in Environment 1 ($E_1 = 4$, $E_2 = 1$), and recollect that success is governed by the equation $W = 20 + 3P + 2E_1 - 2E_2P$. The respective fitnesses for $P = 1$ and $P = 2$ individuals are 29 and 30 respectively. Supposing we begin with equal numbers of each type, the corresponding response to selection (given by the difference between mean phenotype among parents and offspring) is then $89/59 - 30/20 = 0.008$. Consider an alternative environment in which E_2 remains 1 but $E_1 = 0$. Now the fitnesses are 21 and 22, respectively, and the selection differential is $65/43 - 30/20 = 0.012$. In the second case, evolution proceeds somewhat more quickly, and this yields a higher estimate of the strength of selection. But the net effect of E_2, the only adapting condition, is the same in the two cases.

A second way of measuring the strength of selection is to consider the differences in type-specific rates of survival and reproductive success generated by (or on some views constitutive of) the selection process. Selection coefficients and fitness differences are standard examples. But such measures suffer from exactly the same flaws. As measures of the strength of selection simpliciter they may be fine, but as measures of the strength of selection driving adaptation to particular conditions, they are confounded. This can be seen in the example above. When $E_1 = 4$ and $E_2 = 1$, the type fitnesses are 29 and 30, yielding relative fitnesses (dividing through by the maximal fitness) of $w_1 = 0.967$ and $w_2 = 1$, for a selection coefficient $S = 0.033$. Although the effect of E_2 on reproductive success remains unchanged in the alternative environment $E_1 = 0$, $E_2 = 1$, the relative fitnesses and selection differentials have changed. The fitnesses are now 21 and 22, with relative fitnesses $w_1 = 0.955$, $w_2 = 1$, and the selection coefficient is $S = 0.045$. Again, our measures of the strength of selection are responsive to changes in E_1, when to measure the strength of selection driving the fixation of $P = 1$ as an adaptation to E_2 they ought not be.

A third kind of measure of the strength of selection is closer to what we require. This third way of tracking the strength of selection identifies the strength of selection with some measure of the association between phenotype and (components of) fitness, where the measure of association is in turn interpretable as the strength of the causal influence of the phenotypic variable on fitness. The use of linear regression methods in evolutionary ecology (c.f. Roughgarden 1979, though the methods have been employed at least since the 1960s), and of selection gradients (partial regression coefficients) in population genetic treatments (c.f. Lande and Arnold 1983), are illustrations. Prospects here are more promising, but there are substantive methodological problems to be resolved. I will illustrate just one.[10]

---

[10]The interactive causal connection between adapting conditions and fitness is of particular concern, but space prevents any useful elaboration here.

Fig. 6 Two phenotypic
causes of fitness with a shared
common cause. P_1 and P_2
both cause W, and share a
measured common cause G;
α, β, χ and δ are path
coefficients measuring the
conditional association
between pairs of variables

Fig. 7 Complex structure
with a common cause. P_2
causes P_1, and both cause W;
in consequence P_2 influences
W directly, and indirectly
through its influence on P_1

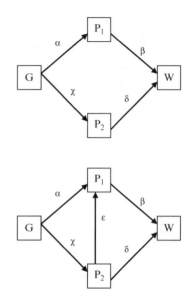

The idea behind selection gradients was to sort out the extent to which different phenotypes that might share a common genetic cause were individually influencing fitness, and hence evolutionary outcomes. Put another way, given that fitnesses vary among phenotypic classes as defined by values of P_2, as in Fig. 6 below, how much of that difference in fitness is explained by selection on P_2, and how much by selection on the associated trait variable P_1? Given the relatively simple structure in Fig. 6 (and assuming linear dependencies), one could simply consider the total association (the correlation or regression coefficient) between P_2 and W. But that total association will be proportional to $\chi\alpha\beta + \delta$, i.e. it will confound the effect of P_2 with an association induced by the effect of P_1 and the fact that P_1 and P_2 share a common genetic cause. In such simple structures, one can produce an unconfounded estimate of the path coefficient δ by taking the partial regression coefficient between P_2 and W, conditioning on P_1. If we identify the strength of selection on P_2 with its causal effect on fitness, and measure that effect by estimating δ, we seem to be on safe ground.

We might then try something similar with respect to environmental conditions. If we have the structure represented in Fig. 4 above, we could measure the strength of selection driving adaptation to E_2 by the association between E_2 and W conditioning on E_1 (a non-interactive cause of fitness, and therefore not an adapting condition). Problems arise however. Consider the structure represented in Fig. 7. Here, P_2 actually causes P_1, and so the total causal influence of P_2 on fitness is really best represented by $\delta + \varepsilon\beta$; here δ represents the direct influence of P_2 on W, and $\varepsilon\beta$ the indirect influence of P_2 on W through P_1. Here, the partial regression of P_2 on W controlling for P_1 will yield an estimate of δ. Thus, the selection gradient is a *biased* estimate of the causal influence of P_2 on W, and hence of the strength of selection on W.

Fig. 8 Complex structure relating adapting condition to fitness. E_2 causes W directly and interactively with P, and indirectly but not interactively through its influence on E_1

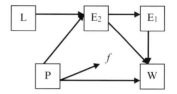

These worries, and others, beset the extension of these measures to environmental conditions. If, as in Fig. 8, the adapting condition influences W directly and indirectly by way of some further environmental variable, conditioning on that variable will lead to biases in estimates of the strength of selection. There are therefore methodological problems associated with the idea that the strength of selection to particular environmental conditions should be measured by the causal influence of those conditions on fitness.

These difficulties are, nonetheless, less pressing than those confronting alternatives, and hence provide reasons for preferring a causal measure of the strength of selection driving adaptation to over the much less sensitive statistical measures using the response to selection or fitness differences.

7   Summary

The arguments above show, I believe, that if 'adaptation to' is to carry more explanatory power than 'adaptation,' something like the Interactive Causal Condition will have to be endorsed. They show further that any such endorsement of the Interactive Causal Condition carries with it a commitment to explicit measures of environmental conditions as against measures of habitat quality. Finally, I hope the subsequent sections have suggested the range of quite intricate statistical and causal relationships between environment, phenotype, and fitness to which we might wish to attend. Failure to be clear about which structures we intend to denote when speaking of 'adaptation to' will lead to explanatory incoherence. And failure to explicitly model those structures will lead to biased or confounded estimates of the strength of selection.

References

Brandon, Robert N. 1990. *Adaptation and environment*. Princeton: Princeton University Press.
Brandon, Robert N., and Janis Antonovics. 1996. The coevolution of organism and environment. In *Concepts and methods in evolutionary biology*, ed. Robert N. Brandon, 161–178. Cambridge: Cambridge University Press.
Caswell, Hal. 2001. *Matrix population models*. Sunderland: Sinauer.
Dawkins, Richard. 1999. *The extended phenotype*. Oxford: Oxford University Press.

Eldredge, Niles. 1986. Information, economics, and evolution. *Annual Review of Ecology and Systematics* 17: 351–369.

Glymour, Bruce. 1998. Contrastive, non-probabilistic statistical explanations. *Philosophy of Science* 65: 448–471.

Glymour, Bruce. 2007. In defense of explanatory deductivism. In *Causation and explanation*, ed. Joseph Campbell, Michael O'Rourke, and Harry Silverstein, 133–154. Cambridge, MA: MIT Press.

Glymour, Bruce. 2011. Modeling environments: Interactive causation and adaptation to environmental conditions. *Philosophy of Science* 78: 448–471.

Grant, Peter R. 1991. Natural selection and Darwin's finches. *Scientific American* 265: 82–87.

Guisan, Antione, and Wilfried Thuiller. 2005. Predicting species distribution: Offering more than simple habitat models. *Ecology Letters* 8: 993–1009.

Heywood, James. 2010. Explaining patterns in modern ruminant diversity: Contingency or constraint? *Biological Journal of the Linnean Society* 99: 657–672.

Hunt, Kevin D. 1994. The evolution of human bipedality: Ecology and functional morphology. *Journal of Human Evolution* 23: 183–202.

Jaenike, John. 1978. An hypothesis to account for the maintenance of sex within populations. *Evolutionary Theory* 3: 191–194.

Johnson, Matthew D. 2007. Measuring habitat quality: A review. *The Condor* 109: 489–504.

Kerr, Benjamin, and Peter Godfrey-Smith. 2002. Individualist and multi-level perspectives on selection in structured populations. *Biology and Philosophy* 17: 477–517.

Lande, Russell, and Stevan J. Arnold. 1983. The measurement of selection on correlated characters. *Evolution* 37: 1210–1226.

Lennox, James G., and Bradley E. Wilson. 1994. Natural selection and the struggle for existence. *Studies in History and Philosophy of Science* 25: 65–80.

Levene, Howard. 1953. Genetic equilibrium when more than one ecological niche is available. *American Naturalist* 87: 331–333.

Levin, Donald. 1975. Pest pressure and recombination systems in plants. *American Naturalist* 109: 437–451.

Lewontin, Richard C. 1970. The units of selection. *Annual Review of Ecology and Systematics* 1: 1–18.

Maynard Smith, John. 1976. What determines the rate of evolution? *American Naturalist* 110: 331–338.

McFadden, Bruce. 1992. *Fossil horses*. Cambridge: Cambridge University Press.

Odling-Smee, F. John, Kevin N. Laland, and Marcus W. Feldman. 2003. *Niche construction: The neglected process in evolution*. Princeton: Princeton University Press.

Pearl, Judea. 2000. *Causality*. Cambridge: Cambridge University Press.

Roughgarden, Jonathan. 1979. *Theory of population genetics and evolutionary ecology*. New York: Macmillan.

Schluter, Dolph. 1988. Estimating the form of natural selection on a quantitative trait. *Evolution* 42: 849–861.

Spirtes, Peter, Clark Glymour, and Richard Scheines. 2000. *Causation, prediction, and search*, 2nd ed. Cambridge, MA: MIT Press.

Toju, Hirokazu, and Teiji Sota. 2006. Imbalance of predator and prey armament: Geographic clines in phenotypic interface and natural selection. *American Naturalist* 167: 105–117.

van Fraassen, Bas. 1980. *The scientific image*. Oxford: Oxford University Press.

Wheeler, Peter E. 1991. The influence of bipedalism on the energy and water budgets of early hominids. *Journal of Human Evolution* 21: 117–136.

Environmental Grain, Organism Fitness, and Type Fitness

Marshall Abrams

Abstract Natural selection is the result of organisms' interactions with their environment, but environments vary in space and time, sometimes in extreme ways. Such variation is generally thought to play an important role in evolution by natural selection, maintaining genetic variation within and between populations, increasing the chance of speciation, selecting for plasticity of responses to the environment, and selecting for behaviors such as habitat selection and niche construction. Are there different roles that environmental variation plays in natural selection? When biologists make choices about how to divide up an environment for the sake of modeling or empirical research, are there any constraints on these choices? Since diverse evolutionary models relativize fitnesses to component environments within a larger environment, it would be useful to understand when such practices capture real aspects of evolutionary processes, and when they count as mere modeling conveniences. In this paper, I try to provide a general framework for thinking about how fitness and natural selection depend on environmental variation. I'll give an account of how the roles of environmental conditions in natural selection differ depending the probability of being experienced repeatedly by organisms, and how environmental conditions combine probabilistically to help determine fitness. My view has implications for what fitness is, and suggests that some authors have misconceived its nature.

1 Introduction

1.1 Overview

Natural selection is the result of differences in fitness, and fitness depends on organisms' interactions with their environment. But environments vary in space and

M. Abrams (✉)
Department of Philosophy, University of Alabama at Birmingham, Birmingham, AL, USA
e-mail: mabrams@uab.edu

G. Barker et al. (eds.), *Entangled Life*, History, Philosophy and Theory
of the Life Sciences 4, DOI 10.1007/978-94-007-7067-6_7,
© Springer Science+Business Media Dordrecht 2014

time, sometimes in extreme ways. Variation in what biologists call patches, habitats, environments, etc.—or what I'll call *subenvironments* of a *whole environment* over which a population ranges—is generally thought to play an important role in evolution by natural selection, maintaining genetic variation within and between populations, increasing the chance of speciation, selecting for plasticity of responses to the environment, and selecting for certain behaviors such as those involved in habitat selection and niche construction.[1] Variation in subenvironments can involve both spatial variation and temporal variation, and can partly result from migration and other kinds of dispersal (since these can expand the variety of environments to which a population is exposed), or from niche construction. Environmental variation goes beyond this, however. The interaction of any organism with its environment will nearly always differ from the interactions of other conspecifics with the same environment. Even if two organisms were genetically, physiologically, and cognitively identical, a consequence of the complexity of most environments is that the organisms' interactions with their surroundings will differ.

There is a wide variety of kinds and dimensions of environmental variation. Do all of these sorts of interaction and other kinds of environmental variation matter for natural selection? Do different kinds of environmental variation play different roles in natural selection? When biologists make choices about how to divide up an environment into subenvironments for the sake of modeling or empirical research, are there any constraints on these choices, or is any way of dividing up the environment legitimate?

In this paper, I try to provide a general framework for thinking about how fitness and natural selection depend on environmental variation. It turns out, I'll argue, that some concepts of fitness popular among philosophers of biology cannot play the role they are thought to play. A concrete example at this point will help to suggest the range of environmental variation that I think must be discussed.

1.2   An Example

In order to illustrate kinds of environmental variation that might be thought to be relevant to interactions affecting fitness, I'll describe some characteristics of house sparrows. Some of the interactions that I want to highlight are not described in detail in the scientific literature; some variations are simply impractical to study systematically. However, my plausible stories are based on a large body of research about house sparrows, primarily that reported by Anderson (2006) where not otherwise noted. I'll use some research on other birds, too, following the common

---

[1]Odling-Smee et al. (2003) treat habitat selection—cases in which organisms "choose" their subenvironment—as a form of niche construction. I use the latter term in a narrower sense requiring modification of the environment by organisms (cf. Sterelny 2005).

scientific practice of making reasonable hypotheses about a species' properties from what's known about closely related species.

A house sparrow that finds food in an area exposed to the sky may give a call to alert other house sparrows, waiting until others arrive—sometimes even foregoing food if they don't. This is not thought to be an altruistic behavior; it allows sparrows to feed more efficiently, as they share the task of visually scanning the sky for raptors. Whether an instance of such food-calling behavior makes survival and reproduction more or less likely than alternative behaviors plausibly depends on various factors: Are there any raptors or other predators who will notice the house sparrow feeding? Did the call bring the calling sparrow to the attention of a predator? How much food is there to share with other house sparrows? How many house sparrows are there nearby? How easy it to hear the call? Are other house sparrows upwind or downwind? Is there intervening foliage that will degrade the sound (Kirschel et al. 2009; Crozier 2010)? Is there noise that will mask it (Hu and Cardoso 2009)? Are nearby house sparrows already satiated? Are they busy with other tasks, such as nest building, courting, sitting on eggs? Would a male who is trying to attract mates to a desirable nest cavity have more offspring if he responds to the food call or remains to guard his nest cavity? Are there other areas with food which are safer, or do they have other dangers, such as nearby feral cats? Note that although a sparrow may be able to eat more if other sparrows share scanning duties, there is also competition for food within a feeding flock, and aggressive encounters use up some potential feeding time. Some sparrows are more aggressive than others, so the benefit of sharing the meal with other sparrows depends partly on which sparrows respond to the call.

There are additional factors that affect the possible benefits of food-calling behavior. What kind of food is available in the exposed area? What nutrients is the calling house sparrow likely to need for energy for activity, for feather production, for maintenance of feather coatings, for fighting parasites, or for pigmentation? Male house sparrows may get substances from food that allow them to synthesize melanin used to create their black chest patches; these may influence attractiveness to females, or play a role in aggressive interactions between males. However, melanin and related substances may also be used for processes which fight parasite infections (Catoni et al. 2009). Whether eating some substances or eating others is more or less beneficial for a given male house sparrow might depend on the season, the number and kinds of parasites that it has or will have, and the cognitive and physical properties of other nearby house sparrows.

The preceding discussion suggests ways in which very small-scale, detailed patterns of variation might affect the survival and reproduction of a house sparrow, though it ignores many aspects of house sparrow life, such as environmental interactions affecting development.[2] Notice, though, that many of the elements

---

[2] For example, the intensity with which a house sparrow nestling begs, influencing parents' feeding behavior and subsequent nestling survival, seems to be the result of an earlier gene-environment interaction (Dor and Lotem 2009).

mentioned above are likely to exhibit large-scale spatial or temporal patterns. What foods are available and where they are located depend on plant growth patterns that vary across the landscape, and over time. Some plants grow best in some soils, at some heights, under certain weather conditions, etc. Kinds of predators vary, and depend on the presence of other prey, which in turn depends on what other plants and animals are present. Weather patterns affect house sparrow survival, but interact with other environmental conditions (Ringsby et al. 2002). Weather patterns of course vary from week to week, from season to season, from year to year, and across different parts of the world: House sparrows are found in Europe, northern Africa, southern Africa, the Middle East, central Asia, the Indian subcontinent, and the Americas. There are morphological and behavioral differences between these different subpopulations, and some of the differences plausibly have to do with differences in environmental conditions. House sparrows are known to live in urban, agricultural, and other rural areas, each of which might favor certain morphologies or behaviors over others (cf. Evans et al. 2009).

Environmental variations of different kinds at different spatial and temporal scales plausibly affect natural selection on house sparrows. Scientists doing research on house sparrow evolution may decide to study some such differences, but no one would try to study all such differences, and different scientists will make different choices about scales of variation to study. Nothing about the house sparrow example suggests that it is particularly unusual; similar points can be made about most species. I note that among humans, social interactions generate particularly complex large-scale and small-scale environmental variations. Although human behavioral plasticity and social institutions may make sustained gene-based evolution in humans less common, I suggest that the framework described below applies equally well to all species (see §5).

1.3 Goals of the Paper

If we are interested in understanding what, in general, fitness and natural selection consist in, the following questions are important. Does *every* subset of conditions that might coexist within a whole environment count as determining fitness relative to it? If not, why not? Since diverse evolutionary models relativize fitnesses to subenvironments, it would be useful to understand when such modeling practices capture real aspects of evolutionary processes, and when structured environments count, instead, as mere modeling conveniences. This will help to clarify relationships between models, the systems they concern, and empirical evidence. Note that a working assumption here is that fitness differences and processes of natural selection are real aspects of the world that are investigated and approximated by empirical methods and modeling strategies. This assumption allows us to make a distinction between pragmatic fitness concepts which are useful for measurement and modeling, and theoretical fitness concepts which are assumed to represent underlying properties approximated by pragmatic concepts (Abrams 2012c).

The following rough question is a reasonable starting point for this investigation:

What level of environmental grain (Levins 1968) captures differences in fitness that can play a role in natural selection?

Speaking of "environmental grain" may suggest that environmental variation always involves clear transitions from one state to another over space or time, but that seems unlikely. So rather than asking about a concept of "grain," we might simply ask about environmental variation itself:

What sorts of environmental variation make a difference to fitness that can play a role in natural selection?

Note that these are *not* empirical questions about how environments affect organisms and populations. They are questions about what fitness is, and about what its relationship is to environmental variation, in general.

With a few exceptions, there has been little attention devoted to giving general characterizations of the relationship between environmental variation and natural selection. Such a characterization would help to clarify the view that the diversity of ways of modeling natural selection approximates a small set of evolutionary processes that can be given unitary characterizations. In (Abrams 2009c), I proposed general ways of characterizing the extent of the *whole environment*—all those conditions determining probabilities relevant to natural selection and biological fitness for a population, providing answers to this question: What conditions constitute the overall, whole environment of a population? That is, what are all the conditions that determine the fitnesses of types competing across the population as a whole (independent of our models, empirical research, etc.)? Earlier, I'd argued that very detailed microenvironments or microhabitats, or *circumstances*, play only a very limited role in determining biological fitness (Abrams 2007). Brandon (1990, Ch. 2) has discussed a kind of intermediate variation like that which is my focus, but I'll argue that his account is incomplete. Wimsatt, drawing on other work such as that due to Lewontin (1966) and Levins (1968) has also discussed differences in intermediate-level variation. My proposals can be viewed as an attempt to extend some of Wimsatt's, Lewontin's, Levins', and Brandon's ideas into more general principles.[3] More specifically, the argument of this essay proceeds along the following lines.

1.4 Outline of the Paper

The role of environmental variation in determining fitness and natural selection depends on what entities are bearers of fitness: The range of environmental variation

---

[3]Smith and Varzi (e.g. (2002)) also discuss a concept of "environment," which they take as equivalent to "niche," but their discussion concerns issues which have little relevance here. More generally, niche concepts bring up issues other than those which are my focus (cf. Abrams 2009c).

encountered by a single organism differs from the range of variation encountered by all instances of a heritable type. Much work in philosophy of biology has assumed that fitness is primarily a property of token organisms. I believe that this assumption has created unnecessary problems. In Sect. 2, I'll argue that because natural selection requires heritable variation in fitness (Lewontin 1970; Godfrey-Smith 2009), only fitness treated as a property of heritable types is directly relevant to natural selection. Fitness as a property of token organisms may play a subsidiary role in natural selection, but if so, this role has usually been misconceived. A crucial step in my account will be to argue that, in part because of the subtlety of an individual's interactions with its surroundings, fitnesses of token organisms depend on extremely specific subenvironments, which I call "circumstances" (Abrams 2007). Circumstances are unlikely to be experienced more than once, and I argue that as a result, any fitnesses which they help determine are not heritable. I will make a fundamental distinction between such circumstances, and broader subenvironments with a significant probability of recurring. (Throughout, I use "recurrence" and related terms to capture this idea.) Heritable biological types differ from token organisms in that the former can be repeatedly realized by different instances of the latter, and thus can enter into interactions with recurrent subenvironments. (I don't mean to suggest that the interaction between a subenvironment and an individual which realizes a type is the same for each individual, as will become clear below.)

The view that fitness must primarily be a property of types will provide a foundation for the discussion in the rest of the paper. Section 3 lays out a general way of thinking about how fitnesses relative to subenvironments combine to generate the overall fitness of a type in a whole environment. Section 4 then fills in the framework suggested by earlier sections. It begins with a summary of Brandon's (1990) distinction between three concepts of environment (§4.1): the external environment, which includes all physical characteristics in a population's surroundings; the ecological environment, which includes only properties that can affect fitnesses; and the selective environment, which includes only properties that affect differences in fitness. Brandon's discussion seems to suggest that only selective environments matter for natural selection, except in certain cases, as when when organisms choose which ecological environment to inhabit. I argue the latter point must be generalized, and that there are many cases in which fitness depends on probabilities of experiencing different ecological environments. I continue by elaborating a central point of Sect. 2: Subenvironments that can play a central role in determining fitnesses will be the ones likely to occur with a systematicity to which natural selection can respond (§4.2). Senses of "fitness" that make it relative to more ephemeral subenvironments are not necessarily unimportant; fitness in this sense can function as a sort of "component" of fitness, but cannot by itself serve in the role that fitness differences play in natural selection (§4.3). Finally, in Sect. 4.4, I present an objection to my view: It appears to make natural selection in response to continuous environmental gradients impossible. I'll argue, however, that there is a way of applying my framework that avoids this problem. Section 5 summarizes my

conclusions and their implications for broader issues. Before continuing, I review
assumptions that I'll use in the rest of the paper.

1.5 Assumptions

In the rest of the paper I'll make the following assumptions, except where noted:

- Natural selection occurs when the frequencies of heritable types (alleles, genotypes, phenotypes) in a population change over time *because* these types have
  different fitnesses, or when frequencies remain the same *because* the types'
  fitnesses are the same.[4]
- Conceptions of fitness must—at least—allow the possibility of natural selection
  over generational time. Natural selection must be capable of providing a causal
  explanation for the distribution of organisms and traits in the world, and for
  understanding how populations change in systematic manners over time. This
  requires it to be able to act in a sustained manner over many generations, which
  usually means that fitnesses change slowly or in systematic ways.
- Fitness can be defined partly in terms of probabilities causally relevant to the
  number of descendants that instances of a biological type have. Call these
  reproductive probabilities (cf. e.g. Brandon 1990; Abrams 2009b).
- In studying evolution, biologists make some choices, at least in a rough sense,
  about what aspects of the world to investigate (Abrams 2009b,c, 2012c). In
  particular, they choose how to delineate populations and what properties of populations to study. This last choice includes a choice about an interval of time over
  which the population might evolve, and with it, what environmental factors might
  be relevant to that evolution. (It would be theoretically convenient to restrict
  populations to sets of organisms experiencing a common environment, and to
  sets of organisms unlikely to experience gene flow to or from other populations.
  However, neither restriction is observed in actual biological practice. What is
  required is that the possibility that subpopulations experience different conditions
  and that the population is not reproductively isolated be taken into account, either
  by incorporating these possibilities into models and measurements, or by having
  reasons to think that their effects can be ignored.[5])

---

[4]This assumption is uncontroversial for many philosophers and biologists, but "statisticalist"
philosophers of biology have challenged it (e.g. Walsh 2010; Matthen and Ariew 2009). Their
arguments are addressed in many other publications, including some of my own.

[5]The constrained arbitrariness of population definitions is illustrated by contemporary research
using the Human Genome Diversity Panel (Li et al. 2008). For example, Thompson et al. (2004)
divides this whole-genome data (from roughly 1,000 individuals) into 52 populations, while
Moreno-Estrada et al. (2009) cluster the same data into 39 populations for some analyses, and
seven populations for others.

- Nevertheless, for a specified population in a specified environment in a particular period of time, natural selection takes place independent of our decisions, modeling, empirical studies, etc. (Abrams 2009b,c, 2012c).
- An environment of a population can be viewed as corresponding to a range of variation in conditions that might be experienced by members of the population over a specified interval of time, along with probabilities of such conditions being experienced (Abrams 2009c).
- All probabilities mentioned below can be understood as objective probabilities, perhaps even with causal implications.

There is some ambiguity in how "organism" is used in practice. In what follows, I'll use "organism" exclusively to refer to individual, particular members of a population, rather than to a species.

2 Token Fitness and Fine-Grained Variation

In this part of the paper, I'll distinguish different concepts of fitness in order to clarify what aspects of these concepts will be included in my focus (§2.1). Among other things, I'll make a distinction between token fitness concepts and type fitness concepts. This will then allow me to argue that certain token fitness concepts are problematic because of the way that they make fitness depend on environmental circumstances (§2.2). Most of the rest of the paper will then focus on type fitness: Sect. 3 lays out a framework for thinking about how type fitness depends on environmental variation, and Sect. 4 clarifies various points about the relationship between type fitness, token fitness, and environmental variation.

2.1 Dimensions of Fitness

This section will delineate several classes of fitness concepts in order to clarify which ones are and are not the focus of this essay.

In evolutionary biology, "fitness," related terms such as "adaptive value," "growth rate," and "selection coefficient," as well as related parameters and variables in models, are all defined and used in diverse ways. I'll use "fitness" as a blanket term for all such related concepts. Ambiguity in these terms generally causes no problem: Context, area-specific traditions, and researchers' explicit definitions make intended senses clear enough for practical use. To avoid unnecessary complications or confusion, I'll begin by setting aside certain dimensions of variation in fitness concepts that I'll ignore in this essay.

It will be useful to distinguish between several classes of fitness concepts. Some fitness concepts are essentially tied to methods of empirical measurement; others might be thought to characterize underlying processes of evolution. My focus here will be on the latter. I distinguish elsewhere (Abrams 2012c), first, between measurable and tendential token fitnesses.[6] A *measurable token fitness* concept is one, such as actual number of offspring, which allows one to measure some property of an individual, where that property is relevant to evolutionary success.[7] A *tendential token fitness* concept, on the other hand, attempts to capture the idea that a particular individual in its particular circumstances has one or more tendencies to realize properties relevant to evolutionary success. Some versions of the propensity interpretation of fitness describe tendential token fitness concepts (e.g. Beatty and Finsen 1989; Brandon 1990, chapter 1; Ramsey 2006). Bouchard and Rosenberg's (2004) characterization of fitness in terms of solving design problems can also be viewed as a tendential concept. I also distinguish (Abrams 2012c) between statistical type fitness concepts and parametric type fitness concepts: A *statistical type fitness* concept is one that defines fitness as a property of a heritable type, in such a way that fitness is a mathematical function of measurable token fitnesses. For example, if we measure the fitness of a trait as the average of the number of offspring that (actual) individuals with that trait have in a certain generation, we are treating the trait's fitness as a statistical type fitness.[8] A *parametric type fitness* concept, by contrast, is one which treats the fitness of a type as an underlying property of the type which might be estimated by one or more statistical type fitnesses; this is a concept of fitness as something potentially entering into processes in the world.[9] Finally, a *purely mathematical fitness* concept is a mathematical concept, defined for use in certain mathematical models, which might usefully be interpreted as one of the other kinds of fitnesses in particular research contexts. For example, de Jong (1994) seems to treat fitnesses in Price's equation as type fitnesses, whereas Price (1970) himself seemed to treat the fitnesses in his model as token fitnesses. In the rest of the paper, my focus will be on tendential token fitnesses and parametric type fitnesses, which I will usually refer to simply as token and type fitnesses.

Warren Ewens, a well-known population geneticist, writes:

> First, while it is universally agreed that fitness is a property of the entire genome of an individual, it is also apparently agreed, with Wright (1931), that to a first approximation, for a short time, a constant net selection value of any allele may usefully be defined. (Ewens 2004, 277)

---

[6] I use "token fitness" rather than "individual fitness" because some biologists use the latter for a property of heritable types. For example, Michod (1999, 9). writes that "...fitness is often defined as the expected reproductive success of a type .... I refer to this notion of fitness as individual fitness." I avoid "organism fitness" for related reasons, although it made sense to include it in the title of the paper.

[7] I intend "evolutionary success" to be vague, capturing the idea of increase in frequency in future generations, or at least maintenance of a type in the population; this vague notion will be sufficient for my purposes here. See (Abrams 2009b) for relevant discussion.

[8] Stearns (1976) and de Jong (1994) survey a variety of statistical type fitness concepts.

[9] The "statistical"/"parametric" terminology is derived from the use of "statistic" and "parameter" in statistics, and is not directly related to the distinction between "statisticalist" and "causalist" views about evolutionary "forces."

This remark occurs in a book which surveys a broad range of population genetics models, typically representing fitness as a property of alleles or simple genotypes, rather than of whole genomes. While I doubt Ewens' claim about universal agreement, he is no doubt correct that fitness is sometimes viewed (1) as a property of a whole genome. Researchers also sometimes view fitness (2) as a property of the organism as a whole—not just its genome—or (3) as a property of an organism and its *particular*, detailed environmental circumstances. Note that all three of these kinds of concepts take fitness to depend on the environment in some sense. What's different about the third set of concepts is that they make it explicit that very specific, particular environmental circumstances can make a difference to fitness. Such fitness concepts would allow even genetically and developmentally identical organisms to have different fitnesses as soon as they are placed in different circumstances within an environment. (The first and second concepts, by contrast, might instead refer to a complex type which might be realized in different, particular environmental circumstances.)

The third conception of fitness seems to be popular among philosophers of biology (e.g. Brandon 1990, ch. 1; Mills and Beatty 1979; Bouchard and Rosenberg 2004; Ramsey 2006), and it has been used as a basis for some attacks on causal conceptions of natural selection (e.g. Ariew and Ernst 2009; Walsh 2007). It's not clear to me how widespread this conception of fitness is among biologists; I believe that biologists' focus is usually on conceptions of fitness with obvious practical utility: measurable token fitnesses, statistical type fitnesses, and purely mathematical fitnesses. As noted above, statistical type fitnesses are defined in terms of measurable token fitnesses, and similar constructions have sometimes been given in terms of tendential token fitnesses: Mills and Beatty (1979) and Sober (1984) defined the fitness of a type in a population as an average of tendential token fitnesses for actual individuals with that type in that population. (See Sect. 4.3 for criticism of this strategy.)

2.2   *Inadequacy of Token Fitness and Environmental Circumstances*

I'll argue now that (tendential) token fitnesses do not play a direct role in natural selection because they are not heritable.[10] I'll argue that type fitnesses, by contrast, can play the role required by natural selection, because they depend on recurrent environmental conditions. In later sections of the paper, I'll explain in more

---

[10]My argument is related to some given by Sober (1984) and Hodge (1987), who argue that overall individual fitness is not causal, but my argument is different. My argument is also related to arguments in Ariew and Ernst (2009) but is more general, and makes it clear that it is not the propensity interpretation of fitness per se that is the problem.

detail how fitness depends on organisms' interactions with varying environmental conditions (§3, §4). This discussion will include a description of a possible role for token fitness in natural selection (§4.3).

Consider the role of fitness in natural selection, taking as our starting point Lewontin's formulation of the conditions required for natural selection by Darwin:

> As seen by present-day evolutionists, Darwin's scheme embodies three principles...:
>
> 1. Different individuals in a population have different morphologies, physiologies, and behaviors (phenotypic variation).
> 2. Different phenotypes have different rates of survival and reproduction in different environments (differential fitness).
> 3. There is a correlation between parents and offspring in the contribution of each to future generations (fitness is heritable). (Lewontin 1970)

Lewontin's formulation spells out fundamental conditions for natural selection in a perspicuous manner, and similar formulations have been given by others (cf. Godfrey-Smith 2009).[11] Now, if fitness is attributed to an entire genome, as Ewens (2004) suggested, the heritability of fitness across more than a few generations will often be very low in those species that undergo significant recombination. Thus if natural selection were to be understood as the result of differences in whole-genome fitness, it would be hard to understand how it could act in a sustained way over many generations, except in special cases (as required in Sect. 1.5). A similar point could be made about the view that fitness attaches to the whole organism. Thus a concept of fitness which is to fill a role like that specified by Lewontin's conditions must be a concept of type fitness—a fitness of either an allele, a genotype, or a phenotype.

Though this conclusion concerns heritable type fitnesses in general, the argument for it is analogous to a well-known argument by Williams (1966), later championed by Dawkins (1976), usually described as an argument that only alleles are units of selection.[12] However, in the Williams/Dawkins account, alleles usually just function as types which are realized by individual organisms. This can be seen from the fact that Williams and Dawkins measure the effects of natural selection by counting token organisms which bear particular alleles, rather than, for example, counting all of the tokens of a given allele which might be found in an organism's cells (cf. Sterelny and Kitcher 1988). Thus what Williams and Dawkins argued for was that natural selection only acts on the distribution of organism types defined by alleles. The mistake that Williams and Dawkins made was to assume that only types which can be replicated nearly perfectly are subject to natural selection. My argument

---

[11]Lewontin suggests that these three conditions are necessary and sufficient for evolutionary change by natural selection. In fact they are neither necessary nor sufficient for natural selection (Godfrey-Smith 2009). However, they capture the core of the notion of natural selection sufficiently well for my purposes here.

[12]Actually, the notion of "allele" that Williams and Dawkins used was unusual, but this subtlety needn't concern us.

generalizes Williams' by assuming that merely heritable properties associated with fitness are all that natural selection requires.[13]

There is a related, deeper problem with token fitness if it's defined so as to depend not just on an individual's genome, but also on the way in which the genes interact with *particular* environmental circumstances, either during development or during mature life stages. (I pointed out above that some philosophers of biology seem to view fitness in this way (e.g. Ramsey 2006).) Then variations that can affect the reproductive success of an individual are not limited to the kinds of patterns explicitly referenced in models and empirical studies of environmental variation, and labeled with terms such as "patch," "habitat," "niche," and "environment"— "micro-habitats" or circumstances might matter as well. As suggested by the house sparrow example in Sect. 1.2, survival and reproduction of an individual can be affected by variation in wind direction in the presence of a predator, variation in numbers and kinds of viruses in nearby conspecifics, or variations in activities of potential mates and potential prey (Abrams 2007). There is no obvious limit to the sorts of minute variations that might affect such individualized fitnesses. For example, the fitness of a prey might be affected by fact that a leaf is blown in such a way as to allow the prey to be noticed by a predator because the predator's gaze followed an unusual movement perceived as that of a potential mate. It's plausible that there are real situations in which any variation in such conditions could make the difference between survival, injury, or death. Even two clones beginning life in (merely) measurably identical circumstances might have very different token fitnesses due to the different circumstances they experience during development and later life.[14] I have argued elsewhere (Abrams 2007) that interactions of individual organisms with their circumstances are effectively deterministic, and that this means that circumstance-relative token fitnesses are equivalent to actual reproductive successes. However, even if we allow that for any token organism in particular circumstances, there is some sort of objective probability distribution over possible outcomes, situations like those described above appear to be ones in which small differences in circumstances would make large differences in fitnesses.

The problem with making fitness depend on circumstances is that survival and reproductive success in response to environmental variation of such a fine-grained

---

[13]Wimsatt (1980b, 1981) argued that when there are nonlinear interactions between alleles, it's inappropriate to treat alleles as units of selection. Analogously, one might argue that when there are nonlinear interactions between heritable types of any sort, its inappropriate to assign fitness values to each type as such. However, given a probability distribution over possible combinations of types, fitness values for any one type can in principle be computed (Abrams 2009b). This is in effect to treat those alternative types, which might be combined with a particular type whose fitness is to be calculated, as the alternative environmental states discussed in Sect. 3 (cf. Dawkins 1976; Sterelny and Kitcher 1988).

[14]In Brandon's (1990, chapter 2) terms, I am arguing that his assumption that there are broad regions of the space of environmental conditions which are objectively homogeneous or which vary only gradually with respect to probabilities relevant to fitness is incorrect, when we consider environmental variation in sufficient detail.

kind is not heritable: The sum of those particular circumstances affecting an individual's fitness is unlikely to recur among its descendants, so any effects of circumstances on organisms' types are, likewise, unlikely to recur. Survival or reproductive success in this sense is therefore not a kind of fitness in the sense required by concept of natural selection. The point holds even for organisms that undergo little or no genetic recombination, as in many asexual species. In the general case then, differences in token fitnesses are not the kind of fitness differences that are essential to the concept of natural selection: No heritability; no natural selection.

Note that Ramsey's (2006) concept of a *fitness environment* is defined in terms of probable conditions that descendants of an individual will encounter. However, Ramsey's fitness environment is defined relative to a particular individual. Different individuals in the same population can have different fitness environments, even if they are genotypically and phenotypically identical. This means, though, that fitnesses relative to such environments need not be heritable.[15]

Now in most species, there will have been selection for robustness of patterns of survival, reproduction, etc., in the face of environmental variation (Wagner 2005; cf. Wimsatt 2007). Complete robustness to environmental circumstances would mean that a given type of organism would have the same number of descendants for any circumstance included in the range of circumstances possible in the population's environment. This seems unlikely in general, and if the members of a population did achieve this sort of robustness, it's likely that fitter variants would eventually arise which allowed the exploitation of new, varying resources despite greater risk of failure. Moreover, even if there were a population which was robust to all variation in circumstances within its environment, a general understanding of fitness could not depend on such cases, since most populations do not have this characteristic.

Thus fitnesses capable of playing the sort of role outlined by Lewontin must attach not to individuals, but to heritable types. In Sect. 4.3, I argue that there may be a sense in which type fitnesses are derived from token fitnesses, but in a way that token fitness advocates have not discussed.

3 How Do Subenvironment-Relative Fitnesses Combine?

We saw a problem with token fitness concepts in the preceding section. This section will lay out a general framework for thinking about how type fitnesses that are relative to subenvironments combine to determine overall fitness.[16] This framework will provide a foundation for Sect. 4, in which I relate Brandon's views about fitness and subenvironments to the framework described in this Sect. (§4.1), discuss how the distinction between circumstances and other subenvironments might be drawn

---

[15]There are some similarities between aspects of Ramsey's (2006) concept of a fitness environment and some of my own ideas (Abrams 2009a,b,c), but the latter focus on type fitnesses.

[16]Glymour (2006, 2011) seems to focus on different questions about environments than I do, but his approach seems broadly complementary to and compatible with mine.

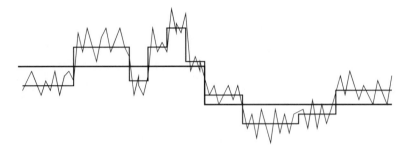

Fig. 1 Schematic representation of variation in fitness in subenvironments at different levels of grain. The horizontal axis represents positions in a space, time, or one or more other continuously varying properties. Heights of lines represent fitnesses of a single heritable type. The line with the most variation in height represents changes in fitness resulting from environmental differences between small regions. Lines with less variation in height represent fitnesses relative to larger regions, treating these fitnesses roughly as averages over fitnesses in smaller regions

(§4.2), explain what role circumstances and token fitnesses might have in natural selection (§4.3), and discuss a problem that environmental gradients pose for my view (§4.4).

For a given partitioning of the whole environment into subenvironments E_j, the overall fitness of a type a is a function of subenvironment-relative fitnesses $F(a|E_j)$ (Abrams 2009a) and probabilities that instances of a will experience each subenvironment (Levins 1968; Wimsatt 1980a) over an interval of time. If each organism experiences only a single subenvironment for its entire life, this function is an average over m subenvironments:

$$F(a) = E(F(a|E_{\bullet})) = \sum_{j}^{m} F(a|E_j) P(E_j)$$

$$= F(a|E_1) \times P(E_1) + F(a|E_2) \times P(E_2) + \cdots + F(a|E_m) \times P(E_m).$$

cf. e.g. (Gillespie 2004; Roughgarden 1979).[17] This is represented schematically in Fig. 1. Where the fitnesses are viabilities, and subenvironments are experienced for short periods of time with independent probabilities, fitnesses combine multiplicatively (Levins 1968; Wimsatt 1980a; Nagylaki 1992):

$$F(a) = \prod_{j} F(a|E_j)^{P(E_j)}$$

$$= F(a|E_1)^{P(E_1)} \times F(a|E_2)^{P(E_2)} \times \cdots \times F(a|E_m)^{P(E_m)}.$$

---

[17]For example, suppose fitness is expected number of offspring O_a for type a, i.e. $F(a) = E(O_a) = \sum_k k\, P(O_a = k)$. The probability of having k offspring is the average across subenvironments E_j, weighted by probability of E_j: $P(O_a = k) = \sum_j P(O_a = k|E_j)P(E_j)$. Together these equations imply that $F(a) = E(F(a|E_{\bullet}))$, as in the text.

However, in general overall fitness is a more complex function of subenvironment-relative fitnesses (Levins 1968; cf. Abrams 2009a,b).

This framework applies recursively all the way down to the level of circumstances: Subenvironment-relative fitnesses of types are a function of fitnesses in, and probabilities of, narrower subenvironments, which in turn are a function of fitnesses relative to even narrower subenvironments, and so on, all the way down to the level of circumstances (cf. §4.3). This conception allows partitioning the environment into subenvironments in different ways for different purposes. For example, one study of a given population of organisms might focus on the effects of variation in amount of rainfall over time, another might focus on variation in rainfall between different regions, while a third might focus the differences between forests and fields.

4   Type Fitness and Coarse-Grained Environmental Variation

The preceding section provided a framework for thinking about how fitness depends on environmental variation. In this section, I'll discuss in more detail my claim that parts of the environment with a significant probability of recurring—of being experienced repeatedly by members of a population, over generational time—give rise to fitness in a sense different than do circumstances that are idiosyncratic to a particular time, place, configuration of organisms, etc. I begin by relating Brandon's previous discussion of environmental variation to my view (§4.1). I then discuss the how one might draw a formal distinction between circumstances and other subenvironments, thus placing a lower limit on what kinds of subenvironments are worth modeling or studying empirically (§4.2). I explain how token fitnesses and environmental circumstances might contribute to type fitnesses relative to larger subenvironments (§4.3), and respond to a challenge that environmental gradients pose for my view (§4.4).

4.1   Brandon's Three Concepts of Environment

Brandon (1978, 1990) sometimes seems to treats fitness as token fitness, but chapter 2 of Brandon (1990) is relevant to type fitness.[18] There Brandon defined three concepts of environment. The *external environment* of a set of organisms consists of any properties of the world external to the organisms. The subset of properties

---

[18]I read parts of chapter 1 of Brandon (1990) as concerned with token fitness, and chapter 2 seems to allude to token fitness, e.g. on page 47, when it mentions the environment of an individual. However, the primary focus of chapter 2 is on environments of populations of organisms and these environments' effects on the fitnesses of types.

in the external environment whose variations can affect these organisms' future contribution to the population count as components of the *ecological environment*. Not all variation in the ecological environment differentiates between distinct heritable types, however. For example, if two genotypically different strains of the roundworm *Caenorhabditis briggsae* in a population respond to differences in climate in ways that affect fecundity (cf. Prasad et al. 2011), but the *relative* fitness between them is constant, then variation in climate counts as variation in ecological environment but not variation in *selective environment*. The selective environment varies only when relative fitnesses of competing types vary. Brandon also defined a concept of selective environment "neighborhoods": roughly, regions of similar relative fitness between competing types. Also note that Brandon often spoke of environmental variation as spatial, using plants as illustrations, but it's natural to extend his notions to complex combinations of variation in many properties over space and time.

Although Brandon initially proposed that it is selective environments that matter to natural selection, his discussion showed that factors *other than* fitness differences relative to selective environments also matter for natural selection. Suppose that the relative fitness relations between two heritable types a and b are the same in two subenvironments E_1 and E_2, while the absolute fitness of both types is greater in E_1 than in E_2. In other words, both a and b individuals are likely to produce more descendants starting from E_1 than E_2, although a is likely to produce more descendants than b in E_j, whether E_j is E_1 or E_2. In this case, the overall fitness of a and b depends on the probabilities of each type being found in E_1 and E_2, and not just on their relative fitnesses. If the difference in absolute fitnesses between E_1 and E_2 is great enough, then b can be fitter than a even if a is fitter than b in each subenvironment. All that's necessary is that b have a sufficiently greater probability than a of landing in E_1.[19]

The significance of Brandon's discussion for this paper should be apparent, but from my point of view his explicit statements, at least, do not go far enough. First, Brandon's own example shows that the concept of a selective environment is relevant to overall selection only in certain cases, since it shows that a type b can be fitter than a overall[20] even though a is fitter than b in each selective (sub)environment. Second, Brandon illustrates the case just described using an example of habitat choice, in which insects have heritable preferences for laying eggs on one kind of plant rather than others. The significance of Brandon's point goes beyond habitat choice, however. For example, if plants have heritable variation

---

[19]Using the first, additive model in Sect. 3, b is fitter than a overall if

$$\mathsf{F}(a|E_1)P(a \text{ in } E_1) + \mathsf{F}(a|E_2)P(a \text{ in } E_2) \quad < \quad \mathsf{F}(b|E_1)P(b \text{ in } E_1) + \mathsf{F}(b|E_2)P(b \text{ in } E_2).$$

Suppose $F(a|E_1) = 10$, $F(a|E_2) = 2$, $P(a \text{ in } E_1) = .1$, $F(b|E_1) = 5$, $F(b|E_2) = 1$, and $P(b \text{ in } E_1) = .9$. Then $\mathsf{F}(a) = 10 \times .1 + 2 \times .9 = 2.8$ and $\mathsf{F}(b) = 5 \times .9 + 1 \times .1 = 4.6$.

[20]That is, b could be fitter in the sense that it has a greater probability of evolutionary success, increased frequency, etc., in either/both the short term and/or the long term.

affecting the probability of seeds being blown by the wind, some types in a population may be more likely to encounter certain environmental variants than others; this is not usually considered habitat choice.[21] Also note that parts of Brandon's discussion assume that ecological environments do not vary in extreme ways over large portions of a whole environment. However, I argued above that minute variations in circumstances within an environment can produce extreme variations in probabilities of reproductive success.[22]

4.2   What Sort of Grain Is Relevant to Selection?

According to our discussion so far, natural selection occurs only when heritable types in a population are differentially reproduced because of differences in type fitness. However, as we saw in Sect. 2, fitnesses are heritable only when environmental conditions are repeatedly encountered by the same type, as members of the population reproduce over time. Thus, there is a lower limit on the grain of environmental variations relative to which heritable fitnesses are environmentally determined. What is that limit? That is the topic of this section.

In order for organisms that are instances of competing heritable types a_i to have fitness relative to a subenvironment E_j, instances of each type a_i must have a significant probability of experiencing the same conditions E_j. The range of conditions corresponding to a subenvironment E_j must be broad enough to allow this. But what is a significant probability of experiencing a particular subenvironment? What is the cutoff value for such probabilities? Any answer would depend on the relative strengths of selection and drift, and more specifically on fitness differences relative to a given subenvironment, as well as on effective population size. What matters is whether selection relative to a subenvironment has a reasonable chance of affecting evolution given the other evolutionary forces acting on the population.

First, the interval of time over which evolutionary change might take place is relevant to what counts as a significant probability of recurrence (cf. Sect. 1.5). Longer intervals may produce greater probabilities of recurrence, since they allow more time for a set of environmental conditions to recur. Thus probabilities of recurrence depend on what kind of time period we want to investigate (Abrams 2009c,b). Second, if fitness differences relative to subenvironment E_j are small, then E_j must be encountered more often for these fitness differences to have a non-negligible impact on selection. Similarly, if fitness differences relative to E_j

---

[21]To be precise, Brandon gives an example in which organisms of different types always choose specific subenvironments, thus in effect creating a uniform "selective environment"—in that each type competes with the other relative to constant environmental conditions. However, it's not difficult to see how to extend this generalization of the concept of a selective environment to cases in which types have different non-extremal probabilities of encountering various ecological environments.

[22]Note that much of Brandon's discussion is driven by concerns other than those that are my focus.

are large, then E_j can have a probable impact on evolution even if its probability of recurrence is relatively low. Third, cutoff probabilities depend on the strength of other evolutionary forces. Consider drift, for example. Effective population size determines the strength of drift. If a population is small, then the effect of minor differences in fitness will be swamped by drift. Thus for a subenvironment to be relevant to natural selection in a small population, it must encompass a broader range of circumstances, *ceteris paribus*. These relationships could be formalized in terms of a particular model, producing a formula allowing calculation of probabilities of recurrence necessary to produce a noticeable effect of selection given values for the parameters just mentioned. For a given population and environment, and a specified level of likely effect of selection, such a formula could in principle be used to estimate a *minimal environmental grain*: a specification of how fine-grained a partition of the environment can be while still making subenvironment-relative fitness differences themselves relevant to selection.

I'm not sure how useful such an estimation project would be, however. The main point is that modeling practices and empirical research are consistent with a vague boundary between subenvironments that have a significant probability of recurrence, and those that don't. Above this vague limit, researchers are free to choose a way of partitioning environmental conditions into subenvironments that is useful for their research goals.

4.3   A Role for Token Fitness?

I think that many philosophers of biology, at least, feel that (tendential) token fitnesses must play a role in natural selection, somehow. I'm not sure that this is on the right track. However, in this section I'll explain what legitimate role token fitnesses might play in natural selection. Since token fitnesses are usually thought to depend on circumstances, we'll also see what role circumstances might have in natural selection.

Suppose that a house sparrow happens to fly across a small opening in the forest cover when a hawk happens to fly overhead. As a result the hawk chases and injures the house sparrow, causing an infection contracted when the house sparrow escapes by diving into a pond. The pond happens to contain bacteria to which the house sparrow has low resistance. The house sparrow's resistance is low because its diet has been limited to a few foods; this is in part a consequence of a recent landslide that caused rainfall to be diverted to other areas. The infection leads to the house sparrow's subsequent death.

This set of circumstances does not give this particular house sparrow low fitness in a sense that's relevant to natural selection, for this set of circumstances will never recur. That hawks are present in the region and are sometimes overhead matters to natural selection. That injuries of various sorts occur matters to natural selection. That infections of various sorts occur matters. That the water supply to nearby plants is sometimes low matters. These things matter because they recur, and they matter to a degree that's weighted by their probability of occurring in various combinations, and by the probable effects on reproductive success of each set of conditions. What matters to natural selection is not this or that organism's particular circumstances and particular fate, but the sorts of conditions that individuals in the population are likely to encounter repeatedly. Nevertheless, if outcomes relative to circumstances were probabilistically combined, they could determine fitnesses relative to more inclusive subenvironments which included those circumstances as possibilities.

Now, if the fitness of a type is defined as a simple arithmetic average of the token fitnesses of those *actual* organisms which realize the type in a particular population, as proposed in Mills and Beatty (1979) and Sober (1984), type fitnesses may fluctuate in odd ways from one time period to the next, as individuals in the population happen to experience this or that "lucky" or "unlucky" micro-habitat (Abrams 2007). Thus it's a mistake to treat type fitness as an average of token fitness of those organisms actually existing in a population in a given period of time.[23] However, a token organism can be viewed as a realization of a complex type including a whole genome, a phenotype produced by whatever factors contribute to development, and a set of environmental circumstances (Abrams 2009a, 2012c). Suppose for each such complex type $O_i E_j$ consistent with a particular heritable type a, there was an objective probability of $O_i E_j$ occurring. Then the overall fitness of a would follow from calculations like those discussed in Sect. 3. In this sense, token fitnesses can play a role in natural selection. Note, however, that the relevant token fitnesses are fitnesses of merely possible (i.e. probable) token organisms and circumstances. Moreover, *differences* in fitnesses relative to circumstances themselves play no meaningful role in natural selection. It is only as contributors to fitness differences in larger subenvironments that these token fitnesses play a role in natural selection per se (cf. Abrams 2009a).

Since concepts of propensity have played a large role in philosophical discussions of fitness, it's worth noting that Abrams (2007) argued that "circumstance probabilities," such as the probabilities of $O_i E_j$ in the preceding paragraph, are unlikely to be propensities. To my knowledge, no one has published an argument that they are propensities. On the other hand, since it is fitnesses of types, only, that are directly involved in natural selection, it may be that token fitness as such plays no role in natural selection. One can think of a population as a whole, in its environment, as a complex causal system in which organism types are realized repeatedly in response to inheritance relations between parts of the system (Abrams 2009a,b, 2012c). The fitness of a type at a time t then corresponds to a function of probabilities of distributions of types in the population at later times t'. No reference to fitnesses of particular organisms in particular circumstances need be

---

[23]Recall that my focus in this paper is on tendential token fitness and parametric type fitness. It's *not* necessarily a mistake to *estimate* parametric type fitness using the average of measurable token fitnesses of actual organisms (Abrams 2012c).

made. Individuals merely function as realizers of types within the entire population-environment system. Inheritance can also be conceived as a relationship between type realizations as such, rather than between concrete individuals.[24]

4.4   Gradients and Recurrent Properties

In many cases, conditions affecting fitness vary continuously over a whole environment. The fact that conditions relevant to fitness can vary continuously raises a potential problem for my characterization of minimal environmental grain. In this section, I present this problem and give a response to it.

Consider an example from the literature. Thompson et al. (2004) found that the frequency of the CYP3A*5 allele of the CYP3A gene in whole-genome data from human populations increases with distance from the equator. Based on statistical patterns in the genetic data, computer simulations, and known differences in phenotypic effects of CYP3A5*3 and its allele CYP3A5*1, Thompson et al. (2004) argued that there was probably selection on these alleles which varied with temperature and humidity.

It's certainly reasonable to assume that an environmental gradient can produce different fitness values at every point along the gradient. Some models do assume this, and hypotheses about fitness gradients are not difficult to test or estimate. However, if the regions between which fitnesses have significant differences are small, it could be that no such region has a significant probability of recurring. My view seems to imply that differences in fitness at every point along a gradient do not contribute to natural selection. My proposal thus seems to rule out an idea that has clear biological sense. Here is an outline of a response. (I'll switch examples; the effect of latitudinal variation in climate on vertebrates may not be the best illustration of this possibility, since population ranges are relatively large.)

Consider a population of plants with wind-blown seeds, sparsely distributed along the side of a mountain. Though each altitude corresponds to a range of environmental circumstances, suppose that the factors correlated with altitude—sunlight, temperature, and atmospheric density—make a large contribution to fitness of two competing heritable types. However, if the probability of seeds of each type growing at any *particular* altitude is very low, then none of the altitude-defined subenvironments are recurrent.

Note, however, that subenvironments corresponding to larger ranges of altitudes may nevertheless be recurrent. And for each point on the environmental gradient

---

[24]For those interested in pursuing alternative interpretations of probability that may be relevant to evolutionary processes, I suggest if the complex system constituted by a biological population and its environment satisfied conditions required for what are known as mechanistic, microconstant, or natural range probabilities (Rosenthal 2010, 2012; Strevens 2011; Abrams 2012a,b) it could turn out that type fitnesses would not derive from token fitnesses. This is a topic better left for later work, but I mention the possibility here for interested readers.

Fig. 2 Schematic representation of overlapping subenvironments along an environmental gradient

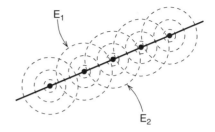

there is a series of larger and smaller regions around it, each of which overlaps similar regions centered on other nearby points (Fig. 2). These are just subenvironments in the usual sense, even if they overlap with other subenvironments with similar ranges of variation. Moreover, fitness differences relative to overlapping recurrent subenvironments can be considered to play a direct role in natural selection. Thus fitness differences in both of the overlapping subenvironments E_1 and E_2 schematically represented in Fig. 2 can be viewed as relevant to natural selection. The fact of the gradient is nevertheless captured by the series of overlapping, recurrent subenvironments.

5 Conclusion

I've argued, first, that fitnesses of types are fundamental to natural selection, and that views that treat differences of fitness for actual token organisms as fundamental cannot make sense of selection's basic character. In part this is because the fully-detailed circumstances of each individual's life can generate significant variation in probabilities concerning reproduction and persistence of descendants. As a result, token fitnesses are often not heritable to the degree required for natural selection.

Second, I've presented a view of fitness as a possibly complex function of reproductive probabilities in subenvironments, and probabilities of an organism experiencing each subenvironment. This scheme allows an environment to be partitioned into subenvironments in various ways, and allows subenvironment-relative fitnesses to be derived from narrower component subenvironments.

Third, I argued that subenvironments that are unlikely to be experienced repeatedly don't determine a sense of fitness that allows comparisons between competing types. That a is fitter than b in a nonrecurrent subenvironment is by itself irrelevant to selection, since this relationship is not heritable. Only fitness differences relative to recurrent subenvironments matter, in the end. I noted that environmental gradients which produce fitness gradients seem to conflict with this picture, since a point on a gradient might be unlikely to recur—in the sense that no two organisms are likely to experience it. I argued, however, that fitness gradients can be understood in terms of overlapping recurrent subenvironments.

My view is that researchers have various constrained choices about what aspects of an evolving population to ask questions about, but given those choices, there are objective answers determined by the world (Abrams 2012c). Researchers can choose what competing heritable types to focus on, what set of organisms to treat as a population, and over what period of time evolution matters. These choices implicitly specify a whole environment relevant to evolution in that population during that period of time (Abrams 2009c) and specify what fitness consists in (Abrams 2009b). The choices delimit a part of the world, with respect to which the facts about whether and how natural selection takes place are objective. Researchers also have constrained choices about how to divide up a population into subpopulations, and about how to divide the environment into subenvironments. These choices will usually be designed to capture different kinds of variation thought to be causally relevant to the evolution of the whole population or of subpopulations.

I see environments as defined by probabilities of the occurrence of subenvironments over time as a result of interactions involving members of a population and other natural processes (Abrams 2007, 2009c,b,a). As a result, I see the views presented in this paper as consistent with interactions between organisms affecting each others' fitnesses, as the house sparrow example in Sect. 1.2 suggested; with organisms affecting the environment, as in niche construction (Odling-Smee et al. 2003); and with a focus on the role of organism-environment interactions in development (cf. Oyama et al. 2001).

I believe that my perspective applies equally well to the biological evolution of humans and their ancestors. There is a common view that the biological evolution of humans has stopped because culture and individual learning allow humans to adapt to new environmental conditions without natural selection on biologically heritable traits (e.g. Dawkins 1976; Barkow et al. 1992). What the present view suggests is a way of framing this claim. I see no reason to think that social and cultural variation, combined with individual differences, do not also affect probabilities of survival and reproduction. Thus there is a sense in which humans experience complex environmental variation that could be relevant to natural selection. If human biological evolution has stopped because of the utility of individual learning and culture, it is because sociocultural environmental variation experienced by humans is insufficiently recurrent: Even if there are patterns to social and cultural conditions, what is stable does not last long enough for selection on genetically-influenced types to act in a consistent manner over many generations. It may be, however, that despite the appearance of rapid sociocultural change, there are higher-order patterns in sociocultural variation which recur sufficiently often that biological natural selection can sometimes respond to it (cf. Richerson and Boyd 2005). Natural selection for genes affecting lactase digestion in response to cattle husbandry has often been mentioned as an instance of natural selection in response to cultural conditions (e.g. Richerson and Boyd 2005). However, there is a growing body of evidence from whole-genome data suggesting that natural selection has had numerous effects on human populations in the last few tens of thousands of years (e.g. Thompson et al. 2004; Moreno-Estrada et al. 2009; Klimentidis et al. 2011; Scheinfeldt et al. 2011). Perhaps some of these effects are the result of higher-order stable patterns in social and cultural conditions.

Acknowledgements I'm very grateful for helpful feedback from the editors of this volume: Trevor Pearce, Gillian Barker, and Eric Desjardins; and for feedback, on related presentations in Groningen (2009), Toronto (2009), and London, Ontario (2010), from Denis Walsh, Jan-Willem Romeijn, Robert Makowsky, Yann Klimentidis, Bruce Glymour, Greg Cooper, John Beatty, André Ariew, and others who gave equally helpful comments. Some ideas in this paper grew from seeds planted by discussions with Bill Wimsatt many years ago. None of these individuals should be assumed to agree with my claims. Olivia Fanizza made early versions of the figures.

References

Abrams, Marshall. 2007. Fitness and propensity's annulment? *Biology and Philosophy* 22(1):115–130.

Abrams, Marshall. 2009a. Fitness "kinematics": Altruism, biological function, and organism-environment histories. *Biology and Philosophy* 24(4):487–504.

Abrams, Marshall. 2009b. The unity of fitness. *Philosophy of Science* 76(5):750–761.

Abrams, Marshall. 2009c. What determines fitness? The problem of the reference environment. *Synthese* 166(1):21–40.

Abrams, Marshall. 2012a. Mechanistic probability. *Synthese* 187(2):343–375.

Abrams, Marshall. 2012b. Chapter 9. Mechanistic social probability: How individual choices and varying circumstances produce stable social patterns. In *Oxford handbook of philosophy of social science*, ed. Harold Kincaid, 184–226. Oxford: Oxford University Press.

Abrams, Marshall. 2012c. Measured, modeled, and causal conceptions of fitness. *Frontiers in Genetics* 3(196):1–12.

Anderson, Ted R. 2006. *Biology of the ubiquitous house sparrow*. Oxford: Oxford University Press.

Ariew, André, and Zachary Ernst. 2009. What fitness can't be. *Erkenntnis* 71(3):289–301.

Barkow, Jerome, Leda Cosmides, and John Tooby (eds.). 1992. *The adapted mind: Evolutionary psychology and the generation of culture*. New York: Oxford University Press.

Beatty, John, and Susan Finsen. 1989. Rethinking the propensity interpretation: A peek inside Pandora's box. In *What the philosophy of biology is*, ed. Michael Ruse, 17–30. Dordrecht: Kluwer.

Bouchard, Frédéric, and Alexander Rosenberg. 2004. Fitness, probability, and the principles of natural selection. *British Journal for the Philosophy of Science* 55(4):693–712.

Brandon, Robert N. 1978. Adaptation and evolutionary theory. *Studies in the History and Philosophy of Science* 9(3):181–206.

Brandon, Robert N. 1990. *Adaptation and environment*. Princeton: Princeton University Press.

Catoni, C., A. Peters, and H.M. Schaefer. 2009. Dietary flavonoids enhance conspicuousness of a melanin-based trait in male blackcaps but not of the female homologous trait or of sexually monochromatic traits. *Journal of Evolutionary Biology* 22:1649–1657.

Crozier, Gillian. 2010. A formal investigation of cultural selection theory: Acoustic adaptation in bird song. *Biology and Philosophy* 25(5):781–801.

Dawkins, Richard. 1976. *The selfish gene*, 1st ed. Oxford: Oxford University Press.

de Jong, Gerdien. 1994. The fitness of fitness concepts and the description of natural selection. *Quarterly Review of Biology* 69(1):3–29.

Dor, Roi, and Arnon Lotem. 2009. Heritability of nestling begging intensity in the house sparrow (Passer domesticus). *Evolution* 63(3):738–748.

Evans, Karl L., Kevin J. Gaston, Stuart P. Sharp, Andrew McGowan, and Ben J. Hatchwell. 2009. The effect of urbanisation on avian morphology and latitudinal gradients in body size. *Oikos* 118:251–259.

Ewens, Warren J. 2004. *Mathematical population genetics, I. Theoretical introduction*, 2nd ed. New York: Springer.

Gillespie, John H. 2004. *Population genetics: A concise guide*, 2nd ed. Baltimore: The Johns Hopkins University Press.

Glymour, Bruce. 2006. Wayward modeling: Population genetics and natural selection. *Philosophy of Science* 73:369–389.

Glymour, Bruce. 2011. Modeling environments: Interactive causation and adaptations to environmental conditions. *Philosophy of Science* 78(3):448–471.

Godfrey-Smith, Peter. 2009. *Darwinian populations and natural selection*. New York: Oxford University Press.

Hodge, M.J.S. 1987. Natural selection as a causal, empirical, and probabilistic theory. In *The probabilistic revolution*, vol. 2, ed. Lorenz Krüger, Gerd Gigerenzer, and Mary S. Morgan, 233–270. Cambridge, MA: MIT Press.

Hu, Yang, and Goncalo C. Cardoso. 2009. Are bird species that vocalize at higher frequencies preadapted to inhabit noisy urban areas? *Behavioral Ecology* 20(6):1268–1273.

Kirschel, Alexander N.G., Daniel T. Blumstein, Rachel E. Cohen, Wolfgang Buermann, Thomas B. Smith, and Hans Slabbekoornc. 2009. Birdsong tuned to the environment: Green hylia song varies with elevation, tree cover, and noise. *Behavioral Ecology* 20(5):1089–1095.

Klimentidis, Yann C., Marshall Abrams, Jelai Wang, Jose R. Fernandez, and David B. Allison. 2011. Natural selection at genomic regions associated with obesity and type-2 diabetes: East Asians and sub-Saharan Africans exhibit high levels of differentiation at type-2 diabetes regions. *Human Genetics* 129:407–418.

Levins, Richard. 1968. *Evolution in changing environments*. Princeton: Princeton University Press.

Lewontin, Richard C. 1966. Is nature probable or capricious? *BioScience* 16(1):25–27.

Lewontin, Richard C. 1970. The units of selection. *Annual Review of Ecology and Systematics* 1:1–18.

Li, Jun Z., Devin M. Absher, Hua Tang, Audrey M. Southwick, Amanda M. Casto, Sohini Ramachandran, Howard M. Cann, Gregory S. Barsh, Marcus Feldman, Luigi L. Cavalli-Sforza, and Richard M. Myers. 2008. Worldwide human relationships inferred from genome-wide patterns of variation. *Science* 319(5866):1100–1104.

Matthen, Mohan, and André Ariew. 2009. Selection and causation. *Philosophy of Science* 76(2):201–224.

Michod, Richard E. 1999. *Darwinian dynamics*. Princeton: Princeton University Press.

Mills, Susan, and John Beatty. 1979. The propensity interpretation of fitness. *Philosophy of Science* 46(2):263–286.

Moreno-Estrada, Andrés, Kun Tang, Martin Sikora, Toms Marquès-Bonet, Ferran Casals, Arcadi Navarro, Francesc Calafell, Jaume Bertranpetit, Mark Stoneking, and Elena Bosch. 2009. Interrogating 11 fast-evolving genes for signatures of recent positive selection in worldwide human populations. *Molecular Biology and Evolution* 26(10):2285–2297.

Nagylaki, Thomas. 1992. *Introduction to theoretical population genetics*. Berlin: Springer.

Odling-Smee, F. John, Kevin N. Laland, and Marcus W. Feldman. 2003. *Niche construction: The neglected process in evolution*. Princeton: Princeton University Press.

Oyama, Susan, Paul E. Griffiths, and Russell D. Gray. 2001. What is developmental systems theory? In *Cycles of contingency: Developmental systems and evolution*, ed. Susan Oyama, Paul E. Griffiths, and Russell D. Gray, 1–11. Cambridge, MA: MIT Press.

Prasad, Anisha, Melanie J.F. Croydon-Sugarman, Rosalind L. Murray, and Asher D. Cutter. 2011. Temperature-dependent fecundity associates with latitude in *Caenorhabditis briggsae*. *Evolution* 65(1):52–63.

Price, George R. 1970. Selection and covariance. *Nature* 227:520–521.

Ramsey, Grant. 2006. Block fitness. *Studies in History and Philosophy of Biological and Biomedical Sciences* 37(3):484–498.

Richerson, Peter J., and Robert Boyd. 2005. *Not by genes alone*. New York: Oxford University Press.

Ringsby, Thor Harald, Bernt-Erik Saether, Jarle Tufto, Henrik Jensen, and Erling Johan Solberg. 2002. Asynchronous spatiotemporal demography of a house sparrow metapopulation in a correlated environment. *Ecology* 83(2):561–569.

Rosenthal, Jacob. 2010. The natural-range conception of probability. In *Time, chance, and reduction: Philosophical aspects of statistical mechanics*, ed. Gerhard Ernst and Andreas Hüttemann, 71–90. Cambridge: Cambridge University Press.

Rosenthal, Jacob. 2012. Probabilities as ratios of ranges in initial-state spaces. *Journal of Logic, Language, and Inference* 21:217–233.

Roughgarden, Jonathan. 1979. *Theory of population genetics and evolutionary ecology: An introduction*. New York: Macmillan.

Scheinfeldt, Laura B., Shameek Biswas, Jennifer Madeoy, Caitlin F. Connelly, and Akey Joshua. 2011. Clusters of adaptive evolution in the human genome. *Frontiers in Genetics* 2:50.

Smith, Barry, and Achille C. Varzi. 2002. Surrounding space: The ontology of organism-environment relations. *Theory in Biosciences* 121(2):139–162.

Sober, Elliott. 1984. *The nature of selection*. Cambridge, MA: MIT Press.

Stearns, Stephen C. 1976. Life-history tactics: A review of the ideas. *Quarterly Review of Biology* 51(1):3–47.

Sterelny, Kim. 2005. Made by each other: Organisms and their environment. *Biology and Philosophy* 20(1):21–36.

Sterelny, Kim, and Philip Kitcher. 1988. The return of the gene. *Journal of Philosophy* 85:339–361.

Strevens, Michael. 2011. Probability out of determinism. In *Probabilities in physics*, ed. Claus Beisbart and Stephann Hartmann, 339–364. Oxford: Oxford University Press.

Thompson, Emma E., Hala Kuttab-Boulos, David Witonsky, Limei Yang, Bruce A. Roe, and Anna Di Rienzo. 2004. CYP3A variation and the evolution of salt-sensitivity variants. *The American Journal of Human Genetics* 75(6):1059–1069.

Wagner, Andreas. 2005. *Robustness and evolvability in living systems*. Princeton: Princeton University Press.

Walsh, Denis M. 2007. The pomp of superfluous causes: The interpretation of evolutionary theory. *Philosophy of Science* 74:281–303.

Walsh, Denis M. 2010. Not a sure thing: Fitness, probability, and causation. *Philosophy of Science* 77(2):141–171.

Williams, George C. 1966. *Adaptation and natural selection*. Princeton: Princeton University Press.

Wimsatt, William C. 1980a. Randomness and perceived randomness in evolutionary biology. *Synthese* 43:287–329.

Wimsatt, William C. 1980b. Reductionistic research strategies and their biases in the units of selection controversy. In *Scientific discovery: Case studies*, ed. Thomas Nickles, 213–259. Dordrecht: Reidel.

Wimsatt, William C. 1981. The units of selection and the structure of the multi-level genome. In *PSA 1980*, vol. 2, 122–183. East Lansing: Philosophy of Science Association, East Lansing, Michigan, University of Chicago.

Wimsatt, William C. 2007. *Re-engineering philosophy for limited beings: Piecewise approximations to reality*. Cambridge, MA: Harvard University Press.

Models in Context: Biological and Epistemological Niches

Jessica A. Bolker

Abstract A model organism's value depends on its biological and epistemological contexts. The biological context of a model species comprises all aspects of its environment in the research setting that may influence its biological characteristics. In contrast, the epistemological context is not a matter of the organism's surroundings, but rather of what question it is supposed to help answer, and the assumptions about its "representativeness" that warrant broader application of results from a unique model. The biological context for model organisms in research is highly controlled and standardized. This strategy has often been productive; however, it risks eliminating essential environmental information and biological mechanisms, including organism-environment interactions that help shape phenotypes. Considering biological context helps us avoid experimental designs that simplify potentially important dimensions out of existence. Clarifying the epistemological context, from background assumptions to the ultimate goal of the research, lets us assess how the research approach we choose—such as employing a particular model—may constrain the range or utility of possible answers. Looking at models in context can enrich understanding of both the history and the practice of biology: how models are selected and evolve to fit questions, and how they in turn influence the direction of future work.

1 Introduction

A model is a representation of, or analogy for, something else. Models come in many forms (e.g. mathematical, statistical, physical, biological) and are used in many ways, for many purposes. Epidemiological and evolutionary models guide

J.A. Bolker (✉)
Department of Biological Sciences, University of New Hampshire, Durham, NH, USA
e-mail: jbolker@unh.edu

G. Barker et al. (eds.), *Entangled Life*, History, Philosophy and Theory
of the Life Sciences 4, DOI 10.1007/978-94-007-7067-6_8,
© Springer Science+Business Media Dordrecht 2014

public health decisions such as the design of flu vaccines; economic models shape both monetary policies and political strategies. In ecology, models play a key role in conservation decisions as well as in understanding complex long-term processes. In developmental biology, a handful of animal and plant models provide a narrow but powerful focus for studying cellular and molecular mechanisms. And in biomedical research, animal models are central both to studying fundamental mechanisms of disease and to developing and testing new drugs and treatments. In all these cases, models are "tools for the job" (Burian 1993): practical and epistemologically useful representations that help us understand less accessible cases, or more general patterns.

I will show in this chapter that a model's value and usefulness in biological research depend on two types of context: biological and epistemological. The biological context of a model species or organism comprises all aspects of the management and maintenance of organisms that may influence their biological characteristics. Culture conditions are critically important for cell lines (particularly stem cells, which are especially sensitive to their environment); for rodent models, housing and breeding protocols can affect biological traits from population genetics to physiology and behavior. The epistemological context is more abstract: it is a matter not of the organism's nature or surroundings, but of what question the model is supposed to help answer. We use some models in experiments designed to shed light on basic biological mechanisms, as with *C. elegans* (Haag 2009); in contrast, purpose-bred rodents serve as a testing ground to assess the safety and efficacy of specific drug candidates.

Because our understanding of context drives decisions (and supports assumptions) about what factors we need to account for, and what we can safely ignore, it's important to recognize both epistemological and biological contexts in model-based science.

For example, nucleocytoplasmic interactions in the oocyte are central to reprogramming stem cells, but obtaining and manipulating human oocytes poses ethical and practical difficulties that impede research in this area. Chimeras comprising a human nucleus in a non-human oocyte have been proposed as an alternative system that avoids the use of human oocytes (Chung et al. 2009). But while this non- (or only partially-) human model simplifies the ethical context of the research, the biological context provided by "foreign" cytoplasm may significantly alter the processes we hope to study. This disparity reduces the ability of the model to represent fully human stem cells, and thus its heuristic value (Robert 2004; Chung et al. 2009).

A species' success as a model depends on the epistemological environment in which it has to function: it may be expected to serve as a surrogate for human patients in an early-stage drug trial, or it may be supposed to represent all vertebrates in studies of gene function during development. In order for a model to do the job we intend it to do, it needs to be well-adapted to its epistemological niche.

Model-based research also needs to include enough biological context, even though a primary reason to use a model in the first place is to reduce the complexity of the experimental material: an inbred mouse housed in a lab cage is a simpler system than a wild rodent at large in its natural environment. But drastically simplifying the environment is a two-edged sword. The obvious advantage, and standard justification, is that doing so eliminates a host of potential variables that are not of interest in a given experiment. We can get much clearer answers to questions about the neurological control of movement in *Aplysia* than in a mammal. But simplifying too far risks eliminating "external" factors that may actually be integral to the phenomenon we seek to study.

2 Biological Contexts: From Ecology to Stem Cells (and Back)

For a model to do its job, it must also be observed in a biological context that includes the features essential to its normal function. Unfortunately, the more "laboratized" (Robinson 1965) a species becomes, the less attention is paid to its natural history, including its original ecological niche.

Determining which aspects of an organism's surroundings are biologically relevant demands close attention both to the intrinsic features of the organism itself, and also to its interactions with its environment. The latter render some aspects of the environment important and others insignificant: for example, the population of zooplankton in a lake may be essential to a carnivorous fish larva, but of no interest to a grazing tadpole swimming in the same water. Lewontin cites the example of a thrush that uses stones as anvils on which to break open snail shells: stones are a key element of a thrush's environment, as a function of its own feeding behavior. The bird determines which aspects of its environment are important, thus effectively helping to construct its own niche (Lewontin 1983). Through such niche construction, organisms actively—and substantially—shape their own environments. Moreover, these interactions are reciprocal: neither organism nor environment can be fully described or understood in isolation from the other (Lewontin 1983; Odling-Smee et al. 2003).

Nevertheless, the core strategy of lab-based, model-centered research is to abstract organisms from the complexity of their natural environments in order to study their traits under highly controlled, standardized conditions. This approach is especially prevalent in cell and developmental biology, where the vast majority of research is based on a handful of species.

In contrast to adult organisms, most embryos do little to shape their environments; but the environment can play a key role in shaping the embryo.[1] The concept

---

[1]Mammals are the exception that proves the rule: the blastocyst manipulates the intrauterine environment to permit implantation, thereby initiating a complex and dynamic maternal-fetal relationship with far-ranging developmental and evolutionary consequences.

of plasticity—the ability of developmental systems to incorporate environmental information into the construction of phenotypes—has been reemerging as a key idea for understanding developmental variation, and the role of such variation in ecological, medical, and evolutionary contexts (West-Eberhard 2003; Tollrian and Harvell 1999; Schlichting and Pigliucci 1998; Gilbert 2001; Gilbert and Epel 2009; Gilbert and Bolker 2003; Perera and Herbstman 2011). Individual organisms adjust their ontogeny according to environmental cues, enabling them to construct a phenotype that maximizes fitness in the context of a specific season, microhabitat, or predation regime.

Despite renewed attention to environmental effects on development, mainstream developmental biology remains concerned primarily with events inside embryos. But ecology reappears on a smaller scale within this more traditional, internalist view of development: cells inside embryos have critical, and often reciprocal, interactions with their surroundings, which are primarily other cells. Effectively, cells compose each other's niches. Analyzing dynamic interactions between individual and environment has thus long been a central theme in developmental biology. Both the physical interactions of morphogenesis (Keller et al. 2003) and the cell-cell communication that guides differentiation in many organisms are characterized by reciprocal interactions. Cells within embryos respond to and shape their immediate developmental context through chemical and physical communication with their neighbors. This is a very local form of niche construction, not in an evolutionary sense, but certainly in an ecological one (Laland et al. 2008).

Back in the lab, ecologists' niche terminology has been enthusiastically adopted by stem cell biologists (Moore and Lemischka 2006; Scadden 2006; Spradling et al. 2001; Calvi et al. 2003; Hackney et al. 2002; Zhang et al. 2003; Xie and Spradling 2000), though the full ramifications of the concept have not necessarily been carried over along with the term. The niche of a stem cell describes its physical location in relation to surrounding cells, which may play a crucial role in maintaining— or even inducing—its broad potency or "stemness." One well-studied example is the essential role of osteoblastic cells in regulating populations of hematopoetic stem cells, by regulating their niche (Calvi et al. 2003; Zhang et al. 2003). Current discussions of the relationship between stem cells and their niches echo classical questions in ecology (Powell 2005).

Ironically, the original ecological context of the models (whether species or cells) studied by most modern biologists concerned with cellular and physiological mechanisms has been largely ignored, or else so drastically simplified in the laboratory that key features of the model itself may be lost. One of the most important is variation, whether in genotype or in environmentally-influenced phenotypic traits. The inability of highly inbred rodent strains to represent (i.e. model) the spectrum of human genetic variation may help explain some of their shortfalls as models in immunology, a discipline focused on understanding cellular and tissue-level mechanisms that mediate our reaction to the environment (Desjardins et al., this volume).

3 Epistemological Contexts

The work a model organism can do and the constraints it may impose depend not just on its biology, but on the epistemological context. What needs to be included in a model—and more generally, what characteristics of the model need to match its target or *representandum*—depends on how we plan to use it (Bolker 2009).

Accessibility and general biological similarity to the target are necessary but not sufficient attributes for a research model. To define a good model in a specific research context we need to articulate additional criteria. In biomedical research the value of a model lies in its ability to represent something of human medical importance: perhaps a disease mechanism, or a set of symptoms that may respond to a new drug. Sometimes we want the model to serve as a surrogate for a patient, for instance a person suffering from asthma or Parkinson's disease. On other occasions, we use models that don't directly substitute for patients, but instead offer experimental access to genes or other factors implicated in disease etiology. Such models serve as tools for basic research whose findings may eventually inform prevention and treatment strategies.

The distinction between surrogate models that directly replace a human patient (e.g. in screening drug candidates) and exemplary models that serve basic research aimed at understanding fundamental, general biological mechanisms has important implications (Bolker 2009). For a model to work well as a surrogate, the traits directly involved in known disease pathways must match. The model need not be a close phylogenetic relative, or have anything else in common with a human patient, as long as some element of the model faithfully represents the focal aspect of the disease—what (Russell and Burch 1959) refer to as "fidelity." Such models serve as causal analogue models (Hesse 1963) within the explicitly defined limits of a focal question: no other aspect of their matching (or lack thereof) with the target is relevant, because there is no intent to generalize beyond the specific question and *representandum*. On the other hand, when a model is intended to exemplify a larger group and provide insights into fundamental biological mechanisms, its phylogenetic position and evolutionary history are important. Here, a good model is one from which we can reasonably generalize about a more inclusive group, rather than one that represents a specific, localized process in a given target species. In basic research, the unique features of a model (such as a disease-causing mutation) that we value in other contexts may become liabilities because they limit generalization, or weaken our ability to make broad inferences.

In addition to clarifying the epistemological role of the model itself, we need to look closely at the question we're trying to address, and the nature of a satisfactory answer. Finding an answer without having thoroughly articulated the question may be of very limited use—and defining the question may be trickier than we first assume (Adams 1979). The history of research into embryonic induction offers one example. The quest for the neural inducing signal in amphibians generated a long list of effective compounds including, oddly, blue jay liver (Wilson and

Hemmati-Brivanlou 1997; Witkowsky 1985). Discovering that the process could be triggered by blue jay liver did not directly advance knowledge of how it normally occurs, but it nevertheless led biologists to reframe their research question. In short, they came to realize that the key to understanding the process was not the identification of the inducer per se, but the response (Wilson and Hemmati-Brivanlou 1997; Witkowsky 1985).

Recognizing after the fact that the original research question was poorly understood undermines the value of any answers we've obtained. On the positive side, carefully considering epistemological context at the outset can help us choose the right models as tools for addressing a given question (Clarke and Fujimura 1992; Burian 1992, 1993; Lederman and Burian 1993).

4 Why Context Matters for Models

The questions we pose create contexts for our use of models. Popular models, in turn, create contexts that can shape or constrain the direction of further research. Some limitations are obvious: we can't study terrestrial locomotion in a fish, or gastrulation in yeast. But other constraints are subtler, and shaped as much by the history as by the biology of particular species. Established models have constituencies of scientists whose perspectives on research, and understanding of biology, have been shaped in part by their experiences with those systems. A scientist often chooses a particular model to address a given question because that species is already established as the "right tool for the job"; retooling is not only technically challenging, but brings the added burden of justifying the use of a non-standard model to peers and grant reviewers.

Over time, scientific communities arrive at consensus about the best way to do things, including the best model to use for particular questions. (Neurobiology may be an exception: Preuss 2000.) This approach has been highly productive, particularly as the communities that form around particular models develop shared questions and resources (Leonelli and Ankeny 2012). It's important to realize, however, that this kind of cultural and practical consensus about how best to do things is part of the epistemological context of scientific research. One effect of that context can be a tendency to canalize inquiry, particularly in the biomedical realm, where nearly all research is now carried out in just a few species, and the lion's share of grant funding supports studies of rodent models. We have both a propensity and a financial incentive to invest in constructing increasingly specific kinds of mice that are designed to address highly targeted questions such as the role of a gene associated with a particular disease (Bedell et al. 1997; Hardouin and Nagy 2000; Thyagarajan et al. 2003). But if diseases such as allergic asthma or diabetes depend strongly on the environment, not just the expression of particular genes carried by standard mouse models (Epstein 2004; Atkinson and Leiter 1999), specialized mouse models reared in standardized lab cages can tell us little about key aspects of their etiology.

There are many good reasons to continue to focus research effort and funding on work with well-established models, from the wealth of existing knowledge and techniques to the opportunities for synergistic interdisciplinary research. But there are also good reasons to think hard about the implications of this approach, and such reflection becomes especially critical when models fail to do the epistemic work we expect of them. As an example, biomedical research in rodents is often intended as a basis for translation to humans—that is, bringing new knowledge from bench to bedside. But there is increasing concern about the failure rate, especially with respect to drug development (Collins 2011), in fields such as immunology, diabetes and asthma research that have come to rely very heavily on highly inbred mouse models (Epstein 2004; Nials and Uddin 2009; Atkinson and Leiter 1999; Mestas and Hughes 2004; Thyagarajan et al. 2003; Desjardins et al., this volume).

Clarifying the scope of the questions we want to address, and the necessary conditions for answering them satisfactorily, is essential to doing good science. Otherwise, the realization that something important was left out—or not recognized as sufficiently significant to be worth controlling—is likely to be retrospective, as part of a post facto analysis of why a study didn't work, couldn't be replicated, or yielded results outside the range of the anticipated possibilities. A classic example is the early search for genes associated with aggressive behavior in mice: careful analyses in different labs yielded opposite results, which were ultimately explained by differences in animal housing and handling (Ginsburg 1966, 1992). Recent work confirms that exactly how mice are picked up—an apparently trivial aspect of routine animal care—may have significant effects on how these models work, and thus the experimental results they yield (Hurst and West 2010). Mouse handling was never intended to be an experimental variable; rather, it is normally an invisible (and unexamined) aspect of lab practice. However, handling techniques turn out to have significant effects on hormone levels and behavior, and thus potentially on the results of studies that use model mice.

5   Implications for Biological Practice

Whether we use models to study ecology, development, or stem cell biology, we have to account for the biological context; there are at least three different ways to do so. One strategy is to control or standardize the environment, and then ignore it. This strategy rests on the assumption that we already know what aspects of the environment require standardization to minimize experimental noise. The problem is that limiting the environment to a set of conditions we can easily control in the lab risks eliminating context that is integral to what we're trying to study. When the refinement of a model reaches a point where standardization outweighs complete representation, it can no longer serve its original epistemological purpose, effectively sacrificing precision for accuracy. Laboratory contexts are rarely intended (or believed) to replicate every aspect of an organism's natural environment.

There are good scientific as well as practical reasons to simplify experimental conditions, and standardize factors ranging from genetics to cage size. But we need to make sure we don't eliminate environmental or contextual information that can affect outcomes. For instance, studies of the genes underlying particular behaviors can yield misleading results when experiments are performed under laboratory conditions (Vanin et al. 2012). Another field in which environmental context is particularly critical is immunology, whose central focus is the interface of organism and environment (see Desjardins et al., this volume; Desjardins et al. 2012). A full accounting of immunological mechanisms represented in mouse models may require attention to the "envirotype" as well as the genotype, recognizing that both— and their interaction—are essential to phenotype (Beckers et al. 2009).

A second strategy is to measure differences between the "natural" environment and the experimental milieu, and factor known discrepancies into the interpretation of experimental results. We may need to do more of this if we want to understand potential ecological impacts of chemicals whose safety has been established only via single-factor lab studies. For example, the insecticide methoprene appears relatively harmless to vertebrates in studies focused exclusively on chemical dosage. However, methoprene's teratogenic effects are magnified dramatically in tadpoles exposed to its breakdown products (La Clair et al. 1998). Similarly, predator stress exacerbates the toxicity of pesticides beyond that observed in single-factor experiments (Davidson and Knapp 2007). Both UV degradation and predator stress are absent in standard laboratory analyses of chemical toxicity, but pervasive in a frog's natural environment; such multifactorial causes may help explain amphibian declines (Davidson and Knapp 2007). The clarity of controlled, single-factor lab experiments rarely fully represents the complexity of the real world—so inference from one to the other requires caution, and attention to what was left out.

A third strategy is to manipulate the environment on purpose, treating it as an independent variable whose effect on outcomes is an explicit focus of study. Ecological developmental biologists use this strategy to analyze developmental mechanisms that underlie the phenotypic plasticity familiar to ecologists (Gilbert and Epel 2009). More generally, it offers a way to learn which factors can be excluded, or what compensations should be made.

Each of these approaches begins by acknowledging the existence and potential significance of context. Considering biological context helps us avoid experimental designs that simplify potentially important dimensions out of existence (e.g. studies of plastic aspects of development in uniform conditions or highly canalized models). It raises the question of how differences in genetic background might affect the function of homologous genes: one potentially significant mismatch between humans and experimental rodents is that we are a genetically diverse and variable species, while key model strains are so inbred that they are effectively a single biological individual (von Herrath and Nepom 2005). Considering context can also suggest (nominally) external factors that might play significant roles in the phenomena we want to understand: for instance, many cancers may be caused by disruptions of tissue-level organization and ecology, not simply a genetic malfunction within an individual cell (Sonnenschein and Soto 1999; Soto and Sonnenschein 2011).

Clarifying the epistemological context, from background assumptions to the nature of the question we hope to answer, lets us assess whether the research approach we choose—such as employing a particular model—might constrain the range or utility of possible answers. Some questions are relatively straightforward, such as whether the results of drug studies in adult males are equally (or at least sufficiently) applicable across a population that includes females and juveniles. But other issues go to the root of entire experimental paradigms: neurobiologists remain deeply divided, for example, about the relative value of primate and rodent models for studies of brain function, in both basic and biomedically-oriented research (Preuss 2000); critiques of mouse-based immunology research raise similar problems (Desjardins et al., this volume).

Epistemological choices about experimental designs, including the selection of models, shape both research questions and the scope of possible answers. We approach any biological problem with some idea of what it's about, what factors we will treat as independent and dependent variables, and what we need to control as part of an experiment. That's how science is normally done, and it works: I am not arguing that we abandon this approach (even if it were possible to do so, which it probably isn't). Rather, I argue that we should pay more attention to those implicit assumptions, especially the ones about aspects we can safely leave out of the picture.

Attending to context may help address two challenges in model-based research. First, it may explain why well-established models are not fulfilling their promise in some areas. Specialized mutant and transgenic rodents used as disease models support research in which all possible "background" factors, from genetic makeup to laboratory environment, are meticulously standardized so that they can then be ignored while we focus on the "intrinsic" biology of the disease. The catch is that a model that omits environmental context that may be essential to disease mechanisms does not adequately represent its target, and thus can't do the epistemic work we expect of it. The use of oversimplified, isolated models may be an obstacle to progress in translational research, especially for immunological disorders in which the environment may play a significant role (Davis 2008; von Herrath and Nepom 2005).

Second, in cases where current models are insufficient, considering context can help us find alternatives. Epistemological and biological context both have to be included in the assessment of a model's utility for a particular task. Beyond its biological suitability, we need to consider how well a candidate model aligns with the question, and what epistemological role we expect it to play (Travis 2006). We must acknowledge the subtler ways in which the tool itself may reshape the question, for example by focusing attention on specific kinds of mechanisms, or eliminating variation in genetic background from functional studies of "disease genes" (Erickson 1996).

6   Biological Models and Epistemological Niches

Why should biologists fuss about the epistemological context? Why not leave that
to the philosophers, and get on with our work? Most successful and productive
biologists do just that. But biological and epistemological contexts are hard to
separate, and can influence one another in scientifically important ways. How one
does science is inseparable from what one is asking—and exactly how one defines
this research question.

In the early days of amphibian embryology, both Wilhelm Roux and Hans
Spemann sought to answer a core question: What is the state of commitment of
embryonic blastomeres at the two-cell stage? (described in (Gilbert 2003)). Roux
killed one blastomere with a hot needle and observed that the other formed a
multicellular half-embryo, equivalent to its original fate. But when Spemann ligated
embryos, separating the two cells along the first cleavage furrow and then culturing
them independently, he found that each blastomere could form a complete larval
body. The first experiment suggests that developmental potency is already limited at
the two-cell stage: the surviving blastomere forms only a half-embryo, and does not
complete the pattern by filling in for the killed cell. The second experiment leads
to the opposite conclusion: each cell can generate an entire embryo even though
normally it would form only half. The explanation for the disparate results lies not
in the cells themselves, but in the context provided for their continued development:
either attached to another (dead) cell, or cultured in isolation.

Depending on the experimental context, cells give disparate answers to what
initially appears to be the same question about their potency. In retrospect, we can
see that the questions were not really the same, because of the different methods
used to answer them. Roux asked about the developmental potential of a single
blastomere adjacent to a (nominally inert) dead cell, while Spemann asked what
a single isolated blastomere could do in the absence of any context other than
the culture medium. In the end, both experimental outcomes point to the same
fundamental principle: a cell's differentiation and fate depend on its context.

The effect of local context on cell differentiation is not merely a problem of
classical embryology: it is a pivotal question in modern stem cell biology and cancer
research. Stem cells are exquisitely sensitive to their surroundings, even to the
stiffness of the substrate and the shape of the well in which they are cultured (Engler
et al. 2006; Kilian et al. 2010); cancer cells become dangerous when they lose the
ability to behave in a tissue-appropriate manner, and disregard environmental signals
that normally stabilize the differentiated state (Sonnenschein and Soto 1999; Soto
and Sonnenschein 2011). For both stem and cancer cells, meticulously cultured cell
lines are primary experimental models, raising the question of how well the in vitro
system represents the complex in vivo environment with which we are ultimately
concerned. Are stem cells in culture enough like stem cells in vivo to support reliable
inferences from one to the other? The differences in their environments add an extra
dimension to the challenge of figuring out what chemical signals influence gene
expression and differentiation pathways in pluripotent cells, and further complicate

the task of drawing inferences about potential clinical applications (which depend on cells' in-vivo behavior) from *in vitro studies*. Ironically, the controllability of the environment that makes in vitro studies so powerful may also limit our ability to apply their results to stem cells in the context of a complex and dynamic living tissue.

7 Conclusion

Ecologists describe the niches of species and organisms in the field, while stem cell biologists study the niches of pluripotent cells in culture in the laboratory. Both observe the active construction and maintenance of niches by their occupants. Model organisms, too, have niches. Some niches are biological and historical; others are constructed by scientists' epistemological practices. Understanding the interactions between models and their contexts can offer practical insights into how model organisms may shape the science they serve. Conversely, ignoring such interactions can blind us to some important issues.

Failing to consider the contexts of model use has several potential costs. We may be unable to resolve apparent conflicts, such as Roux's and Spemann's observations about blastomere potency. We forego a possible explanation for the disappointing record of translational research, namely the epistemic shortcomings of key disease models. We constrain biomedical research by seeking the causes of disease primarily in intrinsic traits such as aberrant genes, and commonly limiting treatment studies to symptoms that are well represented in standard models (e.g. motor symptoms in PD). We may miss potentially critical causes or mechanisms that fall outside either the frame of the question, or the physical limits of the model (whether a gene, a mouse strain, or a cell culture).

In contrast, there is much to be gained by recognizing the epistemic and biological contexts of our models. Such recognition makes it easier to see when it's time to reframe the question and alter research goals accordingly, as in the case of embryonic induction. It helps clarify when we need different models in a particular field (e.g. eco-devo: (Jenner and Wills 2007; Abzhanov et al. 2008; Maher 2009), and how to choose good ones. Considering the epistemic and biological contexts of models lets us see a greater range of possible mechanisms and explanations, and informs the design of studies that include environmental factors (such as the aspects of stem cells' niches that modulate their gene expression). It lets us recognize a broader range of possible research tools and strategies—for example "evolutionary mutant models" (Albertson et al. 2008) that may offer a complementary perspective to what we see in highly-engineered inbred mice. We may develop new insights into the regulation of stem cell populations by looking at their overall ecology, rather than just their gene expression. In the long run, we also stand to gain a richer understanding of the history and the practice of biology: both how models are selected and evolve to fit questions (Kohler 1994; Rader 2004; Burian 1993; Lederman and Burian 1993), and how they in turn influence the direction of future work.

Acknowledgements The author thanks Gillian Barker, Eric Desjardins, and Trevor Pearce for their helpful comments on the manuscript.

References

Abzhanov, Arhat, Cassandra G. Extavour, Andrew Groover, Scott A. Hodges, Hopi E. Hoekstra, Elena M. Kramer, and Antonia Monteiro. 2008. Are we there yet? Tracking the development of new model systems. *Trends in Genetics* 24: 353–360.

Adams, Douglas. 1979. *The Hitchhiker's guide to the galaxy*. New York: Random House.

Albertson, R. Craig, William Cresko, H. William Detrich, and John H. Postlethwait. 2008. Evolutionary mutant models for human disease. *Trends in Genetics* 25: 74–81.

Atkinson, Mark A., and Edward H. Leiter. 1999. The NOD mouse model of type 1 diabetes: As good as it gets? *Nature Medicine* 5: 601–604.

Beckers, Johannes, Wolfgang Wurst, and Martin Hrabé de Angelis. 2009. Towards better mouse models: Enhanced genotypes, systemic phenotyping and envirotype modelling. *Nature Reviews Genetics* 10: 371–380.

Bedell, Mary A., Nancy A. Jenkins, and Neal G. Copeland. 1997. Mouse models of human disease. Part I: Techniques and resources for genetic analysis in mice. *Genes & Development* 11: 1–10.

Bolker, Jessica A. 2009. Exemplary and surrogate models: Two modes of representation in biology. *Perspectives in Biology and Medicine* 52: 485–499.

Burian, Richard M. 1992. How the choice of experimental organism matters: Biological practices and discipline boundaries. *Synthese* 92: 151–166.

Burian, Richard M. 1993. How the choice of experimental organism matters: Epistemological reflections on an aspect of biological practice. *Journal of the History of Biology* 26: 351–368.

Calvi, L.M., G.B. Adams, K.W. Weibrecht, J.M. Weber, D.P. Olson, M.C. Knight, R.P. Martin, E. Schipani, P. Divieti, F. Bringhurst, L. Milner, H. Kronenberg, and D. Scadden. 2003. Osteoblastic cells regulate the haematopoietic stem cell niche. *Nature* 425: 841–846.

Chung, Young, Colin E. Bishop, Nathan R. Treff, Stephen J. Walker, Vladislav M. Sandler, Sandy Becker, Irina Klimanskaya, et al. 2009. Reprogramming of human somatic cells using human and animal oocytes. *Cloning and Stem Cells* 11: 213–223.

Clair, La., J. James, John A. Bantle, and James Dumont. 1998. Photoproducts and metabolites of a common insect growth regulator produce developmental deformities in Xenopus. *Environmental Science and Technology* 32: 1453–1461.

Clarke, Adele E. 1992. *The right tools for the job: At work in twentieth-century life sciences.* Princeton: Princeton University Press.

Collins, Francis S. 2011. Reengineering translational science: The time is right. *Science Translational Medicine* 3(90): 1–6.

Davidson, Carlos, and Roland A. Knapp. 2007. Multiple stressors and amphibian declines: Dual impacts of pesticides and fish on yellow-legged frogs. *Ecological Applications* 17: 587–597.

Davis, Mark M. 2008. A prescription for human immunology. *Immunity* 29: 835–838.

Engler, Adam J., Shamik Sen, H. Lee Sweeney, and Dennis E. Discher. 2006. Matrix elasticity directs stem cell lineage specification. *Cell* 126: 677–689.

Epstein, Michelle M. 2004. Do mouse models of allergic asthma mimic clinical disease? *International Archives of Allergy and Immunology* 133: 84–100.

Erickson, Robert P. 1996. Mouse models of human genetic disease: Which mouse is more like a man? *BioEssays* 18: 993–997.

Gilbert, Scott. 2001. Ecological developmental biology: Developmental biology meets the real world. *Developmental Biology* 233: 1–12.

Gilbert, Scott. 2003. *Developmental biology*, 7th ed. Sunderland: Sinauer.

Gilbert, Scott, and Jessica A. Bolker. 2003. Ecological developmental biology: Preface to the symposium. *Evolution and Development* 5: 3–8.

Gilbert, Scott, and David Epel. 2009. *Ecological developmental biology*. Sunderland: Sinauer.

Ginsburg, Benson. 1966. All mice are not created equal: Recent findings on genes and behavior. *The Social Service Review* 40: 121–134.

Ginsburg, Benson E. 1992. Muroid roots of behavior genetic research: A retrospective. In *Techniques for the genetic analysis of brain and behavior: Focus on the mouse*, ed. D. Goldowitz, D. Wahlsten, and R.E. Wimer, 3–16. New York: Elsevier.

Haag, Eric S. 2009. *Caenorhabditis* nematodes as a model for the adaptive evolution of germ cells. *Current Topics in Developmental Biology* 86: 43–66.

Hackney, Jason A., Pierre Charbord, Brian P. Brunk, Christian J. Stoeckert, Ihor R. Lemischka, and Kateri A. Moore. 2002. A molecular profile of a hematopoietic stem cell niche. *Proceedings of the National Academy of Sciences* 99: 13061–13066.

Hardouin, Sylvie, and Andras Nagy. 2000. Mouse models for human disease. *Clinical Genetics* 57: 237–244.

Hesse, Mary B. 1963. *Models and analogies in science*. Notre Dame: University of Notre Dame Press.

Hurst, Jane L., and Rebecca S. West. 2010. Taming anxiety in laboratory mice. *Nature Methods* 7: 825–826.

Jenner, Ronald A., and Matthew A. Wills. 2007. The choice of model organisms in evo-devo. *Nature Reviews Genetics* 8: 311–319.

Keller, Ray, Lance A. Davidson, and David R. Shook. 2003. How we are shaped: The biomechanics of gastrulation. *Differentiation* 71: 171–205.

Kilian, Kristopher A., Branimir Bugarija, Bruce T. Lahn, and Milan Mrksich. 2010. Geometric cues for directing the differentiation of mesenchymal stem cells. *Proceedings of the National Academy of Sciences* 107: 4872–4877.

Kohler, Robert E. 1994. *Lords of the fly: Drosophila genetics and the experimental life*. Chicago: Chicago University Press.

Laland, Kevin N., F. John Odling-Smee, and Scott F. Gilbert. 2008. EvoDevo and niche construction: Building bridges. *Journal of Experimental Zoology. Part B, Molecular and Developmental Evolution* 310: 549–566.

Lederman, Muriel, and Richard M. Burian. 1993. Introduction. *Journal of the History of Biology* 26: 235–237.

Leonelli, Sabina, and Rachel A. Ankeny. 2012. Re-thinking organisms: The impact of databases on model organism biology. *Studies in History and Philosophy of Biological and Biomedical Sciences* 43: 29–36.

Lewontin, Richard. 1983. The organism as the subject and object of evolution. *Scientia* 118: 65–95.

Maher, Brendan. 2009. Evolution: Biology's next top model? *Nature* 458: 695.

Mestas, Javier, and Christopher Hughes. 2004. Of mice and not men: Differences between mouse and human immunology. *Journal of Immunology* 172: 2731–2738.

Moore, Kateri A., and Ihor R. Lemischka. 2006. Stem cells and their niches. *Science* 311: 1880–1885.

Nials, Anthony T., and Sorif Uddin. 2009. Mouse models of allergic asthma: Acute and chronic allergen challenge. *Disease Models & Mechanisms* 1: 213–220.

Odling-Smee, F. John, Kevin N. Laland, and Marcus W. Feldman. 2003. *Niche construction: The neglected process in evolution*. Princeton: Princeton University Press.

Perera, Frederica, and Julie Herbstman. 2011. Prenatal environmental exposures, epigenetics, and disease. *Reproductive Toxicology* 31: 363–373.

Powell, Kendall. 2005. Stem-cell niches: It's the ecology, stupid! *Nature* 435: 268–270.

Preuss, Todd M. 2000. Taking the measure of diversity: Comparative alternatives to the model-animal paradigm in cortical neuroscience. *Brain, Behavior and Evolution* 55: 287–299.

Rader, Karen. 2004. *Making mice: Standardizing animals for American biomedical research*. Princeton: Princeton University Press.

Robert, Jason Scott. 2004. Model systems in stem cell biology. *BioEssays* 26: 1005–1012.

Robinson, Roy. 1965. *Genetics of the Norway rat*. New York: Pergamon Press.

Russell, William, and Rex Leonard Burch. 1959. *The principles of humane experimental technique.* Springfield: Charles C. Thomas.

Scadden, David T. 2006. The stem-cell niche as an entity of action. *Nature* 441: 1075–1079.

Schlichting, Carl D., and Massimo Pigliucci. 1998. *Phenotypic evolution: A reaction norm perspective.* Sunderland: Sinauer.

Sonnenschein, Carlos, and Ana Soto. 1999. *The society of cells: Control of cell proliferation and cancer.* New York: Springer.

Soto, Ana, and Carlos Sonnenschein. 2011. The tissue organization field theory of cancer: A testable replacement for the somatic mutation theory. *BioEssays* 33: 332–340.

Spradling, Allan, Daniela Drummond-Barbosa, and Toshie Kai. 2001. Stem cells find their niche. *Nature* 414: 98–104.

Thyagarajan, Tamizchelvi, Satish Totey, Mary Jo S. Danton, and Ashok B. Kulkarni. 2003. Genetically altered mouse models: The good, the bad, and the ugly. *Critical Reviews in Oral Biology and Medicine* 14: 154–174.

Tollrian, Ralph, and C. Drew Harvell (eds.). 1999. *The ecology and evolution of inducible defenses.* Princeton: Princeton University Press.

Travis, Joseph. 2006. Is it what we know or who we know? Choice of organism and robustness of inference in ecology and evolutionary biology. *American Naturalist* 167: 303–314.

Vanin, Stefano, Supriya Bhutani, Stefano Montelli, Pamela Menegazzi, Edward W. Green, Mirko Pegoraro, Federica Sandrelli, Rodolfo Costa, and Charalambos P. Kyriacou. 2012. Unexpected features of *Drosophila* circadian behavioural rhythms under natural conditions. *Nature* 484: 371–375.

Von Herrath, Matthias G., and Gerald T. Nepom. 2005. Lost in translation: Barriers to implementing clinical immunotherapeutics for autoimmunity. *The Journal of Experimental Medicine* 202: 1159–1162.

West-Eberhard, Mary Jane. 2003. *Developmental plasticity and evolution.* Oxford: Oxford University Press.

Wilson, Paul A., and Ali Hemmati-Brivanlou. 1997. Vertebrate neural induction: Inducers, inhibitors, and a new synthesis. *Neuron* 18: 699–710.

Witkowsky, Jan. 1985. The hunting of the organizer: An episode in biochemical embryology. *Trends in Biochemical Sciences* 10: 378–381.

Xie, Ting, and Alan Spradling. 2000. A niche maintaining germ line stem cells in the *Drosophila* ovary. *Science* 290: 328–330.

Zhang, Jiwang, Chao Niu, Ling Ye, Haiyang Huang, Xi He, Wei-Gang Tong, Jason Ross, et al. 2003. Identification of the haematopoietic stem cell niche and control of the niche size. *Nature* 425: 836–840.

Thinking Outside the Mouse: Organism-Environment Interaction and Human Immunology

Eric Desjardins, Gillian Barker and Joaquín Madrenas

Abstract Several review articles in immunology indicate that while we have an increasing body of knowledge about the immunology of the mouse, this is translating very poorly into clinical outcomes for humans. This raises several issues for the scientific community, including some related to the apparent inadequacy of the mouse as a model for understanding and predicting human immunity. This paper has two purposes. First, we offer an explanation for why the typical approach to animal model research will most likely fail to produce satisfying clinical outcomes for human immunology. The standard approach to this problem focuses on the lack of similarity between the genes and molecular pathways of model and target systems. Our analysis focuses instead on differences in the ways in which model and target organisms interact with and adapt to their respective environments. We argue that in order to find a proper model organism for studying human immunity we need to think outside the mouse. Second, we advocate abandoning purely reductionist, gene-centered research, giving greater importance to observational studies of humans, and using new emerging technologies for information-processing for *in vivo* observation.

E. Desjardins (✉) • G. Barker
Department of Philosophy, Rotman Institute of Philosophy, Western University, London, Ontario, Canada
e-mail: edesjar3@uwo.ca; gbarker5@uwo.ca

J. Madrenas
Department of Microbiology and Immunology, McGill University, Montreal, QC, Canada
e-mail: joaquin.madrenas@mcgill.ca

G. Barker et al. (eds.), *Entangled Life*, History, Philosophy and Theory of the Life Sciences 4, DOI 10.1007/978-94-007-7067-6_9,
© Springer Science+Business Media Dordrecht 2014

1   Introduction

For the last 50 years, immunology has focused on exploring the molecular bases
of immunity by studying animal models (mostly the mouse) in the laboratory.
Despite the enormous progress achieved by this approach, human immunology is
now facing one of its most difficult challenges: bridging the gap between basic
laboratory research and clinical outcomes. A spate of recent review and opinion
articles in major journals reveals that immunologists are increasingly concerned
about what some have called "the crisis in human immunology" (Lage 2008; Leslie
2010; Hayday and Peakman 2008; Germain and Schwartzberg 2011). The main
objective of this paper is to analyze the nature of this crisis, and the challenge that it
reflects. We argue that the challenge is virtually impossible to surmount within the
existing research framework, with its overwhelmingly dominant focus on mouse
immunology research. In brief, we argue that the complexity and context-dependent
nature of the immune system imply that the only way to really learn about the
immune functions of cells and molecules in humans is to study humans directly.
This in turn means that an important shift in the discipline of immunology is needed.
More resources and time should be devoted to projects that aim at the collection of
data and the development of theories from and about humans—and less to controlled
laboratory study on animal models. We will not here address the far-reaching
practical and institutional adjustments that such change will certainly imply. Our
goal instead is to highlight some pervasive ideological elements at the source of the
actual crisis—a gene-centered and reductionist understanding of living organisms
and a lack of attention to the interaction of organisms with their environments.
Although the mouse may share with us many of the basic building blocks involved
in immunity, treating organisms in isolation from their environments will almost
inevitably fail to produce any substantial harvest of immunological knowledge that
can be transferred to clinical contexts.

  Our thesis is not only negative. We also want to begin a discussion about what
it means to go beyond the standard program of drawing inferences from genetic
factors and molecular pathways. How can we make the metaphor of thinking outside
the mouse more concrete? Our focus here is on the core goals that define human
immunology as a research program. Although there will always be a need for the
development of fundamental and general knowledge about immunity, we think that
human immunology research should take a pragmatic turn, *i.e.*, it should identify the
goal of achieving success in *clinical outcomes* as a central priority and recognize that
achieving this goal will require the production of knowledge in contexts relevantly
similar to clinical contexts.

  In Sect. 2, we give a brief overview of the notion of animal models as Causal
Analogical Models (CAMs) and highlight some of the important structural differences that separate the model (the laboratory mouse) from the target system (human
beings). In Sect. 3, we look at the two basic types of immunity, innate and adaptive
immunity. These give different but powerful reasons to believe that animal models
will be of very limited use in investigating human immune function.The former, a

collection of inherited immune capacities, is evolved to fit each species' ecological context and life-cycle—it suggests that animals with evolutionary histories as divergent as those of mice and humans will have deeply different patterns of immune response. The latter, a form of ontogenic Darwinism that characterizes vertebrate immune systems, implies much more radical constraints on what can be learned from animal models or indeed from any laboratory study. We argue that adaptive immunity entails such a degree of sensitivity to context and individual history that it severely limits the usefulness of animal models in immunology research that is directed toward achieving clinical outcomes. Section 4 investigates some of the basic assumptions that have been driving biomedical research and that can be seen as responsible for the current crisis in human immunology. Finally, Sect. 5 begins to sketch a more concrete picture of what it will mean to think outside the mouse.

2 Animal Models

Animal models[1] in biomedical research are often conceived of as Causal Analogical Models (CAMs). The idea behind a CAM is quite intuitive. If two physical systems share a number of causal properties, then researchers can study one system (the model) and infer how another (target) system will respond to similar interventions (taking into consideration identified differences). In order to make this intuition more precise, we can begin by pointing out that CAMs are a special class of Analogical Models (AMs):

> (AM): X (the model) is similar to Y (the subject being modeled) with respect to a range of properties a, b, c, d, e. X has the additional property f. It is likely that Y has the additional property f. (Shanks and Greek 2009, 105)

This reasoning captures the core of the inference made in biomedical research with animal models.[2] It is crucial to keep in mind that analogies are subject to revisions and are not deductive inferences. The similarities between the model and the target system need to be investigated (empirically – in the case of animal models), and the discovery of the property f in the model does not entail its presence in the target system. Moreover, the correlation between a,b,c,d,e and f tells us nothing about the underlying causal relationship and it does not guarantee that the same will hold for both systems.

---

[1]This section deals with model systems and with inference from these to target systems. As such, it could apply to a wide range of physical systems. But our focus is on biomedical research, more specifically on inference from research on animal models to conclusions about humans, in the area of immunology. Unless otherwise specified, you can assume that the expressions "model" and "target system" refer respectively to "animal model" and "human."

[2]Please note that there is a vast and active literature trying to figure out how the inference from animal models to other target systems really works—e.g., Overmier and Carroll (2001), Gachelin (2006), and Volume 26, Number 2 (1993) of the *Journal of the History of Biology*. But the main idea we are drawing on here is quite uncontroversial.

For example, Lord Rutherford once suggested that the solar system could be a good model for understanding the relationship between electrons and the nucleus of atoms. This analogy turned out to be a fruitful heuristic device allowing physicists to formulate all sorts of testable hypotheses about electrons (Wilson 1984). Yet subsequent developments in quantum mechanics revealed that electrons actually have very little in common with planets. In fact, they do not appear to behave at all like macroscopic bodies.

Although analogies can be useful in some contexts, they can also be misleading if one forgets that they are only analogies. Before we get into the specificities of animal models in immunology, let's see how we go from AM to CAM. According to Shanks and Greek (2009, 106), CAMs are AMs that satisfy the following constraints:

1. The common properties a,b,c,d,e must be causal properties, i.e., they must be effects of various causes.
2. The property f to be projected from model to system modelled should stand in a causal relationship to the properties a,b,c,d,e in the model. Ideally, it should be a cause or effect of a,b,c,d,e.
3. There should be no relevant causal disanalogies between the model and the subject modeled.

Biomedical research assumes that if these conditions are met, then it is possible to predict that f will not only be found in the target system, but it will also possess the same causal dispositions as in the model. In other words, the hope is that once we have established that the correlation between the two systems result from the fact that the same type of causal relationships are at play, then it becomes possible to extrapolate with a fair degree of confidence on a variety of physiological and clinical responses.

There are several ways in which analogies can break down when dealing with complex systems such as living organisms. Organisms of different species cannot be perfectly (or even extremely closely) isomorphic. In fact, given that there are often variations among the members of a species, isomorphism is perhaps impossible to achieve in model research. This idea is well captured by Rosenblueth and Wiener (1945): "the best material model for a cat is another, or preferably the same cat." Moreover, the set of common properties a,b,c,d,e relevant for the response to a stimulus obviously don't fully define the organisms. They represent only a subset of properties of both the model and the target system. There is typically another set of properties in the model, relevant for the occurrence and/or the magnitude of a,b,c,d,e that is not common to both systems (Shanks and Greek 2009). Despite the general belief in biomedical research that mice can mirror human biology remarkably well, the list of differences between the mouse and human is long (Mestas and Hughes 2004). This fact does not come as a surprise when we think that the two species diverged around 65 million years ago and have since adapted to quite different environmental and ecological challenges.

Furthermore, even when properties are shared, the relations among them can differ between the two systems, resulting in very different outcomes. With the possible exception of a few severe genetic diseases, very few traits are determined by the presence of a single gene. Imagine for example that a,b,c,d,e is a set of structural genes involved in the expression of the phenotype f. The mere presence of a,b,c,d,e in both the model and the target organism does not mean that the time and/or level of expression of f will be the same. Or take a more concrete example. Both the human and the mouse species have lymphocytes and neutrophils.[3] But the balance of the two types of cells differs significantly between the two species: humans have a neutrophil-rich blood (50–70 % neutrophils vs. 30–50 % lymphocytes), whereas mice have a lymphocyte-rich blood (10–25 % neutrophils vs. 75–90 % lymphocytes) (Mestas and Hughes 2004, 2731). Since these cells are involved in the production of different classes of defense molecules, it is likely that this difference can impact the immunological responses of the two types of organism.

Note however that none of these difficulties yet challenges the basic assumption made in animal model research. One can have a long list of structural and molecular disanalogies between the model and the target system, and yet conclude that the best way to predict human responses from animal model research is to find a better analogue to our target system. We have a vivid example of this kind of response in the development and enthusiastic uptake of so-called "humanized mice" as a new model for immunology research.[4] These are mice into which human genetic or cellular components have been introduced. The hope is that by working with organisms that are biochemically more similar to the human target, we increase the predictive power that we can achieve. This approach is motivated by the belief that in principle, we should be almost certain to find the relevant property f in the target system if we could base our inference on an almost isomorphic model.

As we will see later, this kind of reasoning can be seriously misleading when combined with a commitment to a framework that ignores the complex web of interactions between organisms and their environment. We will see in the next section that the vertebrate immune system is such that even a perfectly homologous model could fail to predict the behavior of its target if the two organisms differ slightly in their environmental context or in the order of events in their life history. In other words, because the immune system of vertebrates is highly context sensitive and path dependent, the model and the target system have to be not only very similar in their internal structure, but also in the ways in which they interact with their environment.

---

[3]Lymphocytes are small white blood cells. There are two main types of lymphocytes, T and B cells. Both are involved in the production of antibodies (immunoglobulins)—B cells by making them and T cells by regulating their production by B cells. Neutrophils are white blood cells that ingest and destroy bacteria. (National Institute of Allergy and Infectious Diseases, http://www.niaid.nih.gov/topics/immunesystem/Pages/default.aspx)

[4]Biodefense Workshop Summary: Humanized Mice, 2005, Clarion Bethesda Park Bethesda, Maryland Abstract; http://www.niaid.nih.gov/topics/immuneSystem/Pages/frontiers.aspx

3   Innate and Adaptive Immunity

As Darwin (1859) so memorably wrote at the close of the *Origin*, if one stops and looks at the complexity of living organisms and how they are adapted to their conditions of life, one will see grandeur in nature. Among the most impressive such adaptations that we have discovered is immunity, i.e., the capacity of an organism to defend itself against invading microorganisms and substances.

In vertebrates, immunity is typically characterized by two lines of defense mechanisms: innate and adaptive. Innate immunity, as the name suggests, is the part of the immune system that organisms possess at birth. It encompasses cells like macrophages, dendritic cells, neutrophils, natural killer cells and mastocytes.[5] This part of the immune system remains active and virtually unchanged throughout the entire life of the organism. The types of mechanisms involved in innate immunity reflect the evolutionary history of the species. Innate immunity thus changes and adapts to new situations, but it does so very slowly over many generations and not during the lifetime of individual organisms. The evolutionary process by which it becomes able to respond to new challenges is much too slow to cope with the rapid evolution of infectious microorganisms. As a consequence, our species would probably not survive if the only kind of immunity we had was innate immunity. Fortunately for us, our ancestors evolved a second line of defense that can adapt to new infectious treats during the lifetime of an organism or indeed during a single episode of infection. The B and T cells mentioned earlier are key factors in this adaptive immunity. With the help of other factors of the innate immune system, B and T cells are responsible for the rapid production of antibodies, which act very specifically against different types of pathogens, or the development of killer T cells. So the two lines of defense act in concert in helping organisms to keep infectious microorganisms in check.

Unlike the innate system, the adaptive immune system constantly changes during the life of an organism as a function of antigen exposure. It does this by means of somatic selective processes that are made possible by the rapid production of variations *via* somatic recombination and somatic hypermutation. Somatic selection can be compared to a Darwinian selection process that happens within organisms in lines of non-reproductive cells. Some refer to this type of process as "ontogenic Darwinism" (Shanks 2004). Basically, when a pathogen invades an organism, certain cells of the innate system present the pathogen to the antibody-producing cells of the adaptive system. B and T cells produce a tremendous variety of antibodies, but only a few of them can bind to a given pathogen. Defense by antibodies is thus much more specific than innate immunity. The synthesis of a given antibody involves multiple genetic components that are shuffled together to form a complete immunoglobulin gene, which in turn specifies the structure of a

---

[5]National Institute of Allergy and Infectious Diseases
http://www.niaid.nih.gov/topics/immunesystem/Pages/default.aspx

given antibody. This *somatic recombination* process allows organisms to produce a great variety of antibodies—the system is capable of recognizing at least 100 billion different types of antigens (Shanks and Greek 2009, 188). This process of recombination also increases the probability that the system will produce at least one antibody capable of binding to any new antigen.

Once an antibody succeeds in binding to the new antigen, the cell in which it happens goes through a complex series of steps involving chemical signals with other cells and then make exact copies of the antibody and of clonal companion of this B cell, i.e., B cell producing that particular antibody. From this progeny, some cells differentiate into plasma cells (i.e., antibody factories) and others become memory B cells. The latter will allow a much faster response to subsequent infections by the same pathogen. The efficiency of the immune response is also known to increase as an infection progresses. This is because antibodies produced later in the response bind more effectively to the pathogen. The mechanism behind this phenomenon is another selection process called "somatic hypermutation." In this process, the immunoglobulin genes involved in the synthesis of the antibodies undergo random mutation at a very high rate[6] thus increasing the chance of creating varieties of antibodies that are an even better fit for the antigen. This selective fine-tuning allows for an increasingly efficient immune response.

This is of course an extremely simplified representation. But it suffices to show that adaptive immunity stands in the way of laboratory animal model research. Although all higher vertebrates have adaptive immunity and many species share the same general mechanisms, and though reductive methods employing animal models can be very valuable in uncovering those mechanisms, the adaptive immune response that they contribute to is such that each organism will develop a unique response to antigen exposures. This fact has been acknowledged for quite a while now, as Stanford immunologist Peter Parham notes:

> At some point [in] this century the experimental biologists, in an echo of Henry Ford, divorced themselves from evolutionary biologists. This artificial and regrettable separation remains with us today. For the immunologists it was always a sham for the very foundations of their subject are built upon stimulation, selection and adaptive change. Now we see clearly the immune system for what it is, a vast laboratory for high speed evolution. By recombination, mutation, insertion and deletion, gene fragments are packaged by lymphocytes, forming populations or receptors that compete to grab hold of antigen. Those that succeed get to reproduce and their progeny, if antibodies, submit to further rounds of mutation and selection. There is no going back and the destiny of each and every immune system is to become unique, the production of its encounters with antigen and the order in which they happen (Parham 1994, 373).

Parham's overview not only highlights one of the great mistakes of biomedical research, namely its disconnection from evolutionary biology,[7] but it also nicely

---

[6]The rate of mutation in somatic hypermutation is approximately 10^{-3}/*bp/cell division*, which is 10^6 fold higher than the average mutation rate of structural genes (Odegard and Schatz 2006).

[7]Parham is not the only one to bring this critique to immunology. Shanks and Greek (2009) is another, much more articulated example of this claim. But it is also interesting to note that the

brings together several of the key aspects of immunity that make animal models so problematic. It reinforces the point that one of the most important factors in the study of immunology is the *history* of antigen exposure. Our immune identity is not something fixed from the beginning; what defines it is not so much the initial genetic conditions, but the series of encounters and changes accumulated along the way. In that sense, the vertebrate immune system is strongly path-dependent (Desjardins 2011). And the dependence is not only on the series of mutations shaping immunoglobulin genes. If it were, then the prospects of human immunology would probably be in much better shape. Immunity, like many other biological capacities, cannot be satisfyingly understood by studying merely what happens inside the organism. The notion of "antigen *exposure*" makes this fairly explicit, for it implies a relationship with something external, an environment. It is through this relation that the immunity of vertebrates is defined. Unfortunately, it is almost impossible to control this factor—unless we deal with laboratory animals bred in pathogen-free environments and very carefully exposed to one type of pathogen before being destroyed. Moreover, Parham is right to say that each organism is unique in this respect. No two humans, even monozygotic twins, are isomorphic with respect to adaptive immunity. So it is clear that trying to predict immune response in humans from laboratory research on mice, even humanized mice, is missing an important part of the picture. The best model for studying human immunology remains the human. As we will discuss in Sect. 5, this entails that in many cases, we will have to turn away from the 'gold standard' of biomedical research (randomized controlled trials) and conduct observational studies on humans. The following section will give a sense of the challenges that such a change would represent. It means that an entire community of researchers has to make a shift towards an approach typically deemed inferior.

4   Behind the Mouse-centric Paradigm

Let us make something clear. We are not arguing that research with animal models has no contribution to make to immunology. On the contrary, such research will continue to have important roles to play. Many aspects of immunity cannot, if only for moral reasons, be investigated directly on humans. Moreover, vital discoveries

---

clonal selection theory discussed by Parham has been interpreted in Darwinian terms since its early articulation in the late 1950s by Sir Frank Macfarlane Burnet (1899–1985). Before that point, immunologists believed that lymphocytes, the antibody producing cells, were instructed what antibodies to produce from the antigen exposure. Burnet, inspired by Jern's 1955 hypothesis, suggested that lymphocytes were not instructed, but rather selected by the organism on the basis of a positive reaction between the antibody that they were carrying on their surface and the antigen presented to them. This somatic selection produces more cells of the same type, i.e., clones, and thus results in an increased capacity of the organisms to fight the pathogen. In the extended version of his theory, Burnet (1959) explicitly characterizes this process as a form of Darwinian selection.

have been made by using them, and more will probably follow—e.g., the study of gene regulation and function using genetic manipulations (transgenesis of gene deletions), the initial testing of novel therapeutics, the functional assessment cell subsets, and interventional imaging studies. Our main critique is directed instead toward what we are inclined to call an infatuation with understanding the underlying molecular mechanisms involved in animal model immunity, an infatuation that persists in the face of mounting evidence of serious limitations on the capacity of animal models to produce clinical outcomes. The limitations presented in the previous sections are so obvious once we acquire a minimum understanding of immunity that one may wonder why human immunology is facing this crisis today. Why were methods for studying human immunity not developed earlier, and why do the majority of publications in immunology still prefer to present results obtained on animal models? Again, we can only provide a partial answer to this question. But we think that the answer has important implications that go beyond the crisis in immunology, for it speaks to sources of bias that are deeply and broadly entrenched in the culture and institutions of medical research.

Part of the story is a relatively simple version of what economists call a "network effect": when a community of researchers adopts a certain model organism, then it becomes possible to make rapid progress in building a large body of knowledge about numerous interconnected features of that organism. When this knowledge is made available to the entire scientific community, it creates a kind of positive feedback, increasing the scientific payoff for further work with the same model. A better understanding of how all the cogs and wheels interact allows for the formulation of better predictions and hypotheses. Funding agencies, too, are drawn into the cycle, becoming more interested in funding research with the model. Increased funding leads to further enhancement of the body of interconnected theory and data about the model, and methodologies and technologies specially tuned to working with it, and these make further research with it even more rewarding scientifically, and even more attractive to funding bodies. But as some economists interested in network effects have argued, all this positive feedback can result in large-scale reinforcement of bad choices (Arthur 1994; David 2001; Pierson 2004). In the case of biomedical research, the real successes that scientists experience as a result of the accumulation of thorough and fine-grained knowledge about a few model organisms creates a lock-in phenomenon such that researchers may find other kinds of research increasingly difficult to fund or publish, even if those kinds of research would be far more productive than the dominant approach in the long run. As with the allegedly inefficient QWERTY keyboard, once we are all committed to a model like the mouse, the fruit fly, or the flatworm, it becomes extremely difficult to escape the dominant pattern, however irrational we recognize it to be. Network effects of this sort are certainly part of the reason why human immunology is now in crisis. A powerful inertia, generated by the past choices of researchers and funding agencies, biases today's research toward a continued focus on the genetic and molecular bases of immune response in laboratory mice, despite the obvious limitations of this sort of research for understanding human immune function and achieving success in clinical applications.

Network effects are only part of the story, however. Behind the mouse-centrism evident in immunology, we believe, lies a continued attachment to a naive and outdated ideal of scientific reductionism. Reduction and reductionism are the focus of a huge body of philosophical literature, and we can only treat very little of it in this article. One very common form of reductionism (mereological reductionism) attempts to explain the features of complex and large-scale phenomena by reference to the features of their simpler, smaller scale parts—perhaps focusing on a special subset of the parts as particularly illuminating. In immunology this could mean, for example, explaining an organism's immunity to infection merely in terms of molecular pathways. Gene-centered research is another very common species of reductionism; it is an attempt to explain the state of phenotypes or behaviors by the (mere) presence or absence of genes. These reductionist approaches have become increasingly common in the life sciences since the advent of techniques allowing the isolation and sequencing of DNA in the 1980s.[8] One common expression of reductionism in the sciences is especially important in the context of immunology: the procedure of taking individual components out of the context of the larger system to which they contribute and studying them in isolation, in an effort to build up a piece-by-piece understanding of how the system as a whole functions. Seeking to understand societies by studying individuals in the lab, seeking to understand ecosystems by studying populations of organisms in isolation, and seeking to understand immune systems by removing individual components such as cells or antibodies and studying their behaviour in a Petri dish—all these are reductionist approaches in the sense we have in mind, and all are widely practised.

A common response to worries about genetic/molecular reductionism in the life sciences is to recognize that life is complex, and to shift accordingly to a fallback position that treats genes and molecules more modestly as contributing factors rather than as determining factors for the phenomena under study. This appears to be a much more realistic stance, one that acknowledges the importance of the context represented by the larger system as a whole. In the case of immunology, this sort of approach would require us to attend to the ways that genes, molecular pathways, or immune cells interact with other components of the organism's immune system as a whole. This response is a step in the right direction, but it fails to recognize a further trap in the mouse-centric approach: the illusion that the context beyond the boundaries of the individual organism—the context represented by the organism's local environment and developmental history—can be ignored. When we assume that the functions of molecules and cells discovered by experimentation on inbred laboratory animals will be the same in members of a different species, or in members of the same species living in their natural habitat, we are making a bet on the irrelevance of development and the environment. To put it crudely, it is almost as if we assume that the laboratory mouse is just a human under a different guise.

---

[8] Another well-known example of reductionism is the attempt to explain consciousness in terms of the capacities of neurons.

As LaFollette and Shanks (1994) nicely argue, one of the earliest and most influential such "gamblers" in medicine was the French experimental physiologist Claude Bernard (1813–1878). Bernard maintained that medicine should be modeled on physics and chemistry, and that a mature medical science would find its true home in the laboratory, not in the hospital:

> We cannot imagine a physicist or a chemist without his laboratory. But as for the physician, we are not yet in the habit of believing that he needs a laboratory; we think that hospitals and books should suffice. This is a mistake; clinical information no more suffices for physicians than knowledge of minerals suffices for chemists and physicists (Bernard 1949, 148).

Bernard's belief that laboratory research was the most valuable part of biomedical science was based in part on his conviction that the functions that we were to discover in the laboratory in animals of one species could be translated to animals of any other species possessing analogous parts, just as in physics the knowledge that we gain about electrons in one context is applicable to any electrons in relevantly similar contexts.

> Experiments on animals, with deleterious substances or in harmful circumstances, are very useful *and entirely conclusive* for the toxicology and hygiene of man ... for as I have shown, the effects of these substances are the same on man as on animals, save for difference in degree (Bernard 1949, 125, emphasis added).

Bernard thus believed in a kind of general principle of interchangeability of species. This does not mean that he believed that any organism would react identically to a given stimulus. But for him the difference between species was a mere matter of degree. The function of the heart in a fish, a pig, a monkey or a human is always the same: pumping blood. The only thing that changes is the extent to which these different organs perform their function or the range of conditions in which they can still be functional. Larger organs could, in principle, resist a greater range of modifications of the internal milieu or a larger dose of different drugs. Thus, Bernard's view of biological kinds implied that it is possible to predict how a human body will react to different stimuli by performing laboratory experiments on animals such as mice, provided that one can find the correct transformation rule. As we have seen in the discussion of animal models as CAMs, this view still prevails in contemporary biomedical research (LaFollette and Shanks 1994, 200–201).

Despite his aspiration to model physiology on physics and chemistry, Bernard was not a radical reductionist seeking to explain everything in terms of physical particles, or even in terms of organs. He accorded immense importance to the internal environment (*milieu intérieur*), *i.e.*, the environment created by the non-dissectible intracellular fluid that bound all the parts of an organism together, apparently often ignored in medicine. According to Bernard, the internal environment had to be stable, or at least within a very limited range.[9] In an important sense, one could

---

[9]This idea of the fixity of the internal milieu was a precursor of the notion of homeostasis still accepted today.

thus say that he was a holist. Yet, Bernard was still a reductionist, albeit in a more subtle sense. His view about physiology implies that we did not need to go beyond this *milieu intérieur*—indeed, the very capacity of the internal milieu to maintain a stable internal state despite extreme changes in the external environment, he thought, meant that we can understand what goes on inside organisms without reference to the particularities of their external environment or their history.

> [O]nly in the physico-chemical conditions of the inner environment can we find the causation of the external phenomena. The life of an organism is simply the resultant of all its inmost workings; it may appear more or less lively, or more or less enfeebled and languishing, without possible explanation by anything in the outer environment, because it is governed by the conditions of the inner environment. We must therefore seek the true foundation of animal physics and chemistry in the physico-chemical properties of the inner environment (Bernard 1949, 99).

This Bernardian philosophy had a tremendous impact on biomedical research in general, and the field of immunology did not (always) escape it. As Pradeu (2009, 7–8) remarks, until the 1960s immunology adopted Metchinkoff's "ecological medical perspective and was mostly interested in understanding the interactions between hosts and micro-organisms." But then, coinciding with the announcement by several physicians of a *quasi* worldwide end of infectious illnesses, immunologists started to worry less about the interactions between the organism and its environment and took the organism as the sole reference point.

The versatile, yet widely accepted hypothesis of the self/non-self distinction developed by Burnet is a very good example of such a return to a Bernardian reductionism in immunology. In *The Integrity of the Body*, Burnet stipulated that except for some rare exceptions, "no ordinary component of the body will provoke an immunological response. Antibody production or any other type of immunological reaction is against foreign material—against something that is not self" (Burnet 1962, 68). In other words, the self is the part of the body that does not trigger an immune response, whereas the non-self is whatever triggers an immune response in the body. So the "self" in Burnet's sense does not refer to the different aspects of an individual that define its person, but to some kind of molecular signature. The self/non-self dichotomy is mostly a matter of recognition, or lack thereof, of certain molecular patterns in cells and body fluids. This conception was also foundational for immunology, for it delimited the range of phenomena that constituted immunity. Soon, Burnet's notion of self became a conceptual foundation and shaped immunologists' conception of the limits and barriers of organisms. But as Pradeu (2009) remarks, this insular conception of the self brought immunology to focus almost exclusively on the organism and on the way in which it isolates itself from its environment in order to maintain its integrity. In *Genes, Dreams and Realities*, Burnet (1971) paints a rather pessimistic portrait of medical research that put too much hope in finding magical cures from observational approaches. It is clear from the opening of the book that he believes that the success of medicine comes from laboratory research and that the main challenge resides in understanding the genetic bases of various diseases.

The success of, and the prestige attaching to medical research came in dealing with what I have come to speak of as the impact of the environment. Physical injury, infection, and malnutrition; these are the types of disability that we have learnt to understand by the systematic use of the scientific method in the laboratory. Many types of infectious disease can be prevented and almost all serious types can be effectively treated if the right methods can be applied early enough. Malnutrition should no longer exist anywhere in the world. No wounded soldier or road victim should die if the resuscitation team arrives before irremediable damage is done. What remains to be prevented and cured has a different set of origins. Broadly speaking, the conditions whose control eludes us have a genetic or a somatic genetic background against which the onset of deterioration can be accelerated by self-indulgence or misfortune. (Burnet 1971, 3)

We see therefore that by taking a gene-and-mouse-centric approach and by giving laboratory research a special status as its favorite method of investigation, immunology (and most biomedical research) is following a Bernardian path. And like the tracks created in the land by the repeated passage of vehicles and travelers, this path has been successfully visited so often since the nineteenth century that it has become extremely well-marked and deep. By now, it is extremely difficult to get out of the ruts of this too-well-travelled path. More concretely, and perhaps paradoxically, getting out of these ruts for immunology today means studying humans in their natural context. But furthering an approach (observational study) that has been so often downplayed and indeed actively devalued in biomedical research for more than a century cannot be straightforward, especially if this means dedicating fewer resources to the gold standard of biomedical research, i.e., controlled experiments.[10] Undertaking this enterprise requires more from researchers than the recognition and rejection of false assumptions. What is required is a profound cultural shift and a corresponding reorientation of research and funding institutions. Some immunologists, at least, have been aware of the limits of animal model research and the lack of clinical outcomes for some time now, and it is clear that such awareness is now widespread in the field. The issue at this point should not be our limited capacity to predict human immune responses from mouse research, but rather the re-orientation of a field of research that has been committed to a certain paradigm for too long.

5   Thinking Outside the Mouse

This paper is an invitation to "think outside the mouse" in two different senses. On the one hand, we have argued that immunity is defined by how certain genes, molecules, and cells are constantly adapting to an organism's environment, which is

---

[10]We are not implying here that experimental research should be left aside. In an ideal world, one could simply bring more resources towards clinical research. But resources are limited, so furthering observational studies would most likely mean a change in the resources allocated to lab research.

in turn uniquely defined by a history of antigen exposure. To understand a mouse's immune system, we must pay attention to the world *outside* that mouse, and *ditto* for humans. On the other hand, we have also argued that immunology, in order to succeed in understanding human immunity and producing clinical outcomes, has to get beyond the mouse as the principal focus of research and investigate humans directly. Combined, these theses imply that immunology needs to focus its attention not just on humans, but on humans in their own larger ecological, historical and social contexts (e.g., how do we gain herd immunity or how can we eradicate infectious agents such as small pox). This section explores this implication by making two general pragmatic recommendations and one observation about the socio-cultural nature of our environments, including our immunological environments but also the environments that shape our research choices.

First, we suggest that the daunting task of overcoming well-established environmental network effects and reorienting a field of research could be facilitated by an explicit redefinition by leading experts and agencies of the goals of human immunology. As we noted in the last section, there is a real sense in which the key problem is not false assumptions but erroneous value-judgments. The gold-standard of controlled experiment makes sense from an epistemic point of view, if your *only* goal is to obtain the fullest and deepest causal understanding of whatever it is you happen to be studying. If your goal is not just knowledge but knowledge that benefits human beings, this evidential standard is quite inappropriate. It is hard not to think of the tale of the man searching for his lost keys under the streetlight, where he can see better, even though he knows he dropped them in the dark of his garden. Bringing to the forefront the ultimate goal of finding clinical outcomes, and not just fundamental general biological functions, can directly counter the entrenched values that continue to reinforce institutional and individual commitment to an unproductive immunological research program.

Second, we recommend a similarly explicit effort to promote the study of immune function in humans in their actual environmental contexts as a central objective of human immunology research. This means that we should seek wherever possible to study the mechanisms of immune function *in vivo*—and preferably in humans, but also that we should engage in observational studies to make the best possible use of data about human immune function in the environments that people actually live in. These two goals are made far more attainable by recent technological advances, the former by advances in imaging technology that permits increasingly rich and non-invasive microscopic observation of the internal functioning of living organisms, and the latter by advances in information technology that permit complex analysis of very large data sets. An interesting example of this kind of work appears in a recent paper entitled "The Human Model: A Genetic Dissection of Immunity to Infection in Natural Conditions." As Casanova and Abel argue:

> Studies of the human genetics of infectious diseases aim to dissect immunity to infection in natural conditions, by elucidating 'experiments of nature.' Observational immunology in humans does not modify the host-environment interaction, unlike experimental immunology in animal models. This makes the human model of interest for studying immunity to infection. (Casanova and Abel 2004, 64.)

Finally, observational human immunology will have to take into consideration the fact that our environment is profoundly and extensively affected by socio-cultural factors. This means that human immunology can benefit from increased interdisciplinary engagement with sociologists, social psychologists, and anthropologists. An interesting and crucial fact about our socio-cultural reality is its heterogeneity. It would be naive to think that we can find the typical/normal history of antigen exposure for all humans. As Casanova and Abel (2004) go on to acknowledge, there might be more than one human model to study.

An interesting example of how different cultural habits can affect immunity is revealed by the hygiene hypothesis. Roughly, the hypothesis suggests that sufficient exposure to microbes early in life allows for a better calibration of adaptive immunity. As such, the hygiene hypothesis offers an explanation of the correlation between the increased sanitation in certain industrialized countries and the incidence of asthma, allergies and other autoimmune diseases. For example, a recent study reports that incidence of allergies is lower for children living on farms than for those raised in urban settings (Kilpelainen et al. 2000). According to the authors: "environmental exposure to immune modulating agents, such as environmental mycobacteria, could explain the finding" (Kilpelainen et al. 2000). In other words, they suggest that living in a microbe-rich environment could, in right proportions, have a protective effect against certain immune-related diseases.

Here we see immune function entangled with environmental factors that are obviously linked to social, economic and cultural aspects of human life. A final twist shows how important and unexpected the effects of such factors can be. As recently argued by the feminist philosopher of science Sharyn Clough, girls seem to be especially affected by the kinds of immune diseases addressed by the hygiene hypothesis. When it comes to hygiene, the developmental environment that girls experience tends to differ systematically from that experienced by boys. As Clough notes: "Girls tend to be dressed more in clothing that is not supposed to get dirty, girls tend to play indoors more than boys, and girls' playtime is more often supervised by parents ... There is a significant difference in the types and amounts of germs that girls and boys are exposed to, and this might explain some of the health differences we find between women and men" (Clough 2011). This suggests that if we are to form different classes of human models in studying immunity, we have to take into consideration differences not only between, but also within cultures. Not doing so could lead to greater health inequalities.

6   Concluding Remarks

We've noted that immunologists are increasingly concerned about the failure of their most successful research to translate into commensurate clinical success. We close with the words of an immunologist who is a colleague of ours at Western University, Steve Kerfoot. "I've recently come to believe," says Steve, "that we need a new principle to guide our research. *Do the science in the most complex natural system that your technology will allow.* That is my rule now." We add to this only that you should, as much as you can, do the science in the system you actually want to know about, and recognize that the complex natural systems that life scientists study do not all end at the skin of the organism, but extend to include the complex interactions between organisms and their environments, particularly where the phenomena of interest are those connected with features, like the immune system and the nervous system, that evolved to allow organisms to respond and adapt to their environments. In studying these systems, especially, it is critical that we learn to think outside the mouse.

References

Adrian, C. Hayday, and Mark Peakman. 2008. The habitual, diverse and surmountable obstacles to human immunology research. *Nature Immunology* 9(6): 575–580.

Arthur, W. Brian. 1994. *Increasing returns and path dependence in the economy.* Ann Arbor: University of Michigan Press.

Bernard, Claude. 1949. *An introduction to the study of experimental medicine.* New York: Henry Schuman.

Burnet, F. Macfarlane. 1962. *The integrity of the body: A discussion of modern immunological ideas*, vol. 3 in Harvard books in biology. Cambridge, MA: Harvard University Press.

Burnet, F. Macfarlane. 1971. *Genes, dreams, and realities.* New York: Basic Books.

Burnet, F. Macfarlane. 1959. *The clonal selection theory of acquired immunity.* Nashville: Vanderbilt University Press.

Casanova, Jean-Laurent, and Laurent Abel. 2004. The human model: A genetic dissection of immunity to infection in natural conditions. *Nature Reviews Immunology* 4(1): 55–66.

Clough, Sharyn. 2011. Gender and the hygiene hypothesis. *Social Science & Medicine* 72(4): 486–493.

Darwin, Charles. 1859. *On the origin of species by Charles Darwin: A facsimile of the first edition with an introduction by Ernst Mayr.* Cambridge, MA: Harvard University Press.

David, Paul A. 2001. Path dependence, its critics and the quest for 'historical economics.' In *Evolution and path dependence in economic ideas: Past and present*, ed. Pierre Garrouste and Stavros Ioannides, 15–40. Cheltenham: Edward Elgar.

Desjardins, Eric. 2011. Historicity and experimental evolution. *Biology & philosophy* 26(3): 339–364.

Gachelin, Gabriel. 2006. *Les organismes modèles dans la recherche médicale.* Science, histoire et société. Paris: Presses universitaires de France.

Ronald, N. Germain, and Pamela L. Schwartzberg. 2011. The human condition: An immunological perspective. *Nature Immunology* 12(5): 369–372.

Kilpelainen, M., E.O. Terho, H. Helenius, and M. Koskenvuo. 2000. Farm environment in childhood prevents the development of allergies. *Clinical and Experimental Allergy* 30(2): 201–208.

LaFollette Hugh, and Niall Shanks. 1994. Animal experimentation: The Legacy of Claude Bernard. *International Studies in the Philosophy of Science* 8(3): 195–210.

Lage, Austin. 2008. Connecting immunology research to public health: Cuban biotechnology. *Nature Immunology* 9(2): 109–112.

Leslie, Mitch. 2010. Immunology uncaged. *Science* 327(5973): 1573.

Mestas, Javier, and Christopher C. W. Hughes. 2004. Of mice and not men: Differences between mouse and human immunology. *The Journal of Immunology* 172(5): 2731–2738.

Odegard, Valerie H., and David G. Schatz. 2006. Targeting of somatic hypermutation. *Nature Reviews Immunology* 6(8): 573–583.

Overmier, J. Bruce, and Carroll, Marilyn E. 2001. Basic issues in the use of animals in health research. In *Animal research and human health: Advancing human welfare through behavioral science*, ed. M.E. Carroll and J.B. Overmier. Washington, DC: American Psychological Association.

Parham, Peter. 1994. The rise and fall of great class I genes. *Seminars in Immunology* 6(6): 373–382.

Pierson, Paul. 2004. *Politics in time: History, institutions, and social analysis*. Princeton: Princeton University Press.

Pradeu, Thomas. 2009. Darwinisme, évolution et immunologie. In *Les mondes darwiniens*, ed. T. Heams, P. Huneman, G. Lecointre, and M. Silberstein, 759–788. Paris: Syllepses.

Rosenblueth, Arturo, and Norbert Wiener. 1945. The role of models in science. *Philosophy of Science* 12(4): 316–321.

Shanks, Nail. 2004. *God, the devil, and Darwin: A critique of intelligent design theory*. New York: Oxford University Press.

Shanks, Nial, and Ray C. Greek. 2009. *Animal models in light of evolution*. Boca Raton: Brown Walker.

Wilson, David. 1984. *Rutherford: Simple genius*. Cambridge, MA: MIT Press.

Part III
Emerging Frameworks

Integrating Ecology and Evolution: Niche Construction and Ecological Engineering

Gillian Barker and John Odling-Smee

Abstract Ecology and evolution remain poorly integrated despite their obvious mutual relevance. Such integration poses serious challenges: evolutionary biologists' and ecologists' conceptualizations of the organic world—and the models and theories based upon them—are conceptually incompatible. Work on organism-environment interaction by both evolutionary theorists (niche construction theory) and ecologists (ecosystem engineering theory) has begun to bridge the gap separating the two conceptual frameworks, but the integration achieved has so far been limited. An emerging extension of niche construction theory—*ecological niche construction*—now promises to achieve a richer integration of evolutionary and ecological conceptual frameworks. This work raises broader philosophical problems about how to choose and combine idealized models of complex phenomena, which can be addressed with the aid of ideas developed by biologists (such as Richard Levins) and philosophers (such as Sandra Mitchell) on the goals and strategies of model-building in the complex sciences. The result is an opening up of new pathways for conceptual change, empirical investigation, and reconsideration of the familiar that has only just begun. Ecological niche construction combines with new developments in evolutionary developmental biology to reveal the need for a deep transformation of the conceptual framework of evolution and the emergence of an integrative biology re-uniting development, evolution and ecology.

G. Barker (✉)
Department of Philosophy, Western University, London, ON, N6A 3K7, Canada
e-mail: gbarker5@uwo.ca

J. Odling-Smee
Mansfield College, University of Oxford, Oxford, UK
e-mail: john.odling-smee@mansfield.ox.ac.uk

G. Barker et al. (eds.), *Entangled Life*, History, Philosophy and Theory
of the Life Sciences 4, DOI 10.1007/978-94-007-7067-6_10,
© Springer Science+Business Media Dordrecht 2014

1 Ecology and Evolution

The importance of integrating ecological and evolutionary thinking has been discussed for decades now (Hutchinson 1965; Levins 1966), yet this "newest synthesis" (Schoener 2011) remains more notional than real. The Modern Synthesis of the 1930s and 1940s succeeded in integrating genetics with Darwinian evolutionary theory in a framework that combined intuitive appeal with mathematical rigor, generating simple models that could be elaborated to fit increasingly complex evolutionary scenarios (Mayr and Provine 1998). The complexities of organismal development and of organisms' ecological relationships to their environments were for the most part put aside in the construction and early elaboration of that first synthesis, as inessential to the main story of genetic replication, assortment, and selection. Criticism of both of these omissions became prominent in the 1960s and 1970s (Levins 1968; Gould 1977; Lewontin 1978; Gould and Lewontin 1979). Positive work integrating evolutionary biology with developmental biology has recently moved forward rapidly in the wake of progress in the understanding of genetic regulatory mechanisms and their role in development (Hall 1992; Gilbert et al. 1996; Carroll 2005; Laubichler and Maienschein 2007). Ecology and evolution, however, remain poorly integrated at a theoretical level despite their obvious mutual relevance.

The separate treatment of ecological and evolutionary change was justified, originally, by the presumption that these two kinds of processes take place over time scales so disparate that there is no possibility of significant interaction between them (Slobodkin 1961). According to this presumption, the evolutionary environments that ecosystems constitute are usually stable over evolutionary timescales: the short-term ecological fluctuations that disturb this underlying stability are too ephemeral to have any important effect on evolution. Evolutionary change, on the other hand, is seen as too slow to matter to ecology. Critics of the presumption of separated time-scales have shown it to be false on both counts: evolutionary processes can be both sensitive enough to be influenced by ecological processes and rapid enough to influence them (Thompson 1998; Palumbi 2001; Hairston et al. 2005; Caroll et al. 2007; Pelletier et al. 2009). But there is a more general point to consider. The stability or mutability of ecosystems is itself a complex matter, affected by evolved characteristics of member organisms and perhaps by evolved ecosystem qualities, as well as by ongoing evolutionary processes at various levels. And the tempo of evolution is itself importantly affected by ecosystem functioning—by the stability or change occurring within the ecosystem, and by causal interactions among its component parts. To the extent that the presumption of markedly different evolutionary and ecological time-scales holds, this may itself be an outcome of eco-evolutionary interactions rather than a barrier to them. Understanding those interactions is thus inescapably important for both fields of inquiry.

Another way to think about the disjuncture between ecology and evolution is to see it as an outcome of differences in theoretical perspective. Evolutionary biologists and ecologists conceptualize the world differently enough that the relevance of ecological knowledge to evolutionary questions (and vice versa) is often not easy to see. More problematically, the basic simplifying assumptions with which they work may yield conceptual systems that cannot easily be combined. If this is the case there is clearly work to do: if evolutionists' and ecologists' conceptualizations— and the models and theories based upon them—are really in substantial tension with one another, it is worth investigating why that is, and exploring what to think about it. A central question concerns the best way to bridge the gap separating the two conceptual frameworks. Is it possible, and desirable, to unify the divergent frameworks? Or can the two fields of study be brought into illuminating interaction without such unification? This paper begins by exploring the conceptual disconnection between ecology and evolution, and its implications for thinking about the role of eco-evolutionary interaction in explaining change and stability in both ecological and evolutionary contexts. Sections 2 and 3 provide overviews of ideas about organism-environment interaction that take steps, from both sides, toward bridging the gap between the two fields: *niche construction* and *ecosystem engineering*. Sections 4 and 5 outline an emerging extension of niche construction theory—*ecological niche construction*—that is now beginning to achieve a richer integration of evolutionary and ecological conceptual frameworks. Sections 6 and 7 examine more closely the challenges posed by such an integration and how they may be met, in light of work by biologists and philosophers on the broader issue of how best to choose and combine idealized models of complex phenomena.

The traditional evolutionary picture divides the world into two parts: a population of organisms (whose relevant features are typically taken to be defined by their genes) and their environment. What is to be explained is change (or sometimes stability) in the genetic constitution of the organisms in a population. The organisms of each generation inherit their genes from their parents (perhaps slightly modified by mutation and recombination); the environment then selects among them, determining how many offspring each genetic variant contributes to the next generation. These processes, iterated, result in evolutionary change over time. In the traditional version of this picture, as it appears in the simple population genetics models central to evolutionary theory, the environment is taken to be unchanging (in which case it may be represented simply in the form of fixed fitness values assigned to the various genotypes) or as changing only as a result of causes that are independent of the organisms that inhabit it. A new and important revision to this evolutionary picture[1] has recently become widely accepted, however; it adds to the picture an explicit recognition that organisms affect their environments as well as being affected by them (Lewontin 1978, 1982, 1983, 2000; Dawkins 1982; Laland et al. 1999; Odling-Smee et al. 1996, 2003). This modified evolutionary picture differs from the traditional one in more important ways than is initially obvious; the differences will be explored in the next section. The point here is simply

---

[1]Though one with deep historical roots: see Lewontin (1978), Godfrey-Smith (1996), and Pearce (2010).

that, despite its important innovations, even this enriched picture leaves in place the basic conceptual division between organism and environment that it inherits from the traditional picture.

The traditional and revised evolutionary pictures can be contrasted with equally-simple pictures from two different fields of ecological theory: population ecology and ecosystem ecology. Like population genetics (and evolutionary theory more broadly), population ecology begins by partitioning the world into population and environment.[2] But where the evolutionary theorist sees a population whose members have properties that vary within each generation and can change between generations, the population ecologist sees a collection of interchangeable individuals, whose common and unchanging features play out against the background of a (possibly changing) environment to produce, and explain, the properties and dynamics characterizing the population as a whole. Like evolutionary theory, population ecology has developed models that begin to take account of the two-way causal interplay between organisms and their environments; as in the evolutionary case, these revisions have far-reaching implications but leave the basic conceptual architecture of the picture in place.

The traditional ecosystem ecology picture is quite different from the others considered so far. Instead of beginning with the organism-environment division, it begins by dividing the part of the world falling within the ecosystem of interest into multiple interconnected functional components, both biotic and abiotic, whose interactions are understood in terms of flows of energy and materials. Parts of the world external to the ecosystem appear as sources and sinks for these resources. What is to be explained here is change or stability of features of the ecosystem structure. Though the factors external to the ecosystem are often referred to as "the environment," nothing in this picture corresponds closely to the "organism-environment" division of the evolutionary and population ecology pictures, since there is no single focal population of organisms relative to which "the environment" can be defined. Many of the most important components of any organism's local environment are other organisms, including members of its own population and of other local populations, and every organism is a part of many other organisms' environments. There is also nothing in this picture that corresponds easily to the genetically-defined individuals and populations of evolutionary models: what matters for ecosystem ecology is the functional role that the organisms play, not their genetic constitution. Finally, in sharp contrast to the simple structure of the evolutionary models, in which the fundamental division is into elements "internal" to the organism and those "external" to it, ecosystem ecology pictures natural systems as composed of many interacting elements, linked in hierarchically-structured webs of causal connection.

There are thus two conceptual gaps to be considered, presenting rather different sorts of challenges to the would-be synthesizer. The gap between evolutionary

---

[2]Population genetics and population ecology individuate populations somewhat differently, but this contrast is not important here.

genetics and population ecology is the result of idealizations that leave out different aspects of the relationship between population and environment: the two frameworks parse the world in roughly the same way, but they take different perspectives on the elements that result. As others have noted (Levins 1966; Shavit and Griesemer 2011), the factors that each of these two frameworks treats as variable, the other treats as fixed. The gap separating both of these frameworks from ecosystem ecology appears to pose a much more serious obstacle to the achievement of any substantial integration of ecology and evolution, representing as it does the divide between two fundamentally different representations of the causal structure of the organic world.

2 Niche Construction Theory

Niche construction theory (NCT) was first formulated as a revision of evolutionary theory, one of several different theoretical developments in the mid-to-late twentieth century that began to explore the variety of ways in which organisms interact with their selective environments—others included co-evolution theory (Janzen 1966) and extended phenotype theory (Dawkins 1982, 2004). As Sects. 4 and 5 will show, however, a broader application of the core idea of NCT can now help integrate the conceptual frameworks of evolutionary biology and ecosystem ecology.

Niche construction theory, as initially formulated, made a point about evolution. By modifying their own environments—in diverse ways—organisms modify some of the selective pressures that their environments exert upon them, and thus create reciprocal relationships between their own genetic characteristics and features of their environments. These relationships can affect the course of evolution in certain distinctive ways that are characteristic of causal feedback structures, such as producing rapid evolutionary change (and environmental change) via positive feedback, or ensuring evolutionary (and environmental) stability via negative feedback (Robertson 1991). They can thus, for example, change the tempo of evolutionary change—causing evolutionary time-lags, generating momentum and inertia effects, or precipitating episodes of abrupt evolutionary change. They can also change the equilibrium states of the population—creating conditions that lead to the fixation of genes that would otherwise be deleterious, supporting stable polymorphisms where none would otherwise be expected, eliminating polymorphisms that would otherwise be stable, or influencing the population's linkage disequilibrium (Laland et al. 1996, 1999, 2001a, b; Odling-Smee et al. 2003). Two simple examples illustrate some of these effects. The modern earthworm, despite its terrestrial habitat, retains many features that were important for the survival of its freshwater-dwelling aquatic ancestors. This evolutionary stability is maintained by the interaction between earthworms and their environment: by tunnelling, moving materials in and out of the soil surrounding their tunnels, and secreting mucus to coat tunnel walls, earthworms create an environment to which their quasi-aquatic physiology is well suited (Turner 2000). On the other hand, orb-weaving spiders create environments radically different from those experienced by their non-weaving ancestors: environments containing spider webs. Consequently, orb-weaving spiders have evolved a battery of morphological and behavioural features, fitting them for life in this special kind of environment.

Niche construction consists of two separate causal "steps" or sub-processes: the sub-process by which the organisms of a population modify their environment, and the sub-process by which the modified environment subsequently exerts modified natural selection on a population (Post and Palkovacs 2009).

Relative to the first of these sub-process, several different kinds of niche construction can be distinguished (Odling-Smee et al. 2003). Though cases such as earthworms' tunnels and spiders' webs are the most obvious, very diverse interactions between organisms and their environments can play the distinctive evolutionary role that characterizes niche construction. For example, organisms can affect their evolutionary environments by *perturbation* (by physically chang-ing some properties of the world around them—building structures, consuming resources, or producing waste, for example) or by *relocation* (changing which parts of the world they interact with, by moving or growing into a new location where they confront different environmental properties). Either of these kinds of niche construction can be *inceptive* (producing novel changes in the effective environment) or *counteractive* (responding to externally-produced change in ways that override or limit its effects). Niche construction processes can also vary in their plasticity: *obligate* processes of niche construction are those that organisms cannot avoid (waste-production is an obvious example) while *facultative* niche construction processes are possible but not necessary for the organism. Thirdly, niche-construction processes can be classified according to their current selective effects or their selective histories. *Positive* niche construction enhances the current fitness of the niche-constructing organisms, while *negative* niche construction re-duces it. *Historically-selected* niche-construction has been selected for in the history of the population, while *adventitious* niche construction has not been selected for (adventitious niche-construction is often a side-effect of features or processes that are themselves the result of selection, metabolism, for instance). Facultative niche construction is often historically-selected—the building of structures such as beaver-dams and spider-webs are paradigm cases here. Historically-selected niche-construction, in turn, will of course often be positive in its current effect, but in changing environments this is by no means guaranteed.

The second sub-process of niche construction, the exertion of modified natural selection on a population, is contingent on the capacity of the first sub-process to generate an *ecological inheritance* for a recipient population. The defining characteristic of niche construction is thus not the modification of environments *per se*, but the modification of natural selection pressures in environments, combined with the subsequent transmission of modified natural selection pressures from niche-constructing populations to recipient populations, via ecological inheritances, in ecosystems (Odling-Smee 1988; Odling-Smee et al. 2003).

For an ecological inheritance to become evolutionarily consequential it is then also necessary for whatever selection pressures have been modified by the prior niche construction to persist in their modified form in the environment, and therefore in the ecological inheritance of a recipient population, for long enough to cause an evolutionary response in the recipient population. However, in practice that is not a demanding requirement. The persistence of a modified selection pressure in a given environment over a period of generations in an evolving recipient population can be achieved in a variety of ways. For instance, it can be achieved by the sheer physical endurance of an environmental change caused by prior niche construction. The long term persistence of some changes in soils caused by earthworm niche-construction is an example. Conversely, it can also be achieved by highly transitory modifications of environments, through the constant repetition of the "same" niche-constructing acts by a series of generations of a niche-constructing population, often simply as a function of the "same" genes being inherited by successive generations of that niche-constructing population. Webs—repeatedly constructed and repaired by orb web spiders, generation after generation—are one example (Odling-Smee et al. 2003; Odling-Smee 2010). Therefore, it is not only possible but frequently inevitable for features of environments that are produced or maintained by niche-constructing organisms to be reliably passed on to descendent organisms, in the form of ecological inheritances. When that happens, an evolutionarily significant feedback path is likely to be completed. An ecological inheritance may then enable the prior niche-constructing activities of a population to influence the subsequent development of individual organisms in a population within each generation, and the subsequent evolution of a population between generations.

Thus the main differences distinguishing niche construction theory (NCT) from standard evolutionary theory (SET) are twofold: a changed picture of the causal relationship between organisms and their environments, and a changed conception of inheritance in evolution. First, natural selection, combined with niche construction, results in *reciprocal causation*, both in development and in evolution (Laland et al. 2011). Causal influences flow from environments to organisms, as described by SET, but also return from organisms to environments, as described by NCT. Second, because niche construction cannot be evolutionarily consequential until it generates an ecological inheritance, NCT is a dual-inheritance theory of evolution. It necessarily depends on genetic inheritance, as per SET, and ecological inheritance, as per NCT. (Instances of ecological inheritance are more diverse than those of genetic inheritance in their fidelity, in the classes of organisms that they link, and in the time-spans that they involve, but this difference does not obviate their importance as channels of inheritance.) NCT thus introduces one further novel concept, *niche inheritance*. Niche inheritance in evolution is constituted by interrelated genetic inheritance and ecological inheritance processes. It is not just genetic inheritance, but rather niche inheritance, that allows descendent organisms to inherit viable "start-up" niches from their parents (Odling-Smee 2010).

These differences between NCT and SET also give NCT a new focus: SET is about the evolution of organisms; NCT is about the evolution of organisms *together with* those changes in environments that are caused by the evolution of organisms. Hence NCT sees evolution in the same way that Richard Lewontin once articulated so succinctly: "Organism and environment coevolve, each as a function of the other" (Lewontin 1983, 282).

Last, insofar as genes are involved in both of NCT's two sub-processes (they are not always[3]) they can be connected by different feedback paths. The simplest form of niche construction is one in which the genes responsible for a niche-constructing trait are also the recipient genes that are affected by the changed selection pressures that, via their phenotypes, they themselves induced. Instances of this sort constitute a special case of niche construction, similar to what Richard Dawkins (1982, 2004) has called the "extended phenotype." Recognizing the independence of the two niche-construction sub-processes, however, allows us to take account of more complex reciprocal interactions between organisms and their environments, in which the modified environment exerts new selective pressures on genes other than those responsible for the niche construction (Post and Palkovacs 2009). This possibility can be represented in two-locus population genetics models (Laland et al. 1996; Odling-Smee et al. 2003). The state of some resource R in the environment is dependent on the niche-constructing activity associated with genes at the first locus. The state of R, in turn, influences the pattern and strength of selection acting on the second locus. Niche construction results in a changed environment, and this may affect the course of subsequent evolution for the niche-constructing population in many different ways. Thus beavers are adapted by evolution in numerous ways—morphological, physiological and behavioural—both to create and to thrive in environments containing lodges, dams, and the kinds of ponds and modified woodlands that beaver-dams produce; oaks are similarly adapted in many ways both to create and to thrive in environments containing frequent low-intensity fires. In both cases, the adaptations that fit the niche-constructing organisms to their modified environments extend far beyond the traits involved in the niche-construction itself. The genes involved in producing the niche-constructing traits thus help to create a modified environment that bestows selective benefits on the many other genes involved in producing traits that are adaptive in the modified environment.

3   Ecosystem Engineering

The effects of niche construction modify selection pressures not only for the niche-constructing organisms, but for other organisms as well. Beaver ponds, forest fires, spider webs, and the modified soil structure that earthworms produce, all have important selective consequences for many organisms other than their creators. This wider effect of niche construction connects it with a set of ideas that have been developed to address issues in ecology.

---

[3]In humans, for example, niche construction is typically cultural; it depends primarily on acquired cultural traits, and not directly on inherited human genes (see Laland et al. 2001b, 2010; Odling-Smee and Laland 2012).

In ecosystem ecology, the concept of *ecosystem engineering* (EE) was introduced to make a point about ecological structure: that in modifying their own surroundings organisms change ecosystem features in ways that have effects on other organisms as well; that the features of ecosystems that are affected may be either biotic or abiotic; and that these processes have certain kinds of ecological consequences (Jones et al. 1994, 1997; Moore 2006; Cuddington et al. 2007; Cuddington et al. 2009). Types of ecosystem engineers can be distinguished according to the nature of their effects. Berke (2010) distinguishes four main categories. *Structural engineers* change or create relatively durable structural features of their surroundings: beaver dams, termite mounds, coral reefs, and the woody parts of plants are all examples of this sort of engineering. Structural engineers often reduce disturbance and increase the heterogeneity of their surroundings. *Bioturbators* such as burrowers and excavators disturb and mix materials in their surroundings, often producing an increase in uniformity. *Chemical engineers* modify the chemistry of soil, water, or air through processes such as respiration or photosynthesis, or by moving or depositing materials. *Light engineers* alter the local patterns of light transmission, changing the intensity of light in nearby locations by casting shade or causing light scattering, for example. All of these kinds of ecosystem engineering can be either *allogenic* or *autogenic*, i.e., they can take the form either of effects organisms have on their (external) surroundings, or of aspects of the organisms' own growth and development (Jones et al. 1994). The structural engineering carried out by beavers and termites, for example, is allogenic, while that carried out by trees or giant kelp is autogenic. In either case, ecosystem engineers have effects on ecosystem functioning that may be important for other organisms as well as for themselves. Importantly, ecosystem engineering is defined so as to exclude competitive and trophic interactions, since the ecological roles of these are already accounted for in existing models and theories.

4   Ecological Niche Construction

Many—perhaps all—instances of ecosystem engineering are also instances of the first sub-process of niche construction,[4] and it is easy to see that the two theoretical frameworks can be extended to reveal a further important relationship between them. Both frameworks emphasize the importance of the processes by which organisms modify their environments. Niche construction theory, as originally articulated, focuses on the evolutionary effects these processes have on the organisms that initiate them, while the ecosystem engineering perspective

---

[4]Though the reverse is not the case, since niche-constructing activities that are part of the trophic web would not normally be regarded as ecosystem engineering, and relocational niche construction would also normally be excluded. For a different view of the relationship between ecosystem engineering and niche construction see Pearce (2011).

focuses on their ecological effects on local ecosystems and on other organisms within it. But it is obvious that these basic insights can be brought together to show the possibility—indeed the inevitability—of *organisms' modifying their environments in ways that have effects on ecosystem functioning that in turn affect the evolution of other members of the ecosystem.* Niche construction theorists have thus begun to focus more closely on cases of niche construction in which the two sub-processes of niche construction involve different populations of organisms, so that the genes responsible for the modification of the environment and the genes subject to modified selection pressure as a result are found in organisms belonging to different populations (usually of different species) (Barker 2008; Post and Palkovacs 2009; Laland et al. 2009; Laland and Boogert 2010). Ecologists working with models of ecological engineering processes have meanwhile begun to consider the evolutionary effects of those processes (Moore 2006; Erwin 2008). The result is an emerging framework that some ecologists have called *ecological niche construction* (Loreau and Kylafis 2008). This new framework promises insights into the relationship between evolution and ecology, including a new approach to thinking about the evolution of ecosystems (Odling-Smee et al. 2013). At a different level of analysis, it provides an illuminating example for thinking about the challenges and importance of the integration of different theoretical and conceptual systems in the sciences.

Consider a simple model of ecological niche construction. The first sub-process of niche construction modifies some R, a resource or feature of the local environment that plays a role in natural selection for some population. R may be abiotic (e.g. topsoil, or a water hole), biotic (e.g. another population of organisms), or artifactual (e.g. a beaver dam or termite mound).[5] We can represent the change produced in R by δR, so that the outcome of the first process of niche construction is a new $\mathbf{R'} = [\mathbf{R} + \delta\mathbf{R}]$. Niche construction leaves different δR *ecological signatures* of change in different kinds of R. Typical δR signatures of prior niche construction in abiota include geo-chemical and thermodynamic effects (often simply by-products of biotic processes or activities). In biota, typical δR signatures are ecological (e.g. demographic changes in other populations). In artifacts, typical δR signatures include the features often identified with "design."

Distinguishing the two sub-processes of niche-construction, and their effects, allows us to enrich the very simple original picture of niche construction, and its role in evolution, with which we began (Post and Palkovacs 2009). Environment-altering populations and the recipient populations whose evolution is affected need not be identical, and various kinds of causal pathways linking them are possible. A niche-constructing population can act directly on a recipient population, or indirectly via intermediate biota or abiota. There can be (and often are) both one:many

---

[5]This division is not an exclusive one, since artifacts are usually composed of abiotic and occasionally of biotic components. But artifacts as such have a distinctive role to play as environmental resources for organisms, as is indicated by their typical δR signatures.

and many:one relationships between environment-altering populations and recipient populations in ecosystems.

Certain conditions must be met if the first niche-construction sub-process is to give rise to the second. There are no evolutionary consequences of niche construction if the δR ecological changes caused by organisms are too variable, or if they dissipate too rapidly. To influence evolution, a population's niche construction must generate an ecological inheritance: i.e., it must reliably cause a δR change that modifies at least one natural selection pressure for at least one recipient population (itself or another) in an ecosystem, and that persists for a sufficient span of time as measured in generations of the recipient population for selection to be effective.

Ecologists have identified factors that scale up the consequences of ecosystem engineering in ecosystems (Jones et al. 1994, 1997; Hastings et al. 2007). Since the possible consequences include evolutionary ones, the same factors also enhance the evolutionary role of niche construction. They include factors associated with the nature of the niche-constructing population (the lifetime per-capita niche-constructing activity of individual members of the population, the density of the population, and the length of time that it persists in the same place), factors associated with the nature of the δR modifications the population produces (the durability of the modifications in the relevant environmental context, and the number and types of flows of resources (materials and energy) that they modulate), and factors associated with the ecological role of the modifications (how many other species utilise those flows).

The simple one-population picture of niche-construction showed how genes involved in producing a niche-constructing trait and genes involved in producing traits that are advantageous in the resulting modified environment can come to be associated within a population. The genes that contribute to making oaks prone to experiencing frequent fires are associated with the genes that contribute to making them good at surviving fires. The broader conception of niche construction reveals that such *environmentally-mediated gene-associations* (EMGAs) may cross the boundaries between populations or species within an ecosystem. EMGAs can connect *any* environment-altering phenotypic traits (expressed by any genotypes, in any niche-constructing population) to *any* recipient genotypes (in any recipient populations) via *any* modified natural selection pressures in the niches of the recipient populations (Odling-Smee et al. 2003, 2013).

The linked evolutionary processes that produce trans-species EMGAs can bring about a close coordination between the traits of two closely-associated species to produce impressive instances of co-evolved mutualism between the two species. In these cases, there are often several different niche-construction pathways connecting the two species. In the case of the mutualistic relationship between acacia ants and swollen-thorn acacias in Central America (Janzen 1966), for example, the acacias provide ants with shelter in the form of enlarged hollow thorn-like stipules, and food in the form of nectar and specialized detachable leaf-tip structures (Beltian bodies) rich in fats and proteins. The ants, in turn, protect the trees from herbivore damage (attacking both insects and vertebrates that come in contact with the trees) and from competition (cutting away other nearby plants).

Less obvious, but probably much more common, are EMGAs involving a larger number of species linked by looser but more complex webs of niche-construction. Consider, for example, the web of interrelationships involved in a meadow community, in which numerous species affect each other's environments, but each species' niche-construction affects a different subset of the others in the community, and some effects are felt only indirectly through the activity of a mediating species. The evolution of such webs may produce "facilitation networks" that play important roles both in maintaining the stability of ecosystems (Verdú and Valiente-Banuet 2008; see also Bruno et al. 2003) and in enabling assemblages of introduced organisms to succeed in invading established ecosystems (Simberloff and von Holle 1999; O'Dowd et al. 2003; Simberloff 2006; Lindroth and Madritch 2009).

5   Ecosystem Evolution

Ecosystem evolution occurs when evolutionary change in a population or populations of organisms brings about change to ecosystem properties. Ecologist Michel Loreau distinguishes three ways that such change can come about: classical individual-level selective evolution, evolution involving organism-environment feedback, or ecosystem-level selection (Loreau 2010). The first type of ecosystem evolution is represented well by standard evolutionary theory. It comes about when evolutionary changes in one or more populations within an ecosystem, brought about by simple individual-level selection, result in changes to ecosystem properties. Thus, for example, evolutionary changes in the ability of particular species of plants or decomposers to compete for resources can modify nutrient-cycle functioning in the ecosystem as a whole. Here the ecosystem-level changes are no more than side-effects of organismal evolution. This type of ecosystem evolution is possible whenever at least one population that plays a significant role in the ecosystem undergoes evolutionary change, though whether ecosystem evolution actually occurs depends on the particular traits that are evolving, and their contribution to the organisms' ecological role.

The second type of ecosystem evolution is represented well by niche construction theory. It occurs when there is feedback between the two kinds of change involved in the first type of ecosystem evolution: the evolutionary changes at the organism level, and the ecosystem-level changes that these bring about. The results may ramify far beyond the populations that are most directly involved in starting the process, and can also involve environmentally-mediated coevolutionary interactions linking two or more species.[6] In the most complex cases, Loreau points out, this type of ecosystem evolution involves diffuse coevolution among many interacting

---

[6]Loreau notes that the first type of ecosystem evolution can also involve coevolution between species, but extended coevolutionary networks usually depend on the links provided by niche construction, as in the second type of ecosystem evolution.

populations, together with associated changes in the ecosystem processes that they affect—changes that in turn modulate the coevolutionary selection pressures acting on the populations. In addition to the conditions required for the occurrence of ecosystem evolution of the first type, this second type requires long-lasting interactions between populations and their environments; where coevolution is involved, it also requires long-lasting interactions between different populations. Models suggest that ecosystem evolution of this type is capable of giving rise to tightly integrated networks of interdependent populations, linked both directly and through abiotic resources via the two sub-processes of niche-construction (Loreau 2010).

The third type of ecosystem evolution is the most demanding, and indeed Loreau suggests that it may not occur naturally in a pure form. This is the evolution of ecosystem properties by ecosystem-level selection. Loreau argues that selection at the ecosystem level is best understood on the model of trait-group selection as articulated by Sober and Wilson (1998).[7] In trait-group selection, the fitness of each individual organism is determined in part by the kind of group that it belongs to, which in turn is determined by the nature of the organisms that constitute the group and the interactions among them. Thus, for example, an individual organism belonging to a group containing many altruists is fitter than one that is identical to the first except in belonging to a group dominated by selfish individuals; this is true whether the organism itself is selfish or altruistic. Other successful kinds of groups may be composed of particular combinations of individual-level types, such as the different functional castes in social insect colonies. Group-level selection thus favors individuals that belong to groups composed of the best combinations of individual-level types, and so acts to perpetuate such groups. Conflicts between the selective forces at the individual and group levels are common—the classic example is in the case of altruism, which is selected against at the individual level but may be selected for at the group level—and the overall fitness of an individual is determined jointly by the selective forces acting upon it at all levels of selection.

In ecosystem-level selection, then, the fitness of each individual is determined in part by the kind of ecosystem it belongs to, which in turn is determined by the nature of the organisms and abiota that constitute the ecosystem, and the interactions among them. Ecosystem-level selection favors individuals that belong to ecosystems composed of the best combinations of individual-level types, and so acts to perpetuate such ecosystems. Conflicts between the selective forces at the individual and ecosystem levels, like other inter-level conflicts, are expected to be common. This type of ecosystem evolution can occur only when quite stringent conditions are met. It requires that all the conditions for the second type of ecosystem evolution with coevolution be met, but also that interactions between different species, and between organisms and abiota, be strongly localized so that competition between members of the same evolutionary population that are

---

[7]For an approach to ecosystem evolution that treats ecosystems directly as units of selection, see Swenson et al. (2000) and Goodnight (2000).

members of different local ecosystems can occur. Loreau notes, however, that recent research suggests that this condition may be met more often than has been supposed: most nutrient cycling turns out to be very localized, for example.

6   Evolution and Ecology, Revisited

Far from being separated by their disparate timescales, ecology and evolution are tightly linked through the reciprocal causal relationships connecting organisms to both biotic and abiotic components of their local environments. As Loreau puts it, "It is the web of interactions at the heart of an ecosystem that maintains both species and ecosystems as they are, or (more exactly) as they are evolving." (Loreau et al. 2004, 327). As we noted at the outset of this paper, failure to take account of these links leads both evolutionists and ecologists to ways of conceptualizing the systems they study that can be limiting or actively misleading, and that are also difficult to combine with one another.

It is worth looking more closely at these conceptual frameworks, now that we have a larger context against which to consider them. We've seen already that one of the key features of each framework is the set of simplifying idealizations that it makes: which properties and causal relationships it represents, and which it omits as inessential. Classical evolutionary theory treats the environment as causally *self-contained* (usually simply as static, but possibly as changing via processes that are independent of the evolving population), and usually as *simple* in the sense that its causal structure need not be represented; a population's environment can thus often be represented by a single parameter. The structural complexity of the environment and the causal patterns that follow from that complexity are thus made invisible. The organisms themselves are then treated as passive recipients of the selective pressures exerted by the environment; their active role in responding to and changing features of the environment are omitted from the picture. Abiota, and the causal links connecting them to the biota, are commonly not represented at all in evolutionary models, on the presumption that they simply act as the invariant background against which the phenomena of interest appear. When feedback effects between organisms and their environments must be represented, they are often captured in the form of, for example, simple density-dependent selection; the changing biotic and abiotic components of the environment, and the effects upon them that the organisms produce, appear only in the form of the function linking a trait's fitness to the population density of the organisms under selection.

The central models of population ecology, on the other hand, treat populations of organisms as homogeneous, and so as lacking any internal structure that depends on variation among members of the population—the causal implications of such variation are therefore omitted from the picture. Populations are represented as changing only with respect to population-level properties such as population size or rate of increase; evolutionary changes in the nature of the individuals composing the populations—and the causes and effects of such changes—are not registered.

Ecosystem ecology, as we have already noted, begins with a parsing of the world very different from the organism-environment division shared by evolutionary biology and population ecology. The ecosystem is conceived as comprising both biotic and abiotic components, linked by complex causal relationships that include many reciprocal relationships and the feedback effects that these make possible. These distinctive features of the causal structure of ecosystems are not accidental, but follow from more basic causal considerations. The factors involved in ecological relationships are conceived not simply as properties (which could vary *ad lib*) but as stocks and flows of materials and energy within the ecosystem, and so as subject to conservation principles. From this fundamental presumption there follow three further key features of the conceptual framework. First, changes taking place within ecosystems are taken to be constrained by conservation principles within the limits set by the flow of materials and energy across the ecosystem boundaries. Second, ecosystems are therefore characterized by interdependent—often reciprocal—causal relationships among their components, since any local change in a stock or flow of energy or material must be matched by a corresponding change elsewhere in the system. Third, because the life processes of the organisms within the ecosystem depend on the energy and materials that thus cycle through the system, the components of an ecosystem are understood as bound together in a web of functional interdependencies mediated by the flow of these resources. Organisms are seen therefore as active contributors to the web of functional relationships that enable them to survive. Like population ecology, however, the basic models of ecosystem ecology treat populations as homogeneous and evolutionarily static; indeed they may go further and treat organisms only in terms of their ecological roles such as decomposers or top predators rather than as single-species populations.

Niche construction theory in its original form took several important steps toward bridging the gaps separating the simple conceptual frameworks of evolutionary theory, population ecology, and ecosystem ecology. Its most important contribution was to clarify the implications of the fact, already emphasized by Lewontin and Levins, that both populations and their environments are subject to change, that each is capable of causing change in the other, and that this gives rise to a form of ecological inheritance. From this initial step, which brings together elements of the evolutionary and population ecology pictures, several steps toward the ecosystem ecology picture also follow: that organisms and their environments are in reciprocal causal relationships capable of generating feedback effects; that organisms figure as agents of change rather than merely as passive objects of selection; and that organisms and their local environments must be considered as integrated systems that evolve together. The extended form of niche construction theory that results from unifying it with the insights of ecosystem engineering theory, and recognizing explicitly that the two subprocesses of niche construction may involve different populations, goes much further toward accommodating the key elements of the ecosystem ecology picture. It offers ways of representing and taking account, in an evolutionary context, of the causal links among biotic and abiotic ecosystem components and the complex networks of reciprocal relationships and interdependencies that these create. It provides means of representing the flows of energy and materials through the ecosystem and the crucial constraints that result from them, and reveals the nested hierarchical structure that results from the interplay of ecological relationships at different scales. Perhaps above all, it moves decisively away from treating "the environment" of an evolving population either as a mere background or as an object.

7 Strategies of Idealization

We have seen that the simple ideal models that have been central to evolutionary and ecological theory are unable to capture the complex interrelations between ecology and evolution, and noted some moves toward the enriched models that are needed to bridge the gap between the two theoretical frameworks and to enable both disciplines to develop adequate understanding of the multi-layered interplay between organisms and their environments. But the general observation that more complex and inclusive models are needed gives little guidance about how to develop such models, and about the specific desiderata and constraints that must be considered in choosing a modeling strategy.

Simplifying idealizations are, of course, an essential part of science. The complexities of the world must be tamed by models that omit or simplify many features of the real systems they represent, partly just to make the models tractable enough to work with, but also to enable them to uncover the deeper patterns of similarity that underlie the diversity of particular cases (McMullin 1985; Wimsatt 1987; Weisberg 2007). Idealization is thus an essential means to achieving both generality and explanatory power. But choices among idealizing strategies must always be made. In an influential paper, Richard Levins (1966) argued that, given the practical constraints to which both observation and computation are subject, the idealized models that scientists use must make tradeoffs among three desirable features: precision, generality, and realism.[8] The inevitability of such tradeoffs means that it is insufficient to point out that the basic models of evolutionary theory, population ecology and ecosystem ecology variously oversimplify the systems they represent, and to seek to bridge the gaps that separate these frameworks by reinstating the complexities that they put aside. This response will merely result in models so complex as to be unusable. To evaluate a proposed bridging strategy (and indeed to be sure whether one is really required at all), it is necessary instead to assess the strengths and the failings of the current combination of strategies. What problems should we be aiming to correct? What capacities should we be aiming to preserve?

Niche construction theorists have suggested that several important types of error can result from the simplifying idealizations employed in evolutionary biology and

---

[8]For further discussion of such tradeoffs, see Orzack and Sober (1993), Odenbaugh (2003), Orzack (2005), Justus (2005), Weisberg (2006), and Matthewson and Weisberg (2009).

ecology, and from the theoretical disconnection between the two fields of research that these idealizations foster (Odling-Smee et al. 2003; Laland et al. 2009, 2011). In broad terms, the standard models' reliance on idealization strategies that treat either populations or their environments as fixed makes it impossible to see the feedback loops connecting ecological and evolutionary processes, and so to expect the effects that are typical of causal structures involving feedback (such as otherwise unexpected stabilization or runaway change). Recent and more sophisticated models in both population genetics and population ecology do treat both populations and their environments as variable, and some even build in reciprocal relations between the two (e.g. Laland et al. 1999; Krakauer et al. 2009). But without a means of representing the functional relations among both biotic and abiotic ecosystem components in their relationship to evolutionary change, the more serious gap remains.[9] The conceptual disconnection between evolution and ecosystem ecology continues to make it difficult to take full account of two obvious and important facts: that in real-world natural contexts—as opposed to the simplified contexts provided by lab experiments and computer simulations—evolution always takes place within ecosystems subject to constraints set by the conservation of materials and energy, and ecological change always involves species capable both of developmentally plastic responses to environmental change and of evolutionary responses over the longer term (Loreau 2010).

The historical tendency to overlook these structural connections between ecological and evolutionary processes has been consequential. In an era in which an effective understanding of the effects of human interactions with our own environments is of vital practical importance, some of the most serious environmental and evolutionary "surprises" of recent decades[10] involve predictive failures that appear to stem from exactly this sort of conceptual blind spot. Such cases include the rapid evolution of resistant strains of weeds, pests, and especially pathogens (Spellberg et al. 2008; Choffnes et al. 2010); the effects of both the spread of invasive introduced species (plants, animals and pathogens) (Elton 1958; Mooney and Cleland 2001; Facon et al. 2006; Carroll 2011) and the removal of major niche-constructors or "keystone" species (Rosell et al. 2005; Estes et al. 2011); and the effects of human interventions affecting abiota such as the stocks and flows of carbon, water, and topsoil.

The general problem of how to integrate models that represent different aspects of the same system, and that employ incompatible idealizations to do so, is of course a common one in science. Sandra Mitchell (2002, 2003) has distinguished three ways in which such integration can be achieved. In the simplest case, the models simply capture different and independent causes contributing to the system's behavior; they can initially be handled separately in the interests of tractability,

---

[9]Pearce draws a similar conclusion at the end of his (2011).

[10]Earlier events that have recently been given new and radically-different explanations are also relevant here: see for example work on the role of introduced species in facilitating European colonialism (Crosby 2004; Piper and Sandlos 2007).

generality, and explanatory power, and their outputs can then be combined additively to achieve a more realistic complete picture. This approach to the integration of disparate models by using *mechanical rules* to combine the causes that they variously represent is suitable for cases in which the causes really are independent of one another—classical mechanics offers standard examples of this sort, in which separate forces can be modeled separately and then combined additively. Because of the reciprocal relationship between evolutionary and ecological processes, however, this approach is quite inadequate for dealing with their interaction. In Levins's terms, it achieves generality and precision (preserving the broad scope and mathematical rigor of the various separate models) by sacrificing realism (it fails signally to capture the real causal structure of the organism-environment relationship). The second possibility is to seek *local theoretical unification* within the limits set by the particular pragmatic constraints of the case in question, producing a single model that represents multiple aspects of the system in combination. At its best, this approach can strike a distinctive and useful compromise[11] among various epistemic goals, achieving a high level of generality and realism though at the cost of precision. Here we find models that can be used to frame and test hypotheses about large-scale patterns in ecological evolution and evolutionary ecology, but that are too simple to be capable of giving precise predictions about particular complex situations. For that task, the best approach is Mitchell's third option, *explanatory concrete integration*. This approach combines the various component models piecemeal, in ways that are tailored to and supported by detailed information about the particular circumstances of the case at hand. At its best it achieves very high levels of realism and precision, but at the cost of low generality: the detailed and realistic models of particular complex systems that it produces cannot be applied beyond the bounds of those systems.

Successfully integrating ecology and evolution requires both broader theoretical unification and fine-grained explanatory integration in concrete cases. A good deal of excellent work of the latter sort has been done: the studies that have uncovered the complex interactions between evolutionary and ecological processes underlying the environmental "surprises" noted earlier provide many examples of this sort. What is still barely begun is the sort of conceptual and theoretical synthesis that can help uncover the broad patterns of ecological/evolutionary interaction: precisely the sort of synthesis provided by niche construction theory. Several recent studies give some indication of the kind of work such a synthesis can support. Erwin (2005, 2008; see also Crespi 2004) investigates of the role of niche construction in macroevolution and the evolution of diversity, arguing that some niche-constructing processes produce environmental effects that endure through geologic time, modifying evolutionary trajectories for many species simultaneously over extended periods. Such effects, he argues, may have played an essential role in driving major evolutionary transitions and recoveries from mass extinctions. Two teams explore the interaction between niche construction and regulation or control. Krakauer et al. (2009) show

---

[11]Levins (1966) particularly emphasized the virtues of this balance of desiderata.

that positive niche construction results in selection favoring adaptations that make it possible for organisms to monopolize the benefits of their niche-constructing activities, with important implications for the evolution of development, life-cycle patterns, behavioral plasticity, and sociality. McDonald-Gibson et al. (2008), on the other hand, show that interacting negative niche construction processes can coevolve to produce ecosystem-level "rein-control" systems capable of regulating key resources, with important implications for the evolution of ecosystem-level stability and functional integration. Sterelny (2003, 2011) and others (Kerr 2007; Smith 2007; Jablonka 2011; Kendal et al. 2011; Rendell et al. 2011; Van Dyken and Wade 2012a, b) continue to explore the interactions between human genetic and cultural evolution as mediated by niche construction, with models suggesting that niche construction may have played a central role in the evolution of modern human cognitive capacities, behavior patterns, and social systems. Instances like these begin to show how the niche construction perspective can be extended to help researchers in a wide range of contexts investigate the complex interplay between evolving populations and evolving ecosystems.

8   Conclusion

The Modern Synthesis unified key elements of early-twentieth-century theories of evolution and inheritance to yield a set of ideal models of great generality and precision, but lacking contact with the complexities of life outside the fly-bottle—in particular with what was known about how organisms develop and how they interact with their environments. The decision to ignore these aspects of the biological world, and their implications for both heredity and evolution, was justified by theoretical principles essential to the Synthesis: the *Central Dogma* asserting that information flowed only from genes to phenotypes and never in the reverse direction (so that development, including environmental effects, was irrelevant to evolution and heredity) and the *principle of separated time scales* for evolutionary and ecological processes, implying that ecology and evolution could not interact in any very important way. Since the 1970s, however, researchers in many areas of biology have contributed to an increasingly thoroughgoing reconstruction of the life sciences that both elaborates the mathematical models of the Modern Synthesis and integrates them with the flood of information that has been generated over the last decades about the complex realities of genome function and developmental processes in diverse organisms. The result is what many now see as a sea-change in biology: the rise of a new *integrative biology* that differs from the biology of the Modern Synthesis in its core concepts and assumptions, but also in its methods and in the institutional structures that can best help it to thrive (Wake 2001, 2004, 2008; Rose and Oakley 2007; Schwenk et al. 2009). Where the Modern Synthesis fostered specialization and work with mathematical models and fruit flies, integrative biology calls for transdisciplinary work incorporating the strengths of complex computer simulations as well as analytical models, diverse organisms outside as well as inside the laboratory, and contributions from relevant research in adjacent fields including the physical sciences (for understanding genome functioning at the physical level) and the social sciences (for understanding some aspects of human development).

Some characterizations of integrative biology emphasize the importance of ecology as well as development (Wainwright and Reilly 1994; Wake 2004), but the integration of ecological with evolutionary understanding has lagged behind the integration of evolution and development, and its importance has not yet been as widely appreciated. We have seen reason to believe, however, that it is just as consequential for our understanding of evolution and heredity, and that it is urgently needed for our understanding of human impacts on both ecological and evolutionary processes at the global scale.

Like the integration of evolution and development, work at the interface of ecology and evolution has been moved forward partly by purely conceptual work, partly by new empirical results, and partly by a reassessment of the importance of what everybody has known all along. Niche construction theory and ecosystem engineering offered perspectives on organism-environment interaction that now take on a new importance as they have begun to be combined as ecological niche construction, and as empirical results increasingly challenge the assumption of separated ecological and evolutionary time scales. The result is an opening up of new pathways for conceptual change, empirical investigation, and reconsideration of the familiar that has only just begun. Niche construction theorists have been arguing for decades that attention to the "neglected process in evolution" (Laland et al. 1996; Odling-Smee et al. 2003; Laland et al. 2011) reveals the need for a deep transformation of the conceptual framework of evolution. Steps toward an integrative biology that links development, evolution, and ecology seem to confirm that assessment, and indeed to reveal new horizons for transformation beyond the classical bounds of evolution.

Bibliography

Barker, Gillian. 2008. Biological levers and extended adaptationism. *Biology and Philosophy* 23(1): 1–25.

Berke, Sarah K. 2010. Functional groups of ecosystem engineers: A proposed classification with comments on current issues. *Integrative and Comparative Biology* 50(2): 147–157.

Bruno, John F., John J. Stachowicz, and Mark D. Bertness. 2003. Inclusion of facilitation into ecological theory. *Trends in Ecology & Evolution* 18(3): 119.

Carroll, Sean B. 2005. *Endless forms most beautiful: The new science of evo devo and the making of the animal kingdom.* New York: W.W. Norton.

Carroll, Scott P. 2011. Conciliation biology: The eco-evolutionary management of permanently invaded biotic systems. *Evolutionary Applications* 4(2): 184–199.

Carroll, Scott P., Andrew P. Hendry, David N. Reznick, and Charles W. Fox. 2007. Evolution on ecological time-scales. *Functional Ecology* 21(3): 387–393.

Choffnes, Eileen R., David A. Relman, and Alison Mack. 2010. *Antibiotic resistance: Implications for global health and novel intervention strategies.* Washington, DC: National Academies Press.

Crespi, Bernard J. 2004. Vicious circles: Positive feedback in major evolutionary and ecological transitions. *Trends in Ecology & Evolution* 19(12): 627–633.

Crosby, Alfred W. 2004. *Ecological imperialism: The biological expansion of Europe, 900–1900.* Cambridge: Cambridge University Press.

Cuddington, Kim, James E. Byers, William G. Wilson, and Alan Hastings. 2007. *Ecosystem engineers: Plants to protists.* Burlington: Academic.

Cuddington, Kim, Will G. Wilson, and Alan Hastings. 2009. Ecosystem engineers: Feedback and population dynamics. *American Naturalist* 173(4): 488–498.

Dawkins, Richard. 1982. *The extended phenotype.* New York: Freeman.

Dawkins, Richard. 2004. Extended phenotype – But not too extended. A reply to Laland, Turner and Jablonka. *Biology and Philosophy* 19(3): 377–396.

Elton, Charles S. 1958. *The ecology of invasions by animals and plants.* London: Methuen.

Erwin, Douglas H. 2005. Seeds of diversity. *Science* 308(5729): 1752–1753.

Erwin, Douglas H. 2008. Macroevolution of ecosystem engineering, niche construction and diversity. *Trends in Ecology & Evolution* 23(6): 304–310.

Estes, James A., John Terborgh, Justin S. Brashares, Mary E. Power, Joel Berger, William J. Bond, Stephen R. Carpenter, Timothy E. Essington, Robert D. Holt, B.C. Jeremy, Robert J. Jackson, Lauri Oksanen Marquis, Tarja Oksanen, Robert T. Paine, Ellen K. Pikitch, William J. Ripple, Stuart A. Sandin, Marten Scheffer, Thomas W. Schoener, Jonathan B. Shurin, R.E. Anthony, Michael E. Sinclair, Risto Virtanen Soulé, and David A. Wardle. 2011. Trophic downgrading of planet Earth. *Science* 333(6040): 301–306.

Facon, Benoit, Benjamin J. Genton, Jacqui Shyoff, Philipe Jarne, Arnaud Estoup, and David Patrice. 2006. A general eco-evolutionary framework for understanding bioinvasions. *Trends in Ecology & Evolution* 21(3): 130–135.

Gilbert, Scott F., Rudolph A. Raff, and John M. Opitz. 1996. Resynthesizing evolutionary and developmental biology. *Developmental Biology* 173(2): 357–372.

Godfrey-Smith, Peter. 1996. *Complexity and the function of mind in nature.* Cambridge: Cambridge University Press.

Goodnight, Charles J. 2000. Heritability at the ecosystem level. *Proceedings of the National Academy of Sciences of the United States of America* 97(17): 9365–9366.

Gould, Stephen J. 1977. *Ontogeny and phylogeny.* Cambridge, MA: Harvard University Press.

Gould, Stephen J., and Richard C. Lewontin. 1979. The spandrels of San Marco and the Panglossian paradigm: A critique of the adaptationist programme. *Proceedings of the Royal Society of London Series B* 205(1161): 581–598.

Hairston, Nelson G., Stephen P. Ellner, Monica A. Geber, Takehito Yoshida, and Jennifer A. Fox. 2005. Rapid evolution and the convergence of ecological and evolutionary time. *Ecology Letters* 8(10): 1114–1127.

Hall, Brian K. 1992. *Evolutionary developmental biology.* London: Chapman & Hall.

Hastings, Alan, James E. Byers, Jeffrey A. Crooks, Kim Cuddington, Clive G. Jones, John G. Lambrinos, Theresa S. Talley, and William G. Wilson. 2007. Ecosystem engineering in space and time. *Ecology Letters* 10(2): 153–164.

Hutchinson, George Evelyn. 1965. *The ecological theater and the evolutionary play.* New Haven: Yale University Press.

Jablonka, Eva. 2011. The entangled (and constructed) human bank. *Philosophical Transactions of the Royal Society* 366(1566): 784.

Janzen, Daniel H. 1966. Coevolution of mutualism between ants and acacias in Central America. *Evolution* 20(3): 249–275.

Jones, Clive G., John H. Lawton, and Moshe Shachak. 1994. Organisms as ecosystem engineers. *Oikos* 69(3): 373–386.

Jones, Clive G., John H. Lawton, and Moshe Shachak. 1997. Positive and negative effects of organisms as physical ecosystem engineers. *Ecology* 78(7): 1946–1957.

Justus, James. 2005. Qualitative scientific modeling and loop analysis. *Philosophy of Science* 72(5): 1272–1286.

Kendal, Jeremy, Jamshid J. Tehrani, and John F. Odling-Smee. 2011. Human niche construction in interdisciplinary focus. *Philosophical Transactions of the Royal Society B* 366(1566): 785–792.

Kerr, Benjamin. 2007. Niche construction and cognitive evolution. *Biological Theory* 2(3): 250–262.

Krakauer, David C., Karen M. Page, and Douglas H. Erwin. 2009. Diversity, dilemmas and monopolies of niche construction. *American Naturalist* 173(1): 26–40.

Laland, Kevin N., and Neeltje J. Boogert. 2010. Niche construction, co-evolution and biodiversity. *Ecological Economics* 69(4): 731–736.

Laland, Kevin N., John F. Odling-Smee, and Marcus W. Feldman. 1996. The evolutionary consequences of niche construction: A theoretical investigation using two-locus theory. *Journal of Evolutionary Biology* 9(3): 293–316.

Laland, Kevin N., John F. Odling-Smee, and Marcus W. Feldman. 1999. The evolutionary consequences of niche construction and their implications for ecology. *Proceedings of the National Academy of Sciences of the United States of America* 96(18): 10242–10247.

Laland, Kevin N., John F. Odling-Smee, and Marcus W. Feldman. 2001a. Niche construction, ecological inheritance, and cycles of contingency in evolution. In *Cycles of contingency: Developmental systems and evolution*, ed. Susan Oyama, Russell D. Gray, and Paul E. Griffiths. Cambridge, MA: MIT Press.

Laland, Kevin N., John F. Odling-Smee, and Marcus W. Feldman. 2001b. Cultural niche construction and human evolution. *Journal of Evolutionary Biology* 14: 22–33.

Laland, Kevin N., John F. Odling-Smee, Marcus W. Feldman, and Jeremy Kendal. 2009. Conceptual barriers to progress within evolutionary biology. *Foundations of Science* 14(3): 195–216.

Laland, Kevin N., John F. Odling-Smee, and Sean Myles. 2010. How culture shaped the human genome: Bringing genetics and the human sciences together. *Nature Reviews Genetics* 11(2): 137–148.

Laland, Kevin N., Kim Sterelny, John F. Odling-Smee, William Hoppit, and Tobias Uller. 2011. Cause and effect in biology revisited: Is Mayr's proximate-ultimate dichotomy still useful? *Science* 334(6062): 1512–1516.

Laubichler, Manfred D., and Jane Maienschein (eds.). 2007. *From embryology to evo-devo: A history of developmental evolution*. Cambridge, MA: MIT Press.

Levins, Richard. 1966. The strategy of model building in population biology. *American Scientist* 54(4): 421–431.

Levins, Richard. 1968. *Evolution in changing environments: Some theoretical explanations*. Princeton: Princeton University Press.

Lewontin, Richard C. 1978. Adaptation. *Scientific American* 239: 212–228.

Lewontin, Richard C. 1982. Organism and environment. In *Learning, development and culture: Essays in evolutionary epistemology*, ed. Henry C. Plotkin. New York: Wiley.

Lewontin, Richard C. 1983. Gene, organism and environment. In *Evolution from molecules to men*, ed. Derek S. Bendall. Cambridge: Cambridge University Press.

Lewontin, Richard C. 2000. *The triple helix: Gene, organism and environment*. Cambridge, MA: Harvard University Press.

Lindroth, Richard L., and Michael D. Madritch. 2009. Removal of invasive shrubs reduces exotic earthworm populations. *Biological Invasions* 11(3): 663–671.

Loreau, Michel. 2010. *From populations to ecosystems: Theoretical foundations for a new ecological synthesis*. Princeton: Princeton University Press.

Loreau, Michel, and Grigoris Kylafis. 2008. Ecological and evolutionary consequences of niche construction for its agent. *Ecology Letters* 11(10): 1072–1081.

Loreau, Michel, Claire de Mazancourt, and Robert D. Holt. 2004. Ecosystem evolution and conservation. In *Evolutionary conservation biology*, ed. Denis Couvet, Ulf Dieckman, and Régis Ferrière, 327–343. Cambridge: Cambridge University Press.

Matthewson, John, and Michael Weisberg. 2009. The structure of tradeoffs in model building. *Synthese* 170(1): 169–190.

Mayr, Ernst, and William B. Provine (eds.). 1998. *The evolutionary synthesis: Perspectives on the unification of biology*. Cambridge, MA: Harvard University Press.

McDonald-Gibson, J., James G. Dyke, Ezequiel A. Di Paolo, and I.R. Harvey. 2008. Environmental regulation can arise under minimal assumptions. *Journal of Theoretical Biology* 251(4): 653–666.

McMullin, Ernan. 1985. Galilean idealization. *Studies in History and Philosophy of Science* 16(3): 247–273.

Mitchell, Sandra D. 2002. Integrative pluralism. *Biology and Philosophy* 17(1): 55–70.

Mitchell, Sandra D. 2003. *Biological complexity and integrative pluralism*. Cambridge: Cambridge University Press.

Mooney, Harold A., and Elsa E. Cleland. 2001. The evolutionary impact of invasive species. *Proceedings of the National Academy of Sciences of the United States of America* 98(10): 5446–5451.

Moore, Jonathan W. 2006. Animal ecosystem engineers of streams. *BioScience* 56(3): 237–246.

O'Dowd, Dennis J., Peter T. Green, and P.T. Lake. 2003. Invasional 'meltdown' on an oceanic island. *Ecology Letters* 6(9): 812–817.

Odenbaugh, Jay. 2003. Complex systems, trade-offs and mathematical modeling: Richard Levin's "Strategy of Model Building in Population Biology" revisited. *Philosophy of Science* 70(5): 1496–1507.

Odling-Smee, John F. 1988. Niche-constructing phenotypes. In *The role of behavior in evolution*, ed. Henry C. Plotkin, 73–132. Cambridge, MA: MIT Press.

Odling-Smee, John F. 2010. Niche inheritance. In *Evolution—The extended synthesis*, ed. Massimo Pigliucci and Gerd B. Muller, 175–207. Cambridge, MA: MIT Press.

Odling-Smee, John F., and Kevin N. Laland. 2012. Ecological inheritance and cultural inheritance: What are they and how do they differ? *Biological Theory* 6(3): 220–230.

Odling-Smee, John F., Kevin N. Laland, and Marcus W. Feldman. 1996. Niche construction. *American Naturalist* 147(4): 641–648.

Odling-Smee, John F., Kevin N. Laland, and Marcus W. Feldman. 2003. *Niche construction: The neglected process in evolution*. Princeton: Princeton University Press.

Odling-Smee, John F., Kevin N. Laland, Marcus W. Feldman, Eric P. Palkovacs, and Douglas H. Erwin. 2013. Niche construction theory: A practical guide for ecologists. *Quarterly Review of Biology* 88(1): 4–28.

Orzack, Steven H. 2005. Discussion: What, if anything, is "The Strategy of Model Building in Population Biology"? A comment on Levins (1966) and Odenbaugh (2003). *Philosophy of Science* 72(3): 479–485.

Orzack, Steven H., and Elliot Sober. 1993. A critical assessment of Levins' "The strategy of model building in population biology" (1966). *Quarterly Review of Biology* 68(4): 533–546.

Palumbi, Stephen R. 2001. *The evolution explosion: How humans cause rapid evolution change*. New York: W.W. Norton.

Pearce, Trevor. 2010. From 'circumstances' to 'environment' – Herbert Spencer and the origins of the idea of organism-environment interaction. *Studies in History and Philosophy of Biological and Biomedical Sciences* 41(3): 241–252.

Pearce, Trevor. 2011. Ecosystem engineering, experiment, and evolution. *Biology and Philosophy* 26(6): 793–812.

Pelletier, Fanie, Andrew P. Hendry, and Danny Garant. 2009. Eco-evolutionary dynamics. *Philosophical Transactions of the Royal Society B* 364(1523): 1483–1489.

Piper, Lisa, and John Sandlos. 2007. A broken frontier: Ecological imperialism in the Canadian North. *Environmental History* 12(4): 759–795.

Post, David M., and Eric P. Palkovacs. 2009. Eco-evolutionary feedbacks in community and ecosystem ecology: Interactions between the ecological theatre and the evolutionary play. *Philosophical Transactions of the Royal Society B* 364(1523): 1629–1640.

Rendell, Luke, Laurel Fogarty, and Kevin N. Laland. 2011. Runaway cultural niche construction. *Philosophical Transactions of the Royal Society B* 366(1566): 809–822.

Robertson, Douglas S. 1991. Feedback theory and Darwinian evolution. *Journal of Theoretical Biology* 152(4): 469–484.

Rose, Michael R., and Todd H. Oakley. 2007. The new biology beyond the modern synthesis. *Biology Direct* 2(1): 30. doi:10.1186/1745-6150-2-30.

Rosell, Frank, Orsolya Bozser, Peter Collen, and Howard Parker. 2005. Ecological impact of beavers *Castor fiber* and *Castor canadensis* and their ability to modify ecosystems. *Mammal Review* 35: 248–276.

Schoener, Thomas W. 2011. The newest synthesis: Understanding the interplay of evolutionary and ecological dynamics. *Science* 331(6016): 426–429.

Schwenk, Kurt, Dianna K. Padilla, George S. Bakken, and Robert J. Full. 2009. Grand challenges in organismal biology. *Integrative and Comparative Biology* 49(1): 7–14.

Shavit, Ayelet, and James Griesemer. 2011. Mind the gaps: Why are niche construction processes so rarely used? In *Transformations of Lamarckism: From subtle fluids to molecular biology*, ed. Snait B. Gissis and Eva Jablonka, 307–318. Cambridge, MA: MIT Press.

Simberloff, Daniel. 2006. Invasional meltdown 6 years later: Important phenomenon, unfortunate metaphor, or both? *Ecology Letters* 9(8): 912–919.

Simberloff, Daniel, and Betsy von Holle. 1999. Positive interactions of nonindigenous species: Invasional meltdown? *Biological Invasions* 1(1): 21–32.

Slobodkin, Lawrence B. 1961. *Growth and regulation of animal populations*. New York: Holt, Rinehart and Winston.

Smith, Bruce D. 2007. Human niche construction and the behavioral context of plant and animal domestication. *Evolutionary Anthropology* 16(5): 188–199.

Sober, Elliot, and David S. Wilson. 1998. *Unto others: The evolution and psychology of unselfish behavior*. Cambridge, MA: Harvard University Press.

Spellberg, Brad, Robert Guidos, David Gilbert, John Bradley, Helen W. Boucher, Michael W. Scheld, John G. Bartlett, and John Edwards Jr. 2008. The epidemic of antibiotic-resistant infections: A call to action for the medical community from the Infectious Diseases Society of America. *Clinical Infectious Diseases* 46(2): 155–164.

Sterelny, Kim. 2003. *Thought in a hostile world: The evolution of human cognition*. Oxford: Blackwell.

Sterelny, Kim. 2011. From hominins to humans: How sapiens became behaviorally modern. *Philosophical Transactions of the Royal Society B* 366(1566): 809–822.

Swenson, William, David S. Wilson, and Roberta Elias. 2000. Artificial ecosystem selection. *Proceedings of the National Academy of Sciences of the United States of America* 97(16): 9110–9114.

Thompson, John N. 1998. Rapid evolution as an ecological process. *Trends in Ecology & Evolution* 13(8): 329–332.

Turner, Scott J. 2000. *The extended organism: The physiology of animal-built structures*. Cambridge, MA: Harvard University Press.

Van Dyken, David J., and Michael J. Wade. 2012a. Origins of altruism diversity I: The diverse ecological roles of altruistic strategies and their evolutionary responses. *Evolution* 66(8): 2484–2497.

Van Dyken, David J., and Michael J. Wade. 2012b. Origins of altruism diversity II: Runaway coevolution of altruistic strategies via "reciprocal niche construction". *Evolution* 66(8): 2498–2513.

Verdú, Miguel, and Alfonso Valiente-Banuet. 2008. The nested assembly of plant facilitation networks prevents species extinctions. *American Naturalist* 172(6): 751–760.

Wainwright, Peter C., and Stephen M. Reilly. 1994. *Ecological morphology: Integrative organismal biology*. Chicago: University of Chicago Press.

Wake, Marvalee H. 2001. Integrative biology: Its promise and its perils. *Biology International* 41: 71–74.

Wake, Marvalee H. 2004. Integrative biology: The nexus of development, ecology, and evolution. *Biology International* 46: 3–15.

Wake, Marvalee H. 2008. Integrative biology: Science for the 21st century. *BioScience* 58(4): 349–353.

Weisberg, Michael. 2006. Forty years of 'the strategy': Levins on model building and idealization. *Biology and Philosophy* 21(5): 623–645.

Weisberg, Michael. 2007. Three kinds of idealization. *Journal of Philosophy* 104(12): 639–659.

Wimsatt, William. 1987. False models as a means to truer theories. In *Neutral models in biology*, ed. Antoni Hoffman and Matthew H. Nitecki, 23–55. New York: Oxford University Press.

The Affordance Landscape: The Spatial Metaphors of Evolution

Denis M. Walsh

Abstract The adaptive landscape is a metaphorical device employed to depict the evolutionary change in a population or lineage undergoing natural selection. It is a powerful heuristic and didactic tool. This paper has two objectives. The first is to dig beneath the adaptive landscape in order to expose certain presuppositions about evolution concealed there. The second is to propose and motivate an alternative spatial metaphor, one that embodies a wholly different set of presuppositions. I develop the idea that adaptive evolution occurs on an 'affordance landscape.' The conception of adaptation—both the process and the product—that follows from adopting the affordance landscape metaphor is a significant departure from the conception of adaptation embodied in orthodox Modern Synthesis biology.

Spatial metaphors abound in evolutionary biology. Biologists speak of genotype space, phenotype space, and a 'map' between them. Phylogenetic propinquity is measured as a distance in an abstract space of nucleotide sequences (Nei 1972). Molecular evolution is envisaged as occurring in a 'protein space' (Maynard Smith 1970). Morphological diversity is represented by a manifold of dimensions of morphospace (McGhee 2007). Such spatial metaphors contribute to evolutionary thinking in myriad ways. Like scientific metaphors in general, they make recondite theoretical concepts accessible and tractable. They point toward ways in which our theoretical concepts may be extended, expanded and revised. They suggest to us sometimes surprising implications of our theories, and in turn help generate empirical predictions. But they do not come free of cost. In giving form to inchoate concepts, they may also constrain or bias our use of them in subtle and subliminal ways. Such impositions, in turn, may obscure from our view what might otherwise

D.M. Walsh (✉)
Department of Philosophy, Institute for the History and Philosophy of Science and Technology, Department of Ecology and Evolutionary Biology, University of Toronto, Toronto, ON, Canada
e-mail: denis.walsh@utoronto.ca

G. Barker et al. (eds.), *Entangled Life*, History, Philosophy and Theory of the Life Sciences 4, DOI 10.1007/978-94-007-7067-6_11,
© Springer Science+Business Media Dordrecht 2014

be perfectly obvious interpretations, or empirical consequences, of our theories. The use of metaphors may well be essential to scientific thinking: "Metaphor and simile are the characteristic tropes of scientific thought, not formal validity of argument" (Harré 1986, 7). Nevertheless, it imposes a burden on scientific practice: "The price of metaphor is eternal vigilance" (Lewontin 2001a, 1264).[1]

The adaptive landscape is among the most vivid, pervasive and enduring spatial metaphors in biology. It is a device employed to depict evolutionary change in a population or lineage undergoing natural selection.[2]

> The idea ... has become a standard imagination prosthesis for evolutionary theorists. It has proven its value in literally thousands of applications, including many outside of evolutionary theory. (Dennett 1995, 190)

So powerful is it as a heuristic and didactic tool, that we seldom enquire into the ways it might immure our thinking about adaptive evolution. Nor, for that matter, are we inclined to question the commonly held presuppositions about the process of adaptive evolution that make the adaptive landscape such an apt representation. But the 'price of metaphor' suggests that we should.

This essay has two objectives. The first is to dig beneath the adaptive landscape in order to expose certain presuppositions about evolution that the metaphor conceals. The second is to propose and motivate an alternative spatial metaphor, one that embodies a wholly different set of presuppositions. I outline the idea that adaptive evolution occurs on an 'affordance landscape.' The adaptive landscape and the affordance landscape underwrite strongly divergent conceptions of adaptive evolution. The crucial difference resides in the role that each accords to organisms in the process of evolution. Whereas the adaptive landscape entrenches the Modern Synthesis view that organisms make no substantive contribution to adaptive evolution, the affordance landscape underscores the way that the distinctive capacities of organisms create and constitute the conditions under which evolution occurs.

1   The Adaptive Landscape

The adaptive landscape is a pictorial device used to portray the evolution of populations and lineages. Adaptive evolution is depicted as a trajectory traversing a multi-dimensional surface. This surface resides in a space whose axes represent traits, one dimension for each trait. Each point in the multi-dimensional 'design

---

[1]With characteristic modesty, Lewontin credits this dictum—an allusion to a similar saying about the condition or price of liberty—to Rosenblueth and Wiener. However, I was unable to find it in any of their co-authored papers. It appears unattributed, though enclosed in quotation marks, in Lewontin (1963, 230).

[2]Some productive uses of it can be found in Lande (1976), Flyvbjerg and Lautrup (1992), Niklas (1997), and Sloman (2000).

space' thus corresponds to an individual organism's total phenotype or form.[3] There is a further axis in addition to the trait dimensions; each individual total phenotype has a fitness, represented as an altitude on the landscape. Individuals with higher fitness, so the story goes, generally beget phenotypically similar individuals with comparably high fitness. As evolutionary novelties are introduced into a population, some will confer yet higher fitness on their bearers. So long as the selection coefficients are sufficiently high, that is to say, the slopes are sufficiently steep, a population undergoing selection will be drawn inexorably toward a local fitness optimum. The local fitness optima are 'adaptive peaks,' good locations in 'design space.' Populations at these optima are well adapted to their conditions of existence. A population inhabiting a valley may split, each moiety moving toward a different adaptive peak. The adaptive landscape device thus illustrates the way populations undergoing natural selection become both increasingly well suited to survival and reproduction in their respective environments *and* increasingly diverse. All in all, it is an elegant way to represent adaptive evolution. "The value of an adaptive topography is that it is easily visualized and so makes the evolutionary dynamics of the population intuitively clear" (Lande 1976, 315).

Adaptive evolution, then, is visualized as a process in which a population or lineage traverses a fitness surface under the influence of evolutionary forces; its trajectory is explained exclusively or primarily by the topography of that surface.

> Adaptive evolution is a search process—driven by mutation, recombination, and selection—on fixed or deforming fitness landscapes. An adapting population flows over the landscape under these forces. The structures of such landscapes, smooth or rugged, governs both the evolvability of populations and the sustained fitness of their members. The structure of the fitness landscape inevitably imposes limitations on adaptive search. (Kauffman 1993, 118)

1.1   Evolution on the Adaptive Landscape

Unpacking the adaptive landscape metaphor discloses a number of non-trivial, but widely endorsed, preconceptions about adaptive evolutionary change. The landscape suggests that adaptive evolutionary change is robust, in the sense that it is relatively insensitive to initial or perturbing conditions. A population will move toward its local summit, even if it is deflected by deleterious mutations, or impeded by constraints.

The adaptive landscape places other, less conspicuous, demands upon the relation between evolutionary trajectories and the space they move through. Most importantly, 'adaptive space' is autonomous of form, inert and unchanging. A location in design space has its adaptive value (its fitness) whether or not it is occupied.

---

[3]The first use of this device seems to have appeared in Simpson (1944). It is not to be confused with Sewall Wright's (1932) fitness landscape (although it often is) in which the axes are allele frequencies.

Form and landscape are asymmetrically related; form evolves in response to the landscape, but not vice versa. This asymmetry is necessary if adaptive evolution is to be thought of as form *conforming* to the exigencies of design space.

Further, the adaptive landscape can represent adaptive evolution at any scale, from sub-population to kingdom.[4] That being so, adaptive evolution would appear to be scale-independent. The same processes and dynamics that apply within a population within a generation also apply to the evolution of classes, phyla, and kingdoms, over vast stretches of time. The adaptive evolution of higher taxa is simply the adaptive dynamics of populations scaled up. It consists in the introduction of small random mutations and the gradual progression of populations up adaptive slope, and the divergence of lineages toward adjacent slopes.

1.2   Convergence and Contingency

This conception of adaptive evolution forms the backdrop for a number of disputes about the tempo and mode of large-scale evolution.[5] It is most notably evident in the debate concerning the convergence and contingency of evolution. Simon Conway Morris (2010) has consistently maintained that macro-evolutionary trajectories are convergent (and non-contingent).[6]

> What we know of evolution suggests … [that] … convergence is ubiquitous and the constraints of life make the emergence of the various biological properties very probable, if not inevitable. (Conway Morris 2010, 283–284)

Stephen Jay Gould (2002) steadfastly argued for the contrary view, that evolutionary change is non-convergent and highly contingent (Beatty 1995). As historical processes, evolutionary trajectories are subject to all the vagaries of history, the unpredictable occurrences thrown up by chance. The history of any given lineage might easily have turned out very differently than it did. Gould illustrates the contingency of evolution with the metaphor of rewinding a tape:

> You press the rewind button and …. go back to any such time and place in the past. [A]ny replay of the tape would lead evolution down a pathway radically different to the road actually taken. (Gould 1989, 50)

This is a heated and complicated issue. It is not my intention to resolve it, but merely to point out an interesting feature of the dialectic. These diametrically opposed

---

[4]There may be differences in landscape topology as we investigate different levels of detail (Wilkins and Godfrey-Smith 2009), but the processes are the same at every scale.

[5]Indeed the adaptive landscape metaphor figures explicitly in many discussions of macro-evolutionary change (Simpson 1944; Stanley 1998).

[6]See Beatty (1995) for an extended discussion of contingency in evolution. On convergence and parallelism in evolution, see Powell (2007, 2012) and Pearce (2012).

positions arise from a shared conviction—viz. if evolution is adaptive, then it is convergent and non-contingent. The adaptive landscape provides support to this supposition. The trajectory of a lineage undergoing *adaptive* evolution is determined principally by the contours of the landscape (therefore, it is non-contingent). Given that ex hypothesi there are few peaks (few good locations in design space), over time biological lineages will tend to converge upon them, no matter their starting point.

> The phenomenon of convergent evolution means that there are a limited number of ways of making a living in nature, a limited number of ways of functioning well in any particular environment. ... We can model this reality in an adaptive landscape by specifying the location of adaptive peaks for particular ways of life. (McGhee 2007, 35)

1.3   Niches

If the adaptive landscape illuminates the significant features of adaptation as a *process*, then another standard spatial metaphor, the niche, neatly captures the presumptive nature of adaptation as a *product*. A (non-metaphorical) niche is a space into which something—say, a statue—might fit. An evolutionary niche is a set of properties of an organism's environment, to which organismal form may fit.

The niche concept codifies a particular relation between organism and environment, thought to be integral to a genuine understanding of adaptation. Organismal form and the niches to which it adapts are decoupled and asymmetrically dependent.

> To make the metaphor of adaptation work, environments or ecological niches must exist before the organisms that fill them. Otherwise environments couldn't *cause* organisms to fill those niches. The history of life is then the history of coming into being of new forms that fit more closely into these pre-existing niches. (Lewontin 2001b, 63)

Lewontin claims that the decoupling of form and niche is made obligatory by the concept of adaptation:

> So long as we persist in thinking of evolution as adaptation, we are trapped into an insistence on the autonomous existence of environments independent of living creatures. (Lewontin 2001b, 63)

This conclusion follows from the commonly held conviction that adaptation is adaptation *to* some external feature. Lewontin (reluctantly) reaffirms:

> Adaptation is the process of evolutionary change by which the organism provides a better and better "solution" to the "problem," and the end result is the state of being adapted. (Lewontin 1978, 213)

A nice illustration of this use of the niche concept appears in the very issue of *Scientific American* in which Lewontin first questions its coherence. In explaining the distinctive mode of life of New Zealand's three kiwi species (*Apteryx* spp.), William A. Calder III says,

> I prefer to look on this curious bird as a classic example of convergent evolution. In this view an avian organism has acquired a remarkable set of characteristics that we generally associate not with birds but with mammals. ... When there were no mammals present to lay claim to the niches in this hospitable environment, birds were free to do so. (Calder 1978, 142)[7]

Insofar as a trait is an adaptation, then, it must be identifiable as a response to pressures exerted on form by the niche or external environment. That, in turn, requires that we are able to bracket off the contributions that the environment makes to evolutionary change from those of the internal features of biological form. There is an asymmetric relation between environment and organismal form. As Peter Godfrey-Smith notes, in this standard picture...

> organisms respond to the environment, but the environment is largely autonomous with respect to the organisms. The environment is seen as either stable (as far as the time scale of the evolutionary process in question is concerned) or else as changing according to its own intrinsic dynamics. (Godfrey-Smith 2001, 254)

The separation of organism and environment conjoined with the explanatory primacy of environment over form conspire against any substantive role for organisms in the process of adaptive evolution.

> In this view the organism is the object of evolutionary forces, the passive nexus of independent external and internal forces, one generating "problems" at random with respect to the organism, the other generating "solutions" at random with respect to the environment. (Lewontin 2001b, 47)

The traditional niche concept complements the adaptive landscape metaphor nicely. Niches confer on the landscape its fitness structure. If niches are 'extra-organismal' and are the principal determinants of the fitness structure of the landscape, then it follows that the determinants of the fitnesses of biological form are extrinsic to form itself. This seems to accord with the general usage: the fitness that selection increases is a measure of the ability of organisms to meet the exigencies of the extra-organismal environment.

1.4   The Occupancy of Adaptive Space

Another feature of the adaptive landscape trope is immediately apparent, but less obviously significant—viz. that the conception of space encoded in the adaptive landscape metaphor is a decidedly classical, Newtonian one. Newtonian space is a mere container: non-substantival, inert, and unchanging. Its intrinsic properties are exhaustively described by Euclidean geometry. Because space is a mere container,

---

[7]Notice how naturally convergence falls out of the traditional conception of organism/environment relations.

it does not interact with matter. Consequently the properties of a region of space and the relations between spatial regions—relative extensions, displacements, accessibilities—are completely independent of whether that region is occupied. Moreover, the geometrical properties of Classical space are scale-independent. Take a collection of objects. If they were all to be moved in the same direction by the same distance, their spatial relations (relative proximity, accessibility, the angles subtended by any three etc.) would not change. Nor would these relations change if the region these objects occupy were to be doubled in size. Furthermore, doubling the region of space that an object occupies (i.e. doubling its size) may have no effect on its shape.[8] For example, a trajectory that describes a triangle in Classical space, *of any size*, encloses a sum of internal angles of 180°. Consequently, the dynamics of objects moving through space are scale-free.

There are evident analogies between the classical conception of the occupancy of space and the modern synthesis conception of the occupancy of adaptive space. Form does not influence the properties of fitness space. Locations in adaptive space have their properties—in particular their fitnesses—independently of whether they are occupied by biological form.[9] That a location in adaptive space is occupied has no effect on its relations of access to other locations in adaptive space. Most particularly, the dynamics of evolutionary trajectories through adaptive space are independent of location and scale. Selection causes form to move up fitness gradients, and that applies equally to sub-populations within generations and high-level lineages over large time scales. This scale-independence of evolutionary trajectories is manifested in the supposition that the dynamics of macro-evolutionary processes are simply those of micro-evolution scaled up.

2   The Affordance Landscape

It isn't clear, however, that Newtonian space provides the most salutary analogy for the relation between biological form and the space of adaptations. One significant *dis*analogy is that the fitness consequences of a location in adaptive space are not detached from the properties of the form that fills it in the way that objects are from Newtonian space. For example, the traditional niche concept holds that for each environment there is a location in adaptive space that corresponds to the problem posed by the environment. The properties of this location—and hence what would count as a solution to the adaptive problem—are fixed quite independently of the

---

[8]Nerlich (1991) calls the independence of spatial relations from spatial properties the 'Detachment Thesis': "thing-thing relations are logically independent of thing-space relations" (172).

[9]There is a further, related, Newtonian analogy to be considered. Most philosophers of biology seem to hold that adaptive space is inert. Consequently, extraneous causes or forces, like selection and drift, are required to propel form across the adaptive landscape.

features of biological form. The problem with the analogy is that extra-organismal features radically underdetermine what might count as a solution to an adaptive problem. Consider the case of the adaptive solution to 'the problem' of locomotion in water. Paramecia and porpoises have both solved it, but in very different ways. The differences are due to the way that water is *experienced* by organisms of different sizes. A harbour porpoise experiences water in much the way we do; for a porpoise water flows easily. A porpoise swims by setting up smooth laminar flow across its body. Porpoises have evolved a terete shape, a strong caudal fin and a narrow muscular caudal peduncle to concentrate the propulsive power of the tail stroke, as adaptations to the problem of locomotion in water. At a length of approximately 200 μm, a paramecium experiences the viscosity of water differently, much as we would experience being immersed in corn syrup (Purcell 1977). A paramecium cannot displace water by setting up laminar flow. Instead it possesses helical bands of cilia, whose rhythmic beating serves to 'screw' the organism through its thick medium.[10] These are two radically different 'design solutions' to the same environmental feature. The upshot is that the concept of an adaptation is not simply that of an evolutionary response to an environmental condition.

I don't claim that the adaptive landscape metaphor is incapable of accommodating the form-dependence of adaptation. I merely wish to suggest that the detachment of form from the determinants of fitness that is engendered by the metaphor is not the most perspicuous way to think of biological adaptation. The point is that we cannot specify what would count as a solution to an 'adaptive problem' independently of the features of form that solve the problem. In order to identify an adaptation we must cite the way that the environment is *experienced* by the organism. Any metaphor that draws our gaze away from the importance of the *experienced* environment has the very real potential to lead us astray. Conversely, taking the notion of the *experienced environment* seriously ought to occasion a significant shift in our conception of adaptation.

My objective is to motivate an alternative conception of adaptive evolution, one that accords due significance to the way that organisms experience, constitute, and alter their conditions of existence. The alternative is encapsulated in the slogan that *adaptation is an evolutionary response to affordances*. That, in turn, introduces an alternative spatial metaphor for adaptive evolution—the *affordance landscape*. My hope is that the contrasting spatial metaphors will underscore the differences between these conceptions of adaptive evolution.[11]

---

[10] A *Paramecium* actually has three 'gaits,' only two which involve the asymmetric beating of cilia. See Hamel et al. (2011).

[11] Some of the implications of seeing evolution as a response to affordances are discussed in Walsh (2012).

2.1 Affordances

The leading idea behind the experienced environment is captured nicely in
J.J. Gibson's concept of an affordance:

> The affordances of the environment are what it offers the animal, what it *provides* or
> *furnishes,* for good or ill. ... I mean by it something that refers to both the environment and
> the animal .... It implies the complementarity of the animal and the environment. (Gibson
> 1979, 127)

For an organism to experience an environment, or a condition of existence, is for
that condition to '*provide*' or '*furnish*' something to the organism. That, in turn,
depends heavily on the capacities of the organism. We can explain an adaptation as
a response to a challenge faced by the organism, only once we understand how the
features to which form adapts are *experienced* by organisms 'for good or ill.' One
salutary suggestion, then, is that an adaptation is not so much a response to a niche
or environment, traditionally construed, but to an affordance.

There is a considerable amount of debate about how to precisify this notion of
an affordance. If, as some authors have argued (e.g. Turvey 1992), an affordance is
simply a dispositional property of an organism's environment, then it would be possible to reconcile the orthodox Modern Synthesis account of adaptive evolution with
the idea that an adaptation is a response to an affordance. Porpoises and paramecia
simply respond to different dispositional properties of their shared environments.[12]
Construed this way, an affordance could simply be part of a traditional niche—
something wholly independent of biological form—and no change in our conception
of adaptation would be occasioned by adopting the maxim that an adaptation is a
response to an affordance.

It is becoming evident, however, that if the affordance concept is to do the work
initially required of it by ecological psychology, it must be a much richer notion
(Stoffregen 2003; Chemero 2003).[13] "Affordances are opportunities for action;
they are properties of the animal–environment system that determine what can be
done" (Stoffregen 2003, 124). They may be considered intrinsic emergent properties
of the organism/environment system (Stoffregen 2003) or "relations between the
abilities of organisms and features of the environment" (Chemero 2003, 181; see
also Heft, this volume).[14] It doesn't much matter for my purposes so long as

---

[12]Which dispositional properties are represented as Reynold's Number (Purcell 1977).

[13]A sympathetic reading of Gibson (1979), I believe, suggests the same.

[14]It is interesting that in those sciences in which the niche concept plays a genuine theoretical role,
e.g. community and population ecology, the niche concept is often defined more in the way an
affordance is. The niche concept seems to have originated with Elton (Hutchinson 1978) and was
defined in terms of resource utilization. Odum (1959) likens a niche to an organism's 'profession.'
See Beatty (1995). I thank Sahotra Sarkar for pointing this out to me.

an affordance is jointly constituted by the capacities of the organism and the conditions of existence.[15]

Further, if the concept of an affordance is to play its intended role in ecological psychology, then an affordance must imply something of significance for the organism. To perceive the affordances of one's environment is to perceive the *significance* of the opportunities it presents.

> The perceiving of an affordance is not a process of perceiving value-free physical objects to which meaning is somehow added ... it is a process of perceiving a value-rich ecological object. (Sanders 1993, 290)[16]

To respond to an affordance is to respond to the value or meaning of that feature for the organism.

2.2 The Implications of Affordances

There are two crucial implications of the affordance concept that make it a radical departure from the traditional niche concept. The first concerns the form/affordance relation. Affordances are not 'autonomous' from organisms, nor is there an asymmetrical dependence of organisms on affordances. There is a reciprocity between organisms and their affordances that does not hold between organisms and their niches or environments. What a feature of the environment affords an organism depends (in part) upon the organism's capacities, and the capacities of the organism in turn depend (in part) on the features of the environment. Organisms and their affordances are co-constituting and 'commingled.'[17]

The second implication is that, unlike a niche, an affordance implies a purposive system. A statue can have a niche, but only a goal-directed system can have an affordance. Purpose defines affordances: an affordance is an opportunity for, or an impediment to, the attainment of a goal. Conversely, affordances define purposive systems: a purposive system is an entity for which features of its conditions of existence constitute affordances. A purposive system is a system capable of responding adaptively to affordances.

---

[15]One additional advantage of not seeing affordances as dispositional properties of an organism's environment is that it relieves us of the temptation of thinking that all affordance-presenting features are external to organisms. Inner workings of organisms present affordances too.

[16]In the case of organisms, 'value' may be read, often enough, as 'survival value.'

[17]I borrow the term 'commingled' from Haugeland (1998). The relation between organisms and their conditions of existence I envisage includes but extends beyond what Gillian Barker (2008) calls 'selective interaction.' The principal difference is that selective interaction emphasises the ways in which organisms causally influence their conditions of existence. The 'commingling' of organisms and their affordances underscores the way in which the capacities of organisms partly *constitute* those conditions. See Walsh (2012) for a discussion of the distinction.

In order to respond to an affordance, a system must have two features. Firstly, it must be able to experience its conditions *as affordances*. That is to say that it must generally be capable of responding to propitious conditions *as propitious* by exploiting them, and to unpropitious *as unpropitious*, by ameliorating them. Secondly—and concomitantly—a system must also have an adaptive repertoire. That is to say that on any occasion, there must be a range of possible outcomes or activities that the system or its parts could implement. Which elements of the system's repertoire are actualized on an occasion must generally be biased in favour of those that are conducive to the attainment of the system's goals.

Following the suggestion that adaptation is an evolutionary response to affordances, I would like to explore an alternative metaphor for adaptive evolution. We should think of evolution as occurring on an 'affordance landscape.' An affordance landscape is the complete set of affordances—conditions 'for better or ill'—that impinge on an organism. That is to say, it is the complete set of conditions experienced by an organism as impediments to, or as conducive to, its goals of survival and reproduction.

These implications of affordances have special significance for adaptive evolution. Because the capacities of biological form and affordances are co-constituting, any change in one is a change in the other. Form and the affordances to which it evolves co-evolve. Furthermore, as affordances are reflections of purposiveness, then the adaptive goal-directedness of organisms structures and conditions the affordances on which evolution occurs. I discuss these implications of affordances for evolution in turn.

3   The Co-evolution of Form and Affordance

The relation between form and affordance landscape is very *unlike* the relation between form and the adaptive landscape—or, for that matter, matter and Newtonian space. The affordance landscape is not inert or 'detached' from the properties of form, nor does it have 'its own intrinsic dynamics.' It is constantly shifting with changes in organismal form. Nor is there a relation of asymmetrical dependence of form on the affordance landscape. Form and the affordance landscape affect one another reciprocally; they co-evolve. A couple of examples might help to illustrate this reciprocal dependence and its importance for adaptive evolution.

3.1   The Origin of Hominin Tool Use

It is generally acknowledged that the advent of tool use in hominins has been integral to their evolution, especially in late hominin lineages leading to *Homo sapiens*. The affordances provided by tools have long been thought to have been intimately involved in human cognitive, linguistic and social evolution (Gibson 1993). It is less clear how those affordances first arose. Recent work in evolutionary developmental biology suggests that the initial conditions that permitted the expansion of hominin tool use may have been a contingent byproduct of the evolution of obligate bipedalism (Rolian et al. 2010).

Tool use requires, at minimum, 'precision grip' (Marzke 1997). This is the capacity to oppose the thumb against one or more fingers. Advanced tool use requires the ability to oppose the thumb against all of the fingertips at once ('higher order precision grip'). Changes in the structure of the ancestral hominoid hand were required in order for this higher order precision grip and opposition of the digits to be possible. These involved, crucially, the increased robustness of the thumb, its extension distally, and the shortening of the fingers (Rolian et al. 2010).

Recent morphological studies on primate hand and foot development demonstrate a considerable degree of integration in the development of both (Rolian and Hallgrimsson 2009). Hands and feet are serial homologues. Their respective development is controlled by very similar developmental architectures (Hallgrimsson et al. 2002). As a consequence, evolutionary changes occurring in the foot may influence the development and evolution of the hand (and vice versa). Rolian et al. (2010) demonstrate that the changes required to the hominin foot required for bipedal endurance running include the strengthening and distal extension of the big toe and the shortening of the lateral digits. These structural changes to the foot that facilitate endurance running are just those changes that in their homologous structures in the hand are required for higher order precision grip.

Rolian et al. (2010) hypothesize that changes in the hand are a consequence of the evolutionary changes in foot structure. Given the developmental integration, or coupling, of hand and foot development, changes in foot structure drag the hands along.

> Developmental constraints caused hominin fingers to evolve largely as a by-product of stronger selection pressures acting on the toes. Simply put, the shorter fingers and longer, more robust thumbs of humans likely evolved because of selective pressures on their respective homologues in the foot. (Rolian et al. 2010, 1564)

Nevertheless, these changes in hand structure conferred on hominoid ancestors new capacities to grasp implements and use them as tools. In other words, serendipitous changes in form dramatically altered the affordances of our ancestors' environments, *without changing the environment*. Changes in our ancestors' hands put tools in their environments. These altered affordances, in turn, introduced new opportunities for adaptive evolutionary change.

3.2   The Origin of Metazoans

The changes in form that usher in new affordances do not have to be adaptive in any way, nor do they need to be underwritten by genetic changes. These lessons can be gleaned from recent work on the origin of the Metazoa.

The morphological and developmental complexity of metazoans vastly exceeds that of any unicellular organism. Yet, the entire panoply of basic metazoan structures, and a fair degree of phyletic diversity, appears to have arisen rapidly in the Precambrian. A fascinating picture is beginning to emerge about this sudden arrival of metazoan complexity (Newman and Bhat 2009). The original coalescence of unicellular pre-animals into vast assemblages of cells appears to have been the consequence of a precipitous change in the ionic constitution of the seas. Kaźmierczak and Kempe (2004) report evidence of a sudden rise in Ca^{++} concentrations in the Precambrian seas. Increased Ca^{++} is known to promote cell-cell adhesion.

These massive aggregations of cells—the proto-metazoans—encountered 'meso-scopic' physical conditions that had never previously affected the development or diversity of organic form.

> The consequent change in spatial scale created a context in which other pre-existing molecules were able to mobilize mesoscopic (i.e., "middle-scale") physical processes and effects that were irrelevant to objects of the size and composition of individual cells. (Newman 2011, 339)

Thanks to the newly encountered "middle-scale" physical processes and effects, these aggregations of cells had the capacity to produce all the characteristic structures of the metazoans—lumena, tissue layers, blastocoels, tubes, differentiated tissues—spontaneously.

> The forms of the earliest multicellular organisms ... were more like certain materials of the non-living world than are the forms of their modern, highly evolved counterparts, and that they were therefore almost certainly molded by their physical environment to a much greater extent than contemporary organisms. ... Stated simply, tissue forms emerged early and abruptly because they were inevitable—they were not acquired incrementally through cycles of random genetic change followed by selection. (Newman 2003, 221)

These new biological structures, the foundations of metazoan form, are not solutions to adaptive problems posed by an external environment. Nevertheless, they confer on biological form novel capacities, which in turn open up new vistas: threats to survival, opportunities for change, potential for new forms.

The nearest living non-metazoan relatives of the metazoans appear to be the unicellular (and sometimes colonial) choanoflagellates (King 2004). Choanoflag-ellates possess a basic genetic tool kit comprising (*inter alia*) genes coding for proteins that mediate cell-cell adhesions, genes that regulate growth and shape, and extracellular matrix proteins that—in metazoans at least—mediate cell sorting and tissue formation during development. The unicellular precursors of metazoans, then, carried genes that in the new context of multicellular assemblages played entirely new roles in metazoan function and morphogenesis.

> Some components of the protein machinery that mediates animal cell interactions may have originally played other roles in ancestral unicellular eukaryotes before being co-opted to function in signaling and adhesion. (King 2004, 319)

Each of the preceding examples, hominin tool use and the evolution of the metazoans, takes seriously the ways in which biological form partly constitutes the conditions under which evolution occurs. In each of our examples there are reciprocal cycles of changes in form with concomitant changes in affordances, *without changes to the environment*. This suggests that large-scale evolution is not most perspicuously described as the process of form gradually fitting into independently specified problems in static design space. It is the process of form creating and then responding to an ever-changing system of affordances. The affordance landscape metaphor nicely captures this reciprocity of form and conditions in a way that the traditional adaptive landscape metaphor tends not to. The affordance landscape, unlike the adaptive landscape, does not 'obey its own intrinsic dynamics.'

This is not to say that the adaptive landscape could not be 'adapted' to accommodate at least some aspects of the reciprocity of form and the conditions under which it evolves. It is well known that organisms change and 'construct' their environments. We might allow that the adaptive landscape may deform as organisms alter their external conditions of existence. We could even make the concession that the adaptive landscape represents the conditions of existence as *experienced* by organisms. We could, for example, stipulate that in our swimming example the properties of the adaptive landscape represent organism-indexed parameters, such as Reynold's number, rather than intrinsic properties of the environment, such as viscosity.

But while these amendments to the adaptive landscape metaphor would be significant and salutary, they still would not capture the import of the notion that adaption is an evolutionary response to *affordances*. The reason is that, as discussed, affordances imply purposes; it is purposiveness that turns conditions into affordances. The fact of organisms' being purposive, adaptive entities plays no role in the standard Modern Synthesis conception of adaptive evolution. It is hardly surprising, then, that its principal spatial metaphor should decline to represent organismal purposes either explicitly or implicitly. In taking the conditions to which biological evolution responds to be affordances, the affordance landscape does represent organismal purposiveness as a factor in evolution. Here, I believe, is the watershed. The affordance landscape metaphor earns its keep—and distinguishes itself from the traditional adaptive landscape—only if the purposiveness of organisms makes some decisive contribution to adaptive evolution.

4   Affordance and Organismal Purpose

Organisms are highly robust, goal-directed entities. They are capable of building themselves and maintaining their viability despite the considerable vagaries of their conditions (Gibson 2002). Kirschner and Gerhart label this distinctive property of organisms 'dynamic restoration.'

> The organism is not robust because it is built in such a manner that it does not buckle under stress. Its robustness stems from a physiology that is adaptive. It stays the same, not because it cannot change but because it compensates for change around it. The secret of the phenotype is dynamic restoration. (Kirschner and Gerhart 2005, 108–109).

The hallmark of this ability to adapt to, and to compensate for, conditions of existence is phenotypic plasticity. West-Eberhard identifies plasticity as

> the ability of an organism to react to an environmental input with a change in form, state, movement, or rate of activity. ... The words "responsiveness," "flexibility," "malleability," "deformability," and *developmental plasticity* are all synonyms of phenotypic plasticity as defined here. (2003, 34–45)[18]

4.1 Plasticity and Evolution

Phenotypic plasticity contributes to adaptive evolution in the following way. An organism faces a challenge from an environmental perturbation, or a mutation, and proceeds to make an adaptive, compensatory change in its phenotype. As organisms are highly functionally integrated entities, a change in one feature of its phenotype requires concomitant changes in others. This capacity of an organism to make compensatory changes in one part of its phenotype in order to accommodate changes elsewhere, is called 'phenotypic accommodation': it is simply one aspect of phenotypic plasticity.

> Phenotypic accommodation is adaptive mutual adjustment, without genetic change, among variable aspects of the phenotype, following a novel or unusual input during development. (West-Eberhard 2003, 98)

Adaptive evolutionary change requires a considerable degree of orchestration amongst an organism's various subsystems. For example, the increase in the mass of a muscle in response to the demand for greater force also requires changes in the origin and insertion of the bones. It further requires increased vascularisation, innervation, and changes in associated connective tissue.

On the standard Modern Synthesis conception of adaptive evolution, organisms do not initiate or orchestrate evolutionary change. Each evolutionary novelty is initiated by a random mutation, or by recombination of genes. But given the demand for functional integration, if each phenotypic change required just the right mutation (or combination) in each contributing subsystem, adaptive evolution might never occur.

> In contrast to the rapid response produced by plasticity, if the production of newly favored phenotypes requires new mutations, the waiting time for such mutations can be prohibitively long and the risk of subsequent loss through drift can be high. (Pfennig et al. 2010, 459–460)

Phenotypic accommodation facilitates adaptive change by providing the requisite adaptive 'orchestration.'

---

[18]I would suggest an amendment to West-Eberhard's definition. Plasticity should be seen as the capacity to react to an input from *any* source, not merely an environmental input.

> Phenotypic accommodation finesses the problem of correlated change: a genetically-caused modification in one system need not wait for a genetically-caused change in associated systems, even when both must change for either to be adaptive. (Sterelny 2009, 99)

Phenotypic plasticity enters into the process of adaptive evolution by initiating *adaptive* responses to the organism's conditions of existence and then further making adaptive (accommodating) responses that maintain or promote viability. Often enough, adaptive responses, underwritten as they are by the developmental robustness of organisms, are intergenerationally stable. They can be passed from one generation to the next, and hence are candidates for being evolutionary characters.[19]

The adaptive plasticity of organisms is underwritten by the fantastic phenotypic repertoire immanent in development.

> Through its ancient repertoire of core processes, the current phenotype of the animal determines the kind, amount and viability of phenotypic variation the animal can produce ... the range of possible anatomical and physiological relations is enormous. (Gerhart and Kirschner 2007, 8588)

Phenotypic plasticity is the capacity of an organism to marshal its phenotypic repertoire in response to the challenges and opportunities it encounters. It is a manifestation of organismal purposiveness. Organisms make adaptive changes in form or function to those features of their environments, genomes, or developmental systems that threaten or promote their viability. Thus the plasticity of organisms consists in a goal-directed capacity to respond to, and to create, *affordances*.

4.2   Plasticity and the Affordance Landscape

If adaptive evolution is change in response to conditions of existence, then in altering the conditions of existence the adaptive plasticity of organisms contributes to the process of evolution. When an organism makes an adaptive response to its conditions of existence, it also changes those conditions of existence. These changes, in turn, introduce new evolutionary challenges and opportunities. In this way, organisms are participants in adaptive evolution, not mere objects of it.

Traditional Modern Synthesis biology treats the adaptive plasticity of organisms as, at best, a mere consequence of adaptive evolution (Godfrey-Smith 1996). But it accords organisms no active role in promoting adaptive evolution; organisms are "the passive nexus of independent external and internal forces" (Lewontin 2001b, 47). We encountered the rationale for this view in our discussion of the

---

[19]Confusion persists on this point. See, for example, Sterelny (2009, 101) who claims that novelties generated by phenotypic plasticity are "mere ecological events." "Such novelties have no effects on the germline are not inherited [*sic*]" (2009, 94). He muses on how they can be transformed from "mere ecological events" into evolutionary events. My claim is that no transformation is needed; any ecological event that is intergenerationally stable *is* an evolutionary event.

adaptive landscape metaphor, viz. the conditions under which form evolves are taken to be independent of the capacities of form. The external environment is autonomous from biological form and changes, if at all, "according to its own intrinsic dynamics" (Godfrey-Smith 2001, 254).

It is difficult to accommodate the contribution of organisms to evolution within the constraints of the old adaptive landscape trope. Pfennig et al. (2010), for example, try to incorporate the effects of plasticity into the adaptive landscape.

> Phenotypic plasticity also promotes population divergence by facilitating peak shifts or valley crossing on the adaptive landscape. ... a population can traverse a valley rapidly, potentially in one generation, by facultatively expressing an alternative phenotype closer to the fitness optimum. (Pfennig et al. 2010, 462)

To claim that adaptive evolution can readily cross valleys or 'shift peaks' is to concede that evolution does not follow the contours of the adaptive landscape. But the heuristic value of the adaptive landscape metaphor lies specifically in the notion that in adaptive evolution form follows the contours of the landscape, the topography of which is fixed by extra-organismal conditions. Where this relationship breaks down, so too does the utility of the adaptive landscape metaphor.

The affordance landscape metaphor, in contrast, nicely illuminates the relation between organismal plasticity and the conditions under which form evolves. In responding adaptively to conditions of existence, organisms alter their affordance landscapes. These altered affordances, in turn, redound to organisms. Plasticity, then, amplifies the mutual dependence of the capacities of form and the affordances on which it evolves. The plasticity of organisms is one of a number of factors— including genes and environments—that can alter the affordances upon which adaptive evolution occurs. Indeed, according to the affordance landscape metaphor, the contributions of organisms, genes and environments to evolution are no different in kind. They all contribute to the affordances to which organisms respond. That the affordance landscape metaphor can accommodate the role of plasticity in evolution and the adaptive landscape cannot commend the former over the latter.

4.3   The Occupancy of Affordance Space

If the adaptive landscape metaphor embodies a thoroughly classical, Newtonian conception of the occupancy of space, the affordance landscape suggests a radical alternative. The relation between form and the affordance landscape bears some commonalities with the relation between matter and space (well, spacetime) in General Relativity. In General Relativity, space is thought of as a substance. It interacts causally with the matter that occupies it. Most particularly, the geometrical properties of a region of space are not independent of whether that region is occupied. There is a relation of reciprocal dependence between the local properties of spacetime, and the capacities of matter (say, to attract other bodies). The trajectory of a body is influenced by the structure of spacetime around it, which in turn is altered by the motion of bodies.[20] The relations of access and relative proximity between spatial points are not independent of scale, or of the way that spacetime is occupied (Dainton 2001). In curved space, spatial relations change with scale. Because space has no univocal geometrical structure, there is no guarantee that the behaviour of bodies in spacetime is scale free.

Analogously, the features of the affordance landscape—its fitness, the relations of access and proximity between locations—are all influenced by the nature of the form that occupies it. For example, because of plasticity, a given phenotype may be underwritten by a range of developmental mechanisms. Each of these mechanisms, in turn, may have different phenotypic repertoires. Different instances of the same phenotype may differ in the range of conditions over which each is stable, and the kind of adaptive novelties that each can initiate (Ciliberti et al. 2007a, b; Wagner 2011). These novelties, in turn, vary with respect to *other* stable novelties that may be produced. So whether one phenotype is robustly fit, or is close to or accessible from another, may depend upon the way that form affects the array of affordances.

Because evolutionary trajectories are dependent not just on the independent structure of the landscape, but also upon the changes in form itself, there is no reason to suppose that the evolutionary dynamics are scale-independent. There is no reason to suppose that short-range (micro-evolutionary) changes can be extrapolated to long-range (macro-evolutionary) changes. One implication is that even if micro-evolutionary change is convergent and non-contingent, it doesn't follow that macro-evolutionary change is too. This is not to deny that there are convergences in macro-evolution. There certainly are (McGhee 2007). But when they occur they need special explanations.[21] They may be the results of constraints, or reflections of the fact that the same developmental resources are used in different lineages. The point is that convergences are *not* to be expected or explained simply by the fact that evolution is adaptive.

5 Adaptation and Contingency

The affordance landscape is offered here as an alternative spatial metaphor for adaptive evolution. But there is a problem. So different is the process of evolution on the affordance landscape from that on the adaptive landscape that it is questionable whether the former should rightly be considered *adaptive* at all. There are two causes for scepticism. The first concerns the concept of an adaptation. An adaptation is typically thought of as a solution to a 'design problem,' a self-standing, stable,

---

[20]One salutary implication, for both relativity and its metaphorical extension to evolution, is that it is not necessary to posit an additional metaphysical category of force to propel bodies through space.

[21]Perhaps ironically, Conway Morris (2010) offers a number of these.

external condition. Affordances are not external, self-standing or stable. "If … we abandon the metaphor of adaptation, how can we explain what seems the patent 'fit' of organisms and their external worlds?" (Lewontin 2001b, 63). The second cause for scepticism concerns the tempo and mode of evolution. As we saw, the adaptive landscape metaphor reinforces the intuition that large-scale evolution is convergent and non-contingent. We should expect that evolution on the affordance landscape, in contrast, might be contingent and non-convergent. Again, the question arises: "How can such a process be legitimately considered to be adaptive?"

I intend to address these concerns by means of an analogy to two different kinds of 'adaptive' games. In each game the challenge for the player on any given move is to solve a problem. Each of these games presents a player with an affordance landscape, a set of conditions that are propitious for, or impediments to, success. Yet, the relation between the player and the affordance landscape is different in each game. Consequently, the respective dynamics of the games are radically different.

The challenge faced by a sudoku player is to fill in the correct numbers in an array of spaces. The affordances of the sudoku game are the opportunities to fill the empty squares with numbers, and the existing array of filled-in squares. The affordances of the game are (for the most part) largely independent of the state of the game. For instance, what counts as the correct number in any given blank is independent of whether that blank is filled in, or which of the other squares the player has filled. The game has "its own intrinsic dynamics." Consequently, the affordances of the game and the state of the game are (largely) non-co-evolving. The trajectory of a sudoku game depends strongly on the initial conditions. As a consequence of this, the various trajectories of a Sudoku game are convergent and non-contingent. Reasonably adept players will arrive at the same solution to a given game, even if they do so by different routes.

So, the trajectory of a sudoku game is thus rather like the trajectory of adaptive evolution as suggested by the traditional adaptive landscape metaphor. It advances progressively toward the attainment of a pre-specified, unchanging solution to a self-standing problem, the nature of which can be described independently of the trajectory that approaches it. It is convergent and non-contingent because it is adaptive.

The challenge faced by a chess player is much different. At each move the player must respond to the threats and opportunities presented to her by the locations of the pieces, and by their capacities. The capacities of the player at a time and the affordances presented to her are mutually constituting: one cannot be specified without the other. They are also co-evolving. Each change in the location of the pieces, whether it is initiated by the player or the opponent, is a change in the affordance landscape.

The trajectory of a chess game is highly unpredictable. For most arrangements of the pieces on the board there are an enormous number of possible final outcomes. One reason for this is that a player typically has a broad adaptive repertoire. On any occasion, there are any number of moves available to the player that might promote her goals to some degree. Consequently, the trajectories of a chess game are contingent and non-convergent. It is highly unlikely that any two chess games will end up in precisely the same final configuration. Furthermore, unlike sudoku, the trajectory of a game does not depend very largely upon, nor is it predictable by, the initial conditions.

Exploring the contrast between sudoku and chess serves two purposes. First, the chess analogy suggests that we can consider a process to be adaptive even if it is not a progressive convergence upon a solution to an unchanging, independently specifiable problem. Chess moves are adaptive, even if successive moves do not progress toward a single solution. The 'metaphor' of adaptation may not be as constraining as we usually take it to be. Second, the comparison suggests that where an adaptive process is best characterized as the result of an adaptive entity embedded in an interacting system of affordances, there is little reason to expect that process to be convergent and non-contingent. Some processes are contingent and non-convergent *because* they are adaptive.

My suggestion throughout this paper has been that adaptive evolution is such a process. But if so then evolutionary thinking has been ill-served by its most prominent metaphor. The alternative affordance landscape metaphor underscores the important contribution that organisms, as purposive entities, make to adaptive evolution. Adaptive evolution is not to be conceived of as the moulding of passive form to meet the exigencies of an autonomous, external environment. It is the response of form to a mutually constituted set of affordances. Moreover, the affordance landscape metaphor suggests that the process of adaptive evolution should be contingent and non-convergent—not convergent and non-contingent as the adaptive landscape suggests it should be. This is a genuine empirical possibility that has been obscured from our view for much of the history of evolutionary biology, probably through overreliance on the power of a metaphor. Evolutionary thinking has been strongly conditioned by the adaptive landscape metaphor. It may well correctly disclose to us the kind of process that adaptive evolution may and may not be. I think it more likely that the predominant conception of adaptive evolution is just the penalty we incur for not having properly paid the price of metaphor.

6   Postscript: Forging a New Adaptationism

The adaptationist program—so sharply criticized by Gould and Lewontin (1979)— is predicated on a particular conception of the role of organisms in evolution. The properties and capacities of organisms are mere consequences of evolution. Organismal form is an object of evolutionary change. E. O. Wilson expresses the idea vividly. He says of human cognitive evolution,

> however subtle our minds, however vast our creative powers, the mental process is the product of a brain shaped by the hammer of natural selection upon the anvil of nature. (2004, xii)[22]

---

[22]I thank Chris Haufe for the quotation.

The blacksmith metaphor is telling. It neatly conveys the idea that form is a mere malleable substance, shaped by the extrinsic forces of evolution—the 'hammer of selection.' If organismal form contributes to the process of evolution it does so only by exerting resistance—the 'anvil of nature.'

But organisms are not mere objects of evolution. A new adaptationism, I suggest, must develop an understanding of organisms as *subjects* of evolution. Such an adaptationism would highlight the ways in which organisms "actively participate in their own evolution" (Ingold 2000, 292). Organisms are self-building, self-maintaining, purposive systems actively engaged in, commingled with, their conditions of existence. Adaptive evolution is the consequence of a constant dialectical interplay between organisms and their conditions; organisms change them and are changed by them. Perhaps an alternative blacksmith metaphor might be more germane. Hegel illustrates his theory of human freedom with "the old proverb that says 'Everyone is the smith who forges his own fortune.'"[23] Organisms are smiths who forge their own evolution.

Acknowledgments I would like to thank audiences in London (Ontario), Dubrovnik, and Stockholm for their lively responses. I am particularly grateful to Fermin Fulda and Susan Oyama for suggested improvements to an earlier draft. In addition, Jacob Stegenga, Cory Lewis, Michael Cournoyea, and Alex Djedovic provided a very helpful discussion. The editors of this volume also provided very helpful suggestions.

References

Barker, Gillian A. 2008. Biological levers and extended adaptationism. *Biology and Philosophy* 23: 1–25.

Beatty, John. 1995. The evolutionary contingency thesis. In *Concepts, theories, and rationality in the biological sciences*, ed. Gereon Wolters and James G. Lennox, 45–81. Pittsburgh: University of Pittsburgh Press.

Calder, William A. 1978. The kiwi. *Scientific American* 239: 132–142.

Chemero, Anthony. 2003. An outline of a theory of affordances. *Ecological Psychology* 15: 181–195.

Ciliberti, Stefano, Olivier C. Martin, and Andreas Wagner. 2007a. Robustness can evolve gradually in complex regulatory gene networks with varying topology. *PLoS Computational Biology* 3: e15.

Ciliberti, Stefano, Olivier C. Martin, and Andreas Wagner. 2007b. Innovation and robustness in complex regulatory gene networks. *Proceedings of the National Academy of Sciences of the United States of America* 104: 13591–13596.

Dainton, Barry. 2001. *Time and space*. Ithaca: McGill-Queen's University Press.

Dennett, Daniel C. 1995. *Darwin's dangerous idea: Evolution and the meanings of life*. New York: Touchstone.

Flyvbjerg, Henrik, and Benny Lautrup. 1992. Evolution in a rugged fitness landscape. *Physical Review A* 46: 6714.

---

[23]Hegel (1991, §147Z). Quoted in Yeomans (2012, 163). I thank Sally Sedgwick for the quotation.

Gerhart, John, and Marc Kirschner. 2007. The theory of facilitated variation. *Proceedings of the National Academy of Sciences* 104: 8582–8589.

Gibson, James J. 1979. *The ecological approach to visual perception*. Mahwah: Lawrence Erlbaum.

Gibson, Kathleen R. 1993. Animal minds, human minds. In *Tools, language and cognition in human evolution*, ed. Kathleen R. Gibson and Tim Ingold, 3–19. Cambridge: Cambridge University Press.

Gibson, Greg. 2002. Developmental evolution: Getting robust about robustness. *Current Biology* 12: R347–R349.

Godfrey-Smith, Peter. 1996. *Complexity and the function of mind in nature*. Cambridge: Cambridge University Press.

Godfrey-Smith, Peter. 2001. Organism, environment, and dialectics. In *Thinking about evolution: Historical, philosophical and political perspectives*, ed. Rama S. Singh, Costas B. Krimbas, Diane B. Paul, and John Beatty, 253–266. Cambridge: Cambridge University Press.

Gould, Stephen Jay. 1989. *Wonderful life: The Burgess Shale and the nature of history*. New York: W.W. Norton.

Gould, Stephen Jay. 2002. *The structure of evolutionary theory*. Cambridge, MA: Harvard University Press.

Gould, Stephen Jay, and Richard C. Lewontin. 1979. The spandrels of San Marco and the Panglossian paradigm: A critique of the adaptationist programme. *Proceedings of the Royal Society of London B* 205: 581–598.

Hallgrimsson, Benedikt, Katherine Willmore, and Brian K. Hall. 2002. Canalization, developmental stability, and morphological integration in primate limbs. *American Journal of Physical Anthropology* 119: 131–158.

Hamel, Amandine, Cathy Fisch, Laurent Combettes, Pascale Dupuis-Williams, and Charles N. Baroud. 2011. Transitions between three swimming gaits in Paramecium escape. *Proceedings of the National Academy of Sciences of the United States of America* 108: 7290–7295.

Harré, Rom. 1986. *Varieties of realism: A rationale for the natural sciences*. Oxford: Blackwell.

Haugeland, John. 1998. *Having thought: Essays in the metaphysics of mind*. Chicago: University of Chicago Press.

Hegel, Georg W.F. 1991. *The encyclopaedia logic: Part I of the encyclopaedia of philosophical sciences with the zusätze*. Indianapolis: Hackett.

Hutchinson, George E. 1978. *An introduction to population ecology*. New Haven: Yale University Press.

Ingold, Tim. 2000. *The perception of the environment: Essays on livelihood, dwelling and skill*. London: Routledge.

Kauffman, Stuart. 1993. *The origins of order: Self-organization and selection in evolution*. Oxford: Oxford University Press.

Kaźmierczak, Józef, and S. Kempe. 2004. Calcium build-up in the Precambrian sea: A major promoter in the evolution of eukaryotic life. In *Origins: Genesis, evolution and diversity of life*, ed. Joseph Seckbach, 339–345. Dordrecht: Kluwer.

King, Nicole. 2004. The unicellular ancestry of animal development. *Developmental Cell* 7: 313–325.

Kirschner, Marc, and John Gerhart. 2005. *The plausibility of life: Resolving Darwin's dilemma*. New Haven: Yale University Press.

Lande, Russell. 1976. Natural selection and random genetic drift in phenotypic evolution. *Evolution* 30: 315–334.

Lewontin, Richard C. 1963. Models, mathematics and metaphors. *Synthese* 15: 222–244.

Lewontin, Richard C. 1978. Adaptation. *Scientific American* 239: 212–230.

Lewontin, Richard C. 2001a. In the beginning was the word. *Science* 291: 1263–1264.

Lewontin, Richard C. 2001b. *The triple helix: Gene, organism, and environment*. Cambridge, MA: Harvard University Press.

Marzke, Mary. 1997. Precision grips, hand morphology, and tools. *American Journal of Physical Anthropology* 102: 91–110.

Maynard Smith, John. 1970. Natural selection and the concept of a protein space. *Nature* 225: 563–564.

McGhee, George. 2007. *The geometry of evolution: Adaptive landscapes and theoretical morphospaces*. Cambridge: Cambridge University Press.

Morris, Simon Conway. 2010. Evolution: Like any other science it is predictable. *Philosophical Transactions of the Royal Society B* 365: 133–145.

Nei, Masatoshi. 1972. Genetic distance between populations. *American Naturalist* 106: 283–292.

Nerlich, Graham. 1991. How Euclidean geometry has misled metaphysics. *Journal of Philosophy* 88: 169–189.

Newman, Stuart A. 2003. From physics to development: The evolution of morphogenetic mechanisms. In *Origination of organismal form: Beyond the gene in developmental and evolutionary biology*, ed. Gerd B. Müller and Stuart A. Newman, 221–239. Cambridge, MA: MIT Press.

Newman, Stuart A. 2011. Complexity in organismal evolution. In *Philosophy of complex systems*, ed. Cliff Hooker, 335–354. London: Elsevier.

Newman, Stuart A., and Ramray Bhat. 2009. Dynamical patterning modules: A "pattern language" for development and evolution of multicellular form. *International Journal of Developmental Biology* 53: 693–705.

Niklas, Karl J. 1997. Adaptive walks through fitness landscapes for early vascular land plants. *American Journal of Botany* 84: 16–25.

Odum, Eugene P. 1959. *Fundamentals of ecology*. Philadelphia: W. B. Saunders.

Pearce, Trevor. 2012. Convergence and parallelism in evolution: A neo-Gouldian account. *The British Journal for the Philosophy of Science* 63: 429–448.

Pfennig, David W., Matthew A. Wund, Emilie C. Snell-Rood, Tami Cruickshank, Carl D. Schlichting, and Armin P. Moczek. 2010. Phenotypic plasticity's impacts on diversification and speciation. *Trends in Ecology & Evolution* 25: 459–467.

Powell, Russell. 2007. Is convergence more than an analogy? Homoplasy and its implications for macroevolutionary predictability. *Biology and Philosophy* 22: 565–578.

Powell, Russell. 2012. Convergent evolution and the limits of natural selection. *European Journal for Philosophy of Science* 2: 355–373.

Purcell, Edward M. 1977. Life at low Reynolds number. *American Journal of Physics* 45: 101–111.

Rolian, Campbell, and Benedikt Hallgrímsson. 2009. Integration and evolvability in primate hands and feet. *Evolutionary Biology* 36: 100–117.

Rolian, Campbell, Daniel E. Lieberman, and Benedikt Hallgrímsson. 2010. The co-evolution of hands and feet. *Evolution* 64: 1558–1568.

Sanders, John T. 1993. Merleau-Ponty, Gibson, and the materiality of meaning. *Man and World* 26: 287–302.

Simpson, George G. 1944. *Tempo and mode in evolution*. New York: Columbia University Press.

Sloman, Aaron. 2000. Interacting trajectories in design space and niche space: A philosopher speculates about evolution. In *Parallel problem solving from nature – PPSN VI*, ed. Marc Schoenauer et al., 3–16. Berlin: Springer.

Stanley, Steven. 1998. *Macroevolution: Pattern and process*. Baltimore: Johns Hopkins University Press.

Sterelny, Kim. 2009. Novelty, plasticity and niche construction: The influence of phenotypic variation on evolution. In *Mapping the future of biology: Evolving concepts and theories*, ed. Anuouk, Barberousse, Michel, Morange, and Thomas, Pradeu, 93–110. Dordrecht: Springer.

Stoffregen, Thomas. 2003. Affordances as properties of the animal-environment system. *Ecological Psychology* 15: 115–134.

Turvey, Michael T. 1992. Affordances and prospective control: An outline of the ontology. *Ecological Psychology* 4: 173–187.

Wagner, Andreas. 2011. *The origins of evolutionary innovations: A theory of transformative change of living systems*. Oxford: Oxford University Press.

Walsh, Denis M. 2012. Situated adaptationism. In *The environment: Philosophy, science, ethics*, ed. William P. Kabasensche, Michael O'Rourke, and Matthew H. Slater, 89–116. Cambridge, MA: MIT Press.

West-Eberhard, Mary Jane. 2003. *Developmental plasticity and evolution*. Oxford: Oxford University Press.

Wilkins, Jon F., and Peter Godfrey-Smith. 2009. Adaptationism and the adaptive landscape. *Biology and Philosophy* 24: 199–214.

Wilson, Edward O. 2004. *On human nature. 25th anniversary*. Cambridge, MA: Harvard University Press.

Wright, Sewall. 1932. The roles of mutation, inbreeding, crossbreeding and selection in evolution. *Proceedings of the Sixth International Congress of Genetics* 1: 356–366.

Yeomans, Christopher. 2012. *Freedom and reflection: Hegel and the logic of agency*. Oxford: Oxford University Press.

Rethinking Behavioral Evolution

Rachael Brown

Abstract To date, the impact of the evo-devo "revolution" has been almost entirely restricted to the morphological domain—discussions of the role of contingency and development in the evolution of morphological traits being commonplace. In contrast, very little attention has been paid to contingency and development in the evolution of behavioural traits. This observation leads one to ask if there is any in-principle reason why this is the case. In this chapter, I respond to this question by motivating the application of the conceptual toolkit from evo-devo to the behavioural domain. I argue that there is evidence from inheritance of behaviour through social learning that demonstrates that development plays an important causal role in the evolution of behavioural traits. Furthermore, this evidence is as strong as, if not stronger than, analogous evidence used to motivate the evo-devo approach in the morphological domain. On these grounds, we should be just as willing to engage in the evo-devo research program when considering the evolution of behavioural traits as we are when considering the evolution of morphological traits.

1 Introduction

The emerging discipline of evolutionary developmental biology (also known as evo-devo) has been driven, in part, by evidence demonstrating the existence of non-genetic forms of inheritance that are developmentally derived—in particular

R. Brown (✉)
School of Philosophy, Australian National University, Canberra, Australia
e-mail: rachael.brown@anu.edu.au

G. Barker et al. (eds.), *Entangled Life*, History, Philosophy and Theory
of the Life Sciences 4, DOI 10.1007/978-94-007-7067-6_12,
© Springer Science+Business Media Dordrecht 2014

epigenetics or cell memory (Hall 2003; Müller 2008).[1] Where earlier theorists had supposed that evolution results only from natural selection acting on variation that is both produced and transmitted via genetic mechanisms, proponents of evo-devo argue that non-genetic developmental mechanisms can also contribute importantly to changes in the distribution of phenotypes in populations over time. Non-genetic inheritance provides important support for this claim.

Uptake of this new perspective has been uneven, however. Some areas of evolutionary study have seen vigorous engagement with the broad conceptual framework offered by evo-devo—the study of the evolution of morphological traits is a notable example. In contrast, those who work on the evolution of behavioral traits—especially workers in the fields of animal behavior, behavioral ecology and ethology—have shown scarcely any substantial engagement with the theoretical framework presented by evolutionary developmental biology (Ghalambor et al. 2010, 90; Bertossa 2011, 2056–2057).[2]

One reason for the lack of engagement with evo-devo in behavioral biology may be the perception that evo-devo is concerned solely with understanding the role development plays in morphological evolution. Evo-devo has its roots in the evolutionary embryology of the nineteenth century, a field focused on the embryonic foundations of morphology and body plans (Hall 2000; Müller 2008). These origins are still reflected in the way evo-devo is most often discussed today. Many key proponents of evo-devo still describe its explanatory focus as being the "origins of organismal form," and discuss key evo-devo concepts such as innovation and novelty in terms peculiar to morphology (for example, Müller and Newman 2005; Müller 2007, 2008, 2010). This focus upon morphological evolution gives the appearance that evo-devo could only ever be concerned with morphology, but the appearance is deceptive. Evo-devo at its heart is a science concerned with the relationship between development and evolution in general, not just morphological development and morphological evolution. Though much of the empirical work that has moved the field forward has focused on morphological cases, at a theoretical and conceptual level evo-devo is simply concerned with the influence of development upon phenotypic variation, regardless of the traits in question. Thus, if it can be shown that developmental processes play a role in the evolution of behavioral traits, just as they do for some morphological traits, then the existing conceptual framework offered by evo-devo is the obvious starting point for researchers wishing to understand behavior in light of this evidence.

---

[1] There are other threads of evidence supportive of an evo-devo research program besides evidence for non-genetic inter-generational inheritance that is developmentally derived. For example, the growing body of work on developmental constraint and the origins of body plans. Müller (2008) provides a nice overview of the many conceptual foundations of evo-devo that makes these different threads of evidence clear.

[2] There are some notable exceptions to this general trend. For example; Carroll and Corneli (1999), Gottlieb (2001), Sih et al. (2004a, b), Laland et al. (2008), Dingemanse et al. (2010), Ghalambor et al. (2010), Mery and Burns (2010), Bertossa (2011, 2056–2057).

In this chapter I seek to motivate those working in behavioral biology to engage with evo-devo, by pointing out some evidence demonstrating that developmental processes can play a role in the evolution of behavior. I begin the chapter with an overview of evo-devo, focusing on the relationship between the emerging conceptual framework it represents and the status quo within evolutionary biology—the "received" view or Modern Synthesis (this being the most prevalent account of evolutionary biology within behavioral ecology, ethology and animal behavior). I then show how non-genetic inter-generational forms of inheritance lend support to the evo-devo approach, using chromatin marking (a developmentally-derived form of epigenetic inheritance) as an example. In the second part of the chapter, I argue that a type of behavioral inheritance—social learning—presents a challenge to the Modern Synthesis analogous to that provided by epigenetics. Like chromatin marking, social learning is a non-genetic inheritance channel. It is a developmental process via which behavioral traits acquired during the lifetime of the parent can be transmitted to their offspring and subsequent generations, thus contributing to evolution. This interplay between the evolution and the developmental process of social learning is important to explaining the evolution of behavior in numerous species and thereby justifies the application of the evo-devo research approach to the behavioral domain.

2 Evo-Devo: Moving Away from the Modern Synthesis

2.1 What Is the Modern Synthesis?

Since the mid-twentieth century the dominant theory within evolutionary biology regarding how the requirements for selection are satisfied has been the Modern Synthesis (or "received view"). The Modern Synthesis is a general theory of evolution and includes claims both about the conditions that are necessary in principle for evolution by natural selection to take place, and about how these are actually instantiated.

In the simplest case, three necessary conditions must hold within a population for it to undergo evolution by natural selection—(I) phenotypic variation, (II) heritability of phenotype and (III) differential survival and reproduction. When these three conditions hold in any population of entities, evolution by natural selection is highly likely (Lewontin 1970, 76; Godfrey-Smith 2009). A fourth condition, (IV) cumulative selection, is required for the evolution of complex adaptations (Sterelny and Griffiths 1999). This requires that inheritance be stable over many generations. Without such stability of inheritance, the accumulation of beneficial mutations necessary for complex adaptation is not possible.

One key empirical claim of the Modern Synthesis is that the underlying biological structures enabling evolution by natural selection to occur are predominantly genetic (Huxley 2010; Mayr 1982, 542–546; Dobzhansky 1937, 26;

Fisher and Bennett 1999). In other words, genes are the major channel of inheritance for traits, and it is genetic mutation and recombination that provide heritable phenotypic variation within populations. Furthermore, the supply of variation generated by these mechanisms is taken to be largely isotropic (uniform in all directions) and thus unbiased with respect to adaptive value. A further conclusion is generally accepted along with these claims: that the large-scale evolutionary events in the tree of life (such as the emergence of novel capacities or morphological features) are simply the outcome of the accumulation of a series of small-scale events at the genetic level (Jablonka and Lamb 2005; Bonduriansky and Day 2009). But this is not (as the others are) an empirical claim; it is a simplifying assumption or idealization that allows evolutionary biologists to ignore features of the world which, according to proponents of the Modern Synthesis, are not of evolutionary significance on a large scale (Mayr 1982, 832). To illustrate—the Modern Synthesis is only committed to the empirical claim that the underlying biological structures enabling natural selection to occur are predominantly genetic, not that they are exclusively so. Proponents of the Synthesis do not deny that non-genetic structures capable of sustaining selection exist, nor that evolution via these routes of inheritance may occur. For example, Richard Dawkins (a prominent advocate of the Modern Synthesis) takes seriously the possibility that human culture can evolve by natural selection in his discussion of memetic evolution (Dawkins 1976). That non-genetic structures capable of sustaining selection are widespread enough to contribute to evolution beyond a few special contexts, however, is denied. The transmission of behavior via human culture is thus not included as a source of heritable phenotypic variation in the general account of evolution presented by the Modern Synthesis, because it is considered relevant only to a "special" and restricted class of species (i.e. humans and other higher primates) (Tomasello 1999a, b; Dawkins 2004). This idealization is a very strong one. Advocates of the Modern Synthesis are claiming that we need look only to natural selection and gene frequency change over time in order to explain the vast majority of evolution— nothing else is relevant to this task (Stebbins and Ayala 1981; Charlesworth 1996; Dawkins 2004; Mayr 1993).

One consequence of this idealized view of the evolutionary process is that proponents of the Modern Synthesis "bracket off" the study of development and ontogeny from the study of evolution (Sterelny 2000; Müller 2007; Jablonka and Lamb 2005; Gilbert et al. 1996). This key aspect of the Modern Synthesis is best represented by Ernst Mayr's widely-accepted proximate-ultimate distinction (Mayr 1961; Beatty 1994). According to Mayr (a key architect of the Modern Synthesis) when we look at the types of questions asked in biology we are able to identify two domains of enquiry—functional biology and evolutionary biology. Research in functional biology is fundamentally concerned with "how" questions, such as "how does a bat wing develop?" and "how does the bat wing work?" Mayr claims that answering these questions requires the functional biologist to uncover a particular type of cause—a proximate cause. Proximate causes "govern the responses of the individual (and his organs) to immediate factors of the environment" and include development, ontogeny, and agency (Mayr 1961, 1503). Research in evolutionary biology, on the other hand, concerns an entirely different set of questions—"why" questions, such as "why does the bat have wings?" and "why do bats and birds both have wings?" According to Mayr, answering "why" questions requires that the evolutionary biologist look to a different type of cause—an ultimate cause. Ultimate causes "are responsible for the evolution of a particular DNA code of information with which every individual of every species is endowed" (Mayr 1961, 1503). Natural selection is—unsurprisingly—the central ultimate cause. According to Mayr these two domains of enquiry (functional biology and evolutionary biology) are largely causally independent—they are concerned with answering fundamentally different questions; they answer them by appeal to different sets of causes; and the research within each of those domains can provide little (if any) explanatory traction upon the questions of the other. Thus, biologists who conflate these two classes of questions—for example, by attempting to respond to ultimate questions by considering development, or by attempting to respond to proximate questions by considering adaptation—are making a serious category mistake.

Mayr's view makes excellent sense within the context of the gene-focused view he advocated. If, as the Modern Synthesis assumes, the sole inheritance channel is the transmission of DNA in the germ-line cells, it follows that only traits and variants that arise from genetic mutation and recombination in the germ line can contribute to natural selection. Any other traits or variants that arise (for example those acquired via developmental or environmental processes) are not important in the evolutionary context, as they are assumed not to be heritable: while selection may act upon them, no evolutionary change can result. In this context the bracketing-off of developmental biology from evolutionary biology is entirely reasonable. If proponents of the Modern Synthesis are correct to presume that genetic inheritance is the only kind there is, then developmental processes are indeed causally independent from evolutionary processes and irrelevant to them.

2.2 Enter Evolutionary Developmental Biology

Evolutionary developmental biology (or evo-devo) challenges this aspect of the Modern Synthesis, for its subject matter is precisely the relationship between the developmental processes within individuals and evolutionary processes within populations (Hall 1999; Müller and Newman 2005). Proponents of evo-devo take seriously the potential for so-called proximate causes (in particular, developmental processes) to inform our understanding of evolutionary "why?" questions. They thus challenge Mayr's supposition that functional and evolutionary biology are effectively independent.

In particular, evolutionary developmental biologists investigate the potential for developmental mechanisms to contribute to and influence the supply of variation to natural selection. Key issues within evo-devo such as plasticity, novelty, innovation, and evolvability relate to the mechanisms supplying phenotypic variation to selection and the nature of that supply (Müller and Newman 2005). By incorporating both proximate and ultimate causes into their investigations, evolutionary developmental biologists make it possible to entertain a picture of evolution that includes both developmental processes and natural selection (Hall 1999; Müller and Newman 2005).

Proponents of evo-devo thus reject Mayr's sharp dichotomy, but most of them do so in a reformist rather than a revolutionary spirit, regarding the account of the evolutionary process provided by the Modern Synthesis as overly simplified rather than eschewing it altogether (Laland et al. 2011). In particular, the presumption that genes provide the sole route for phenotypic inheritance, and that development therefore has no place in evolution, is the consequence of a view of the nature of evolution that is overly simplified rather than wholly mistaken.[3] Proponents of evo-devo argue that developmental processes are important in structuring the supply of phenotypic variation to selection and thus can explain disparity in the tree of life: biases in the available variation can drive different populations' evolutionary trajectories in different directions (Müller 2008). They also claim that there are there are extra-genetic channels of inheritance that come into play during development and are not captured if we simply focus upon gene frequency change over time in populations (Jablonka and Lamb 2005). Thus, because it "brackets off" developmental processes from the processes of evolution, the Modern Synthesis cannot adequately capture the true nature of evolution and the actual role of natural selection in shaping the tree of life as we see it today (Pigliucci 2007, 2008, 2009; Pigliucci and Muller 2010). One key piece of evidence supporting this assessment of the Modern Synthesis by advocates of evo-devo comes from the study of epigenetics—the functioning of extra-genetic cellular entities that are heritable and apparently widespread (Hall 2003; Raff 2000; Müller 2007, 2008).

2.3 Epigenetics

The science of epigenetics or "cell memory" is the study of changes in phenotype or gene expression that are not generated by changes in the DNA of the cell (Hall 2011). A simple example of an epigenetic effect is seen during growth in our own bodies. As our bodies develop, the cells in our body (despite carrying the same DNA) change in morphology and physiology according to the role they are playing in the body. For instance, bone cells, brain cells, and skin cells all differ in appearance and action despite containing the same complement of DNA. Variation in the appearance and action of these cells is due to the differential activation of gene expression in the cells during development. Importantly for our purposes here,

---

[3]This is itself an oversimplification. Evo-devo is a broad church. There are many ways in which evo-devo can be said to challenge or disagree with the Modern Synthesis; I present but one here. The full extent and nature of the challenge evo-devo presents to the Modern Synthesis is as yet unresolved (Hall 2000; Laubichler 2009; Minelli 2009; Craig 2009, 2010a, b; Müller and Pigliucci 2010).

when such differentiated cells divide during growth, each daughter cell has the same activation pattern as its parent cell. The inheritance of the activation pattern from parent to daughter cell is not solely caused by the transmission of genetic material from parent cell to daughter cells. Epigenetic material—non-genetic cellular factors—must be transmitted as well. For our purposes what is important is that such mechanisms are widespread and, in some cases, have been shown to be capable of facilitating the type of cell-to-cell inheritance needed to satisfy the heredity requirement for natural selection (Grant-Downton and Dickinson 2006; Richards 2006; Jablonka and Lamb 2008; Jablonka and Raz 2009; Gilbert and Epel 2009; Richards et al. 2010). One example of this is chromatin marking.

Chromatin is the material that makes up chromosomes. In addition to DNA this includes non-genetic molecules such as RNA, proteins, and other molecules. The way that these non-genetic factors within the chromatin are distributed along the DNA making up any given chromosome is known as "chromatin marking." Crucially, this marking influences which genes on each chromosome are expressed and when. In other words, these non-genetic factors determine how the genetic code is read and interpreted. Some of these marks and the gene expression patterns they control have been shown to be heritable. The best studied of these is DNA methylation patterns.

DNA methylation is the attachment of a methyl group to the DNA within the chromosome. It is seen in all vertebrates and plants, and in some invertebrates, fungi, and bacteria (Jablonka and Lamb 2005). Methylation affects gene expression; the presence or absence of these methyl groups, and their density in a region of DNA, alters the likelihood of that region of DNA being transcribed. The transgenerational inheritance of methylation patterns and their phenotypic effects has been shown in asexual plants and single-celled organisms (Chong and Whitelaw 2004; Richards 2006; Hauser et al. 2011; Youngson and Whitelaw 2008) and amongst eukaryotes (Morgan et al. 1999; Rakyan et al. 2001; Crews et al. 2007; Anway and Skinner 2006; Cropley et al. 2006; Cuzin et al. 2008; Youngson and Whitelaw 2008).

In single-celled and asexually budding organisms the primary mode of inheritance for methylation patterns, known as structural inheritance, involves the transfer of elements of the parental cells to offspring during mitotic reproduction or binary fission (Jablonka and Raz 2009). In particular, when cells reproduce via meiosis, mitosis, or fission, the parent cell is cleaved into a number of daughter cells. In this process the parent cell is lost, but elements of its structural properties are conserved in the daughter cells. One such conserved structure is methylation patterns in the DNA. The conservation of these patterns results in the replication of phenotypic effects seen in the parent cells, in the offspring cells.

In sexually-reproducing multi-cellular organisms like mammals, the mechanisms of inheritance are less well understood. In these organisms many of the chromatin markings of cells in the offspring are "reprogrammed" rather than maintaining the parental state during reproduction and early embryonic development (Reik et al. 2001). While originally it was thought that this reprogramming completely ruled out the inheritance of epigenetic factors in sexual organisms, more recent research has shown that the erasure of methylation patterns is incomplete, at least with respect to

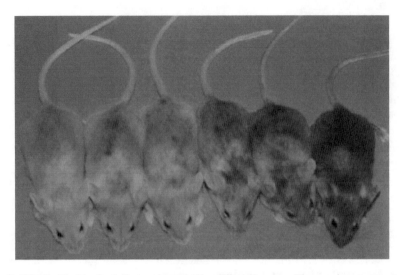

Fig. 1 Genetically identical (but epigenetically different) mice. These mice represent the continuum of phenotypes ranging from yellow on the left to agouti on the right depending on the level of activation of the agouti viable yellow (A^{vy}) allele (Source: Morgan et al. 1999) (Color figure online)

the maternal alleles (Morgan et al. 1999; Chong and Whitelaw 2004; Blewitt et al. 2006). The agouti viable yellow mouse allele provides the most famous example of this type of inheritance in mammals.

The mice in the picture above (Fig. 1) are genetically identical. They all carry two alleles at a genetic locus for wild-type coat colour—A^{vy} and a. What differs between them is the amount of folic acid that their mothers were fed during pregnancy. The smaller brown or "agouti" mice had dams that were fed food with relatively large amounts of folic acid during pregnancy and the other, large yellow mice, had dams that were not fed folic acid (Wolff et al. 1998). The different diets of the dams has led to the generation of distinct methylation patterns on the chromosomes of the embryos developing within them and thus to different phenotypes. In particular, the presence or absence of folate in the maternal environment induces changes to the DNA methylation patterns and this alters the extent to which the dominant "agouti viable yellow" allele (A^{vy}) within that embryo is "activated." The phenotype of A^{vy}/a mice is thus dependent on the level of activity of the A^{vy} allele and ranges from yellow (A^{vy} is strongly activated) through to agouti (A^{vy} is not active).[4]

The phenotypic effects of methylation at the A^{vy} allele are transgenerational. All A^{vy}/a mice are genotypically identical at the A^{vy} allele, yet within this class, agouti

---

[4]Note that A^{vy}/a agouti mice are often described as "pseudoagouti" so as to distinguish them from mice that lack the silent A^{vy} allele but also have the agouti phenotype (i.e. a/a agouti mice). Assume all agouti mice discussed here are A^{vy}/a mice.

dams are more likely than yellow dams to have agouti pups, and vice versa (Wolff et al. 1998; Morgan et al. 1999; Blewitt et al. 2006). It is clear that the inheritance of the phenotype from dam to pup results from the incomplete erasure of the epigenetic modification in the germ line cells (Cropley et al. 2006; Waterland et al. 2007), but the exact mechanism of inheritance is unclear. Although the methylation patterns that influence the expression of the A^{vy} allele in the dam are replicated in the juvenile offspring, those patterns are not seen in the offspring at the blastocyst stage of development, which is good evidence that the methylation patterns themselves are not being retained during cell reprogramming. Some other epigenetic element appears to be underwriting the inheritance here but it is not clear what it is (Blewitt et al. 2006). This problem is not restricted to the agouti viable yellow mouse allele; broader studies suggest that although the DNA methylation patterns are clearly important to the inheritance of the phenotype in a number of cases, some other element is also required for transmission (Daxinger and Whitelaw 2010; Jablonka and Raz 2009).

2.4 Lessons from Chromatin Marking

Chromatin marking challenges some key assumptions of the Modern Synthesis regarding the supply of phenotypic variation to selection—in particular, the assumption that phenotypic variation is exclusively supplied by genetic mechanisms and that the supply itself is isotropic. First, chromatin marking is a route of inheritance that is not genetic—while genes are necessary in such circumstances, they are not sufficient for the expression of the traits in question in the parent and their reiteration in further offspring; the particular chromatin marking patterns are also required. Furthermore, traits underwritten by chromatin marking are stable within lineages over multiple generations (Crews et al. 2007; Anway and Skinner 2006) and potentially widespread—chromatin marking is found in all cells with chromosomes. Chromatin marking thus presents a source of variation that is heritable and potentially capable of contributing to cumulative selection, challenging the assumption that the only evolutionarily significant source of phenotypic variation to selection is via genetic mutation and recombination. In addition, the supply of variation via chromatin marking is not isotropic. Because chromatin marking arises via the interaction between the organism and the environment, it offers a route via which the environment can potentially bias the supply of variation to selection (though not necessarily towards adaptive benefit). Chromatin marking therefore can potentially help to explain the divergent evolutionary trajectories that create disparity in the tree of life.

There are still many questions surrounding chromatin marking, however. First, just how widespread is inter-generational inheritance via the transmission of chromatin marking? Although chromatin marking is clearly heritable across generations within many single-celled organisms, it is not clear how widespread its inheritance is in multicellular organisms.[5] If it turns out to be very rare, then the gene-centrism of the Modern Synthesis could be justified. A second question relates to the stability and fidelity of the inheritance that chromatin marking offers. While it might be capable of maintaining selected traits over generations, can chromatin marking underwrite their persistence over the tens and hundreds of generations required for cumulative evolution and thus for the evolution of complex adaptations? If the evolutionary reach of chromatin marking is only very shallow, perhaps it can still be justifiably partitioned off from general evolutionary theory. These concerns cannot be ignored, but they alone do not threaten the science of evo-devo. Chromatin marking (like other epigenetic mechanisms) presents a potential alternative to genetic inheritance, and thus gives reason to question the idealizations of the Modern Synthesis even though at this time it does not conclusively discredit them. Epigenetic inheritance reveals the need for further research into the role of developmental factors in evolution and thus provides important motivation for the evo-devo research program (Müller 2008).

3   Rethinking Behavioral Evolution

Epigenetic chromatin marking has been shown to underwrite both morphological and behavioral traits, and thus motivates the application of the evo-devo approach across a broad range of contexts (Jablonka and Raz 2009; Bonduriansky and Day 2009). While this alone could be seen as sufficient to motivate an evo-devo of behaviour (Jablonka and Lamb 2005), research in the behavioral domain also offers distinctive motivations of its own. In particular, as I will argue, social learning is a widespread and evolutionarily efficacious route of inheritance for behavioral traits; an extra-genetic channel of inheritance for characters acquired during development. It is also potentially a source of bias in the supply of variation. Thus, I claim, understanding the role that social learning plays in evolution requires us to focus upon the interplay between the developmental processes within individuals (in particular, those that are affected by social learning) and the evolution of populations over many generations— i.e. upon the subject matter of evo-devo. Evidence that social learning acts as a route of behavioral inheritance thereby motivates the extension of the evo-devo research approach to the behavioral domain, while the distinctive features of social learning as a form of inheritance raise new questions for that approach.

I begin by discussing the view of social learning and evolution that is standard within behavioral biology—that inter-generational inheritance of behavioral traits via social learning sufficient to satisfy the requirements for cumulative selection is

---

[5]See Jablonka and Raz (2009) and Bonduriansky and Day (2009) for summaries of the factors found and the species in which they are found.

Fig. 2 Tinbergen's four
problems

1. Causation: How does it work?	2. Ontogeny: How did it develop?
3. Survival value: What is it for?	4. Evolution: Why did it evolve?

limited to only a few special cases in the animal kingdom. I then give an overview of the evidence demonstrating the widespread existence of animal traditions. Such traditions, I argue, represent a feasible source of non-genetic inter-generational inheritance of behavioral traits that is widespread and stable enough to bring behavioral evolution within the ambit of evo-devo.

3.1 Social Learning – The Traditional View

Within behavioral ecology, animal behavior, and ethology there has been little to no engagement with evolutionary developmental biology (Ghalambor et al. 2010, 90; Bertossa 2011, 2056–2057). Rather, the Modern Synthesis, and the approach to development and evolution it advocates, is the norm. This is demonstrated clearly by the central role accorded to Mayr's proximate-ultimate distinction and Niko Tinbergen's related "four questions" of ethology in these fields (Tinbergen 1951, 1963; Griffiths 2008; Manning 2005). Tinbergen argued that behavioral biology is best understood as responding to four interrelated problems or questions—Causation, Ontogeny, Survival Value and Evolution (Fig. 2)—each of which corresponds to a different aspect of the question "Why is behavior *x* as it is?" Tinbergen's taxonomy established a set of principles that have defined research in behavioral biology for more than half a century.

Textbooks and collections of papers on the foundations of animal behavior (Houck and Drickamer 1996); methodology in the studying of animal behavior (Martin and Bateson 2007; Lehner 1998; Manning and Dawkins 1998); behavioral evolution (Slater and Halliday 1994); behavioral ecology (Krebs and Davies 1997); cognitive evolution (Shettleworth 2010); and cognitive ethology (Allen and Bekoff 1999), not to mention hundreds of journal articles published across these fields, all refer to Tinbergen's "four questions." Most of them explicitly combine them with Mayr's distinction as in Fig. 3 (for example, Martin and Bateson 2007).[6]

---

[6]Note that Tinbergen never presented his questions in this manner. This (now prevailing) presentation of the four problems is thought to be originally due to Klopfer and Hailman (1972), Alcock (1975), and Dewsbury (1999).

Proximate	1. Causation:	3. Ontogeny:
"How questions"	How does it work?	How did it develop?
Ultimate	**2. Survival value:**	**4. Evolution:**
"Why questions"	What is it for?	Why did it evolve?

Fig. 3 Tinbergen's four problems with Mayr's proximate-ultimate distinction

Social learning—learning that involves the use of information gained from other individuals such as the location of food sources or successful behavioral patterns[7]—encompasses a range of different learning processes ranging from true imitation to stimulus enhancement (see Brown and Laland 2003, Table 1 for a summary). It is accepted that such learning is widespread in the animal kingdom, in species from insects (Leadbeater and Chittka 2007) to complex vertebrates (Galef Jr and Laland 2005). What is less accepted is that such learning is a route via which behavioral traits can be inherited over multiple generations and thus contribute to the outcomes of evolution. Rather, those interested in the evolution of animal behavior generally assume that the inter-generational inheritance of behavior within animal lineages is solely genetic. This is unsurprising given the widespread acceptance of the Modern Synthesis within behavioral biology.

Yet social learning obviously constitutes a channel through which behavioral patterns can be transmitted from one individual to another; a sort of inheritance. Why omit it from the evolutionary picture? Several widely-accepted assumptions about social learning are commonly invoked to justify this omission. To begin with, it is often argued that most social learning processes lack the fidelity and stability required to underpin the persistence of traits required for the evolution of cumulative and complex behaviors via natural selection; that only true imitation—the explicit copying of the behavior of others—and teaching can support the persistence of traits long enough for cultures or traditions to evolve (Tomasello 1999a, b).[8] Such

---

[7] I am using a very general notion of social learning here. It is worth noting that sometimes imitation (learning about behavior via conspecific observation) is distinguished from social learning (learning information about the environment through social observation) (e.g. Heyes 1993). As I will show, sometimes learning information about the environment through social observation is sufficient for individuals to learn indirectly about the behavior of others.

[8] There is some debate about whether the cases put forward as examples of animal "culture" are truly cultural or something else. In particular, within human societies is generally accepted to that culture is the cumulative product of the transfer of small variations in behavior from one generation to the next via social learning over multiple generations. There is some disagreement as to the extent to which the animal cases satisfy the cumulative aspect of this requirement. For example, the debate as to the extent to which chimpanzees can be considered to have culture (Tomasello 1994, 1999a, b; Boesch 1996, 2003; Whiten et al. 1999; McGrew 1998; Whiten 2005; Langergraber et al.

imitation, it is then claimed, is cognitively demanding and therefore limited to a few "special cases"—the few species that possess what are thought to be the appropriate cognitive capacities, such as humans and primates (e.g. Galef 1992). The assumption is thus that evolutionarily efficacious inheritance of behaviors by social learning is the rare exception, not the rule (Jablonka and Lamb 2005; Laland and Janik 2006; Avital and Jablonka 2000). Recent evidence of social learning and animal traditions beyond primates challenges this assumption, calling into question the exclusive focus upon genetic inheritance and the bracketing-off of development by those interested in the evolution of behavioral traits.

3.2 Social Learning – Challenging the Traditional View

Two types of evidence support the view that the inheritance of behaviors via social learning is both widespread and evolutionarily significant: first, evidence of the existence of culture and traditions beyond the primate lineage; second, evidence that culture and traditions can be generated by relatively simple learning mechanisms.

Evidence of culture and traditions in non-human animals comes from several sources. What is important is demonstrating that behavior is being transferred from one generation to the next via learning rather than via genes. A few different types of evidence are relevant here (in part due to Laland and Janik 2006):

1. The speed at which behaviors infiltrate a population: Behaviors that are produced by genetic mutation are generally slow to invade populations because, even when under very strong selection, they are transferred between individuals only during reproduction. If behaviors are heritable by social learning, however, they can feasibly invade entire populations within a generation.
2. The speed at which behaviors are lost from populations: Socially learned behaviors are more likely to suffer a rapid decline than genetically heritable behaviors. This is because the persistence of socially learned behaviors in populations is much more contingent on the environment and random events. In particular, there is no silent transmission of behavioral traits—if a trait is not expressed it cannot be learned by others. This contrasts with genetic traits, which can be recessive and skip generations.
3. The arbitrariness of persistent behavioral traits: That arbitrary or maladaptive differences in behavior take hold in populations and persist is a good indicator of the social transmission of behavior between individuals, as via genetic inheritance alone we would not expect the traits to reach fixation because they play no adaptive role (Tomasello 1994, 274–275). Persistent variations between

---

2011). As a consequence of these debates some researchers refer to the less complex culture-like animal cases as "traditions." See Avital and Jablonka (2000, 21–23) for a discussion of this.

groups or geographic "clines" in behavior within spatially distributed groups that are not explicable by differences in environment are usually used as an indicator of arbitrariness.[9]

4. The outcomes of "cross-fostering" experiments: This type of experiment gives us information about the extent to which differences in behavior are due to maternal environment or to genetics. To illustrate, imagine two populations of the same species in very similar environments. The majority of individuals in one of those populations exhibit a foraging technique not seen in the other population—in this case, nut cracking. Furthermore the technique is known to have persisted in the population over multiple generations. We might want to test whether the variation in nut cracking capacity between the populations is the product of a genetic difference or a difference in "culture."[10] One way to do this is to look into the inheritance of the nut cracking behavior by doing cross fostering experiments (i.e. fostering the offspring of non-nut-cracking mothers with nut-cracking mothers and vice versa). If the nut cracking behavior is largely genetically inherited, we should not see it in those individuals taken from the non-nut-cracking population and raised in the nut-cracking population. This is because they should lack the appropriate genes. Conversely if the nut cracking behavior is inherited via some social means we should see it only in those individuals raised in the nut-cracking population. Those raised in the non-nut-cracking population will lack the technique because the appropriate social stimulus is unavailable to them in that environment.

It is generally accepted that these types of evidence exist for some primates, particularly chimpanzees, *Pan troglodytes* (Whiten et al. 1999; Whiten 2005). What

---

[9]Looking to human behaviors can help illustrate what is meant by arbitrariness here. Human cultural groups vary in their properties in many ways. For example, cultural groups often differ with respect to their rituals relating to burial of the dead. In some cultures individuals are buried with funerary offerings and other items. In other cultures, burial is very simple, individuals being buried without objects at all. Similarly, in some cultures burial must be soon after death and in others can be delayed. Variation in burial rituals between cultures can, in some cases, be put down to differences in environment. For example, a practice of burying bodies swiftly is unsurprising in a very hot region as bodies present a greater disease threat in warmer climes than colder ones. Burying bodies swiftly is adaptive if it is warm. Not all burial practices, however, are coupled to selection for suitability to the environment; some are arbitrary in this respect. A good example of this is the provision of funerary offerings or absence thereof. Differences between groups in such practices do not seem to be due to differential selective regimes. Rather, they are due to contingent social mores and individual choices. Because they are not adaptive the only explanation for the presence of those behaviors in populations is social choice and social learning. In other words, if burial offerings were genetically derived and inherited we would not expect them to reach fixation in populations as they have no benefit. This is not to say that swift burial must therefore be a product of genetic inheritance (it seems likely that it is not). Arbitrariness in the funerary offerings case simply provides support for the social learning hypothesis not available in the swift burial case.

[10]We can rule out environment here as being the difference maker as the two populations are in similar environments.

is less accepted is the idea that some or all of the characteristics of culture and traditions are seen in other species, even those very far removed phylogenetically from primates. Among the most famous challenges to this view is the tool use of the New Caledonian crow (*Corvus menuloides*).

New Caledonian crows manufacture and use stick- and leaf-based tools for foraging. These tools are complex and their successful use requires skill. There is strong evidence that a significant portion of tool use and manufacture in New Caledonian crows is maintained in populations via social learning and that the complexity of the tools is a product of cultural evolution. First, there is evidence of geographic clines in tool design and manufacture (Hunt and Gray 2003, 2007). Second, there is evidence of the relatively fast transmission of innovative behaviors within groups (Holzhaider et al. 2010). Third, there is evidence that social learning plays an important role in the development of the tool use and manufacture behaviors in juvenile crows similar to that played in the development of stick-fishing and nut-cracking technologies in chimpanzees (Holzhaider et al. 2010; Kenward et al. 2006).

While the crow case is particularly impressive for the cultural complexity involved and demonstrates that the cultural achievements of primates are not unique, it is not a demonstration that animal culture is widespread across different taxa. Simpler animal cultures and traditions have been more widely observed however. The multi-generational transmission of elements of songs and vocalizations via social learning is observed in birds (Kroodsma and Krebs 1980; Podos and Warren 2007; Mundinger 1982) and whales (Deecke et al. 2000; Yurk et al. 2002). Similarly, simple foraging techniques and innovations have been shown to persist within populations and be transmitted between generations via social learning in rodents (Aisner and Terkel 1992), various birds (Lefebvre and Bouchard 2003), and even fish (Laland et al. 2003; Brown and Laland 2003). These cases give grounds for thinking that animal cultures could be widespread. They also give grounds for questioning the supposition that behavioral inheritance via social learning is limited to cases of imitation and teaching. Fish provide a particularly good example of this.

Fish are traditionally thought to be fairly cognitively limited organisms, but recent studies have shown that fish are able to recognise and remember their shoal-mates, foraging and nest locations, and navigational routes. There is also evidence that fish learn via stimulus enhancement and social exposure—relatively cognitively simple processes (see Laland et al. 2003 and Brown and Laland 2003 for a review of the evidence in both cases). In some fish, researchers suggest that this evidence is sufficient to demonstrate the existence of traditions that persist across many generations. A good example of social learning in fish is seen in the bluehead wrasse (*Thalassoma bifasciatum*). In this species, information about the location of arbitrarily determined mating sites and how to get to them is transmitted from older to younger fish via simple learning mechanisms. Individuals learn the location of traditional mating sites by observing and following others rather than via more cognitively demanding means of social transmission such as explicit copying or "true" imitation. The use of such sites is determined not by genes but by the maintenance of information about those sites in the lineage via social learning, and there is evidence that the use of such sites can span several generations (Laland and Janik 2006). Similar mechanisms have been shown to maintain foraging behaviors in other species, such as milk-bottle opening in some birds (Sherry and Galef 1984, 1990; Lefebvre and Bouchard 2003). These cases suggest that social learning is widespread, and can be underpinned by simple mechanisms. This lowers the bar for achieving inheritance via social learning considerably.[11]

3.3 Stable Inheritance via Social Learning

One challenge to the picture I have presented here concerns the fidelity and stability of social learning as a channel of inheritance. It is clear that behavior can be transferred between individuals via social learning over multiple generations, but what evidence is there that this route of inheritance provides sufficient fidelity of transfer or stability over multiple generations to satisfy the requirements for cumulative selection?

Some features of social learning seem likely to undermine its ability to serve as an evolutionarily-significant channel of inheritance. First, unlike genetic inheritance, the inheritance of behavior via social learning cannot be silent: behavioral traits must be expressed in order to be transmitted. This makes the persistence of the traits far more fragile or sensitive to changes in the environment than in the case of genetic inheritance. Second, copying fidelity is crucial. Although copying failure is a potential source of novel behavioral variation in populations, it can also swamp out the effects of selection. This is a potential problem for genetic inheritance as well (a species with a very high mutation rate could not undergo cumulative selection), but simple social learning mechanisms such as stimulus enhancement appear particularly prone to copying error. Although these concerns are legitimate, they are not conclusive. Considering the role of niche construction may reveal mechanisms by which social learning can achieve robustness and high fidelity without the need for cognitively-demanding forms of imitation.

Niche construction refers to the ability of organisms to define, alter and build their own environments (Odling-Smee et al. 2003). The effects of niche construction can endure or accumulate over many generations, so that organisms inherit not just genetic information but features of their selective environment in what is known as "ecological inheritance." Proponents argue that niche construction can have important evolutionary impacts, altering the course of the evolution for niche-constructing species (and for other species with which they interact) by generating long-term changes to the environmental elements of the selective regime. Examples of niche construction include the manufacture of nests, burrows and webs by animals, the alteration of atmospheric gases by plants, and the fixation of nutrients by bacteria.

---

[11] Social learning has even been attested in non-colonial invertebrates, e.g., Coolen et al. (2005) and Mery et al. (2009).

Niche construction has the potential to increase the robustness and fidelity of social inheritance. Niche construction in this context is best thought of as "scaffolding" for inheritance, capable of buffering social inheritance mechanisms from changes in the environment. It aids social inheritance, for example, by reducing the likelihood of losing cues or materials required for the transfer of traits. Niche construction can also increase the likelihood of high-fidelity transfer by adding redundancy to the system. Niche construction coupled with social learning can thus provide a more effective route of inheritance. A good example of both these outcomes is seen in the New Caledonian Crow case.

We've noted already that New Caledonian Crow tool use is a particularly good example of animal culture outside the primate lineage. One way in which juvenile New Caledonian Crows learn to use tools is by interacting with the discarded tools of adult crows. Juvenile crows are naturally interested in the tools. They pick them up and carry them about. They also use them to mimic adults' use of them (Holzhaider et al. 2010; Kenward et al. 2006). By interacting with their environment, New Caledonian Crows have set up a situation (unintentionally, of course) in which juvenile crows are able to gain familiarity with tool structure and manipulation before they even begin to make tools themselves. In this way, the discarded tools of adult crows provide a type of ecological scaffolding for the development of tool use in subsequent generations—i.e. the simple addition of discarded tools to the developmental environment of juveniles makes the effective transmission of tool use and manufacture via social learning more likely, increasing both the fidelity and the robustness of the transfer. It is worth noting that a similar phenomenon is observed in chimpanzee troops that engage in tool use (Tomasello 1994).

The role of niche construction in structuring the environment so as to facilitate the transfer of learned behaviors is not limited to cognitively complex species. Meerkats (*Suricata suricatta*) exhibit a type of niche construction in the transmission of foraging techniques. Meerkats eat scorpions. While they are a good source of protein, scorpions are also a potentially very costly prey because of their sting (which carries enough neurotoxin to kill an adult meerkat). Meerkats use particular predation techniques for scorpions that involve disabling the sting. Interestingly— given the high costs of failure—these foraging techniques are learned. A form of niche construction scaffolds this learning. Adult meerkats modify the juvenile learning environment by presenting their offspring with live scorpions from which the adults have removed the stings. This is enables the naive foragers to learn from adults how to catch and disable scorpions, without the risk of a high cost sting (Thornton and McAuliffe 2006). Once again, niche construction here increases the robustness and fidelity of transmission of behaviors between generations. The stability of the transfer of behaviors via social learning is sensitive to the costs of failure or transfer error—if a behavior is only transferrable via a costly or dangerous learning situation its persistence in a population is fragile. Scaffolding the juvenile learning environment via the provision of "safe" prey items reduces the costs of learning in the meerkat, at least during the initial learning phase, and thus increases the effectiveness of social learning as a form of inheritance.

Though the exact extent of social learning and its evolutionary reach remain unclear, niche construction provides a potential source of stability for social learning as a route of inheritance. Overall, the cases I have described provide good grounds for rethinking the relationship between development and selection in evolution. There is reason to question the standing assumption in behavioral biology that inheritance via social learning plays little role in the evolution of behavioral traits. This evidence, like that of epigenetic inheritance, undermines the "bracketing-off" of development from studies of behavioral evolution. As with epigenetics, inheritance via social learning is developmentally derived—it is the consequence of experience rather than the transfer of genetic material from parent to offspring. It is thus a form of inheritance that is not captured if development is ignored in evolutionary biology. Social learning is also like epigenetics in that it is not totally understood. While there is evidence that social learning could be an important channel of inheritance it is not clear how important it actually is. More research is required to establish how widespread inheritance via social learning is, and to determine its "evolutionary reach." Evo-devo is a science concerned with the relationship between the developmental processes within individuals and evolutionary processes within populations, and thus, research considering the relationship between the developmental process of social learning and evolution falls broadly within its purview. It is in this sense that we should be asking why social learning has not motivated an evo-devo approach to behavior if epigenetic research motivates the application of evo-devo principles elsewhere in evolutionary biology.

4 Conclusion

In this chapter I have argued that, like chromatin marking, social learning presents an important challenge to the bracketing off of development from evolution in the Modern Synthesis by being an example of a non-genetic inheritance route which is active during development. In doing this I have shown that there is clear motivation for the application of the evo-devo research framework in the behavioral domain. We already know that behavioral traits can be transferred over multiple generations via chromatin marking. This alone might be considered sufficient to motivate a reintegration of the study of behavioral development and behavioral evolution. Showing, in addition, that behavioral traits can also be transferred over multiple generations via social learning adds support to the claim that those interested in behavioral evolution should take seriously the interplay between development and evolution. Evo-devo, as a science that does take this interplay seriously, is the obvious place for behavioral biologists to begin the study of the evolutionary developmental biology of behavior.

Moving to an evo-devo of behavior will require openness to change amongst both behavioral biologists and evolutionary developmental biologists. For behavioral biology it means thinking more about the developmental systems underpinning behaviors and the ways in which this could influence evolution. This will require untangling the messy interplay between genetics, development and the environment and a new way of looking at the proximate-ultimate questions of behavioral biology (Laland et al. 2011). This will be challenging, both conceptually and methodologically.

For evo-devo there will also be challenges. The focus of evo-devo research to date has been upon morphological traits and their evolution. Because of this, many key terms and concepts used in evo-devo are tailored to suit this agenda and thus are not directly applicable to behavior. For example, the concepts of novelty and innovation in evo-devo are standardly defined in a manner that explicitly makes reference to variation in morphological features such as the metazoan body plan or anatomy (e.g. Müller 2010, Table 12.1). Such concepts are not directly applicable to behavioral traits. New evo-devo concepts that are either more general or designed specifically with behavior in mind will be needed.

While these challenges are real barriers to the use of the evo-devo research framework in the behavioral domain, the benefits of moving away from the standard approaches to behavioral biology towards an approach that integrates our understanding of development and evolution are potentially very large. For example, not only does social learning present a route of inheritance that is developmental; it is also a route of inheritance that is very often biased to adaptive value. Learning is dependent upon the perceived benefit of the behavior being learned by the organism. Learned behaviors that are beneficial are maintained during the lifetime of the organism and behaviors that fail to be beneficial or lose their value tend to be lost. Unlike genetic inheritance that is blind to adaptive value, social learning is consequently a potential source of bias in the supply of variation to selection. It may increase the rate at which adaptations evolve (for example) and thus drive the emergence of disparity in behavior in the tree of life. Thus, taking into account a role for social learning as a route of inheritance may help to explain many aspects of behavioral evolution, including the persistence of behavioral traits in populations, the rapid loss of behaviors in populations, differences in the rates of behavioral evolution between lineages and differences in extinction rates.

Acknowledgements Thank you to Gillian Barker, Trevor Pearce and Eric Desjardins for their encouragement, thoughtful comments and assistance editing this manuscript. Also to Kim Sterelny, Brett Calcott, Peter Godfrey-Smith, the Sydney-ANU Philosophy Group (particularly Paul Griffiths and Karola Stotz) and participants at the 2010 ISHPSSB Off-Year Workshop "Integrating Complexity: Environment and History" and for valuable discussion and remarks on the ideas presented here.

References

Aisner, Ran, and Joseph Terkel. 1992. Ontogeny of pine cone opening behaviour in the black rat, Rattus rattus. *Animal Behaviour* 44: 327–336.

Alcock, John. 1975. *Animal behavior: An evolutionary approach*, 1st ed. Sunderland: Sinauer Associates Massachusetts.

Allen, Colin, and Marc Bekoff. 1999. *Species of mind: The philosophy and biology of cognitive ethology*. Cambridge, MA: MIT Press.

Anway, Matthew D., and Michael K. Skinner. 2006. Epigenetic transgenerational actions of endocrine disruptors. *Endocrinology* 147: s43–s49.

Avital, Eytan, and Eva Jablonka. 2000. *Animal traditions: Behavioural inheritance in evolution*. Cambridge: Cambridge University Press.

Beatty, John. 1994. The proximate/ultimate distinction in the multiple careers of Ernst Mayr. *Biology and Philosophy* 9: 333–356.

Bertossa, Rinaldo C. 2011. Morphology and behaviour: Functional links in development and evolution. *Philosophical Transactions of the Royal Society B: Biological Sciences* 366: 2056–2068.

Blewitt, Marnie E., Nicola K. Vickaryous, Andras Paldi, Haruhiko Koseki, and Emma Whitelaw. 2006. Dynamic reprogramming of DNA methylation at an epigenetically sensitive allele in mice. *PLoS Genetics* 2: e49.

Boesch, Christophe. 1996. Three approaches for assessing chimpanzee culture. In *Reaching into thought: The minds of the great apes*, ed. Anne E. Russon, Kim A. Bard, and Sue Taylor Parker, 404–429. Cambridge: Cambridge University Press.

Boesch, Christophe. 2003. Is culture a golden barrier between human and chimpanzee? *Evolutionary Anthropology: Issues, News, and Reviews* 12: 82–91.

Bonduriansky, Russell, and Troy Day. 2009. Nongenetic inheritance and its evolutionary implications. *Annual Review of Ecology and Systematics* 40: 103–125.

Brown, Culum, and Kevin N. Laland. 2003. Social learning in fishes: A review. *Fish and Fisheries* 4: 280–288.

Carroll, Scott P., and Patrice S. Corneli. 1999. The evolution of behavioral norms of reaction as a problem in ecological genetics. In *Geographic variation in behaviour: Perspectives on evolutionary mechanisms*, ed. Susan A. Foster and John A. Endler, 52–68. New York: Oxford University Press.

Charlesworth, Brian. 1996. The good fairy godmother of evolutionary genetics. *Current Biology* 6: 220.

Chong, Suyinn, and Emma Whitelaw. 2004. Epigenetic germline inheritance. *Current Opinion in Genetics & Development* 14: 692–696.

Coolen, Isabelle, Olivier Dangles, and Jérôme Casas. 2005. Social learning in noncolonial insects? *Current Biology* 15: 1931–1935.

Craig, Lindsay R. 2009. Defending evo-devo: A response to Hoekstra and Coyne. *Philosophy of Science* 76: 335–344.

Craig, Lindsay R. 2010a. The so-called extended synthesis and population genetics. *Biological Theory* 5: 117–123.

Craig, Lindsay R. 2010b. Criticism of the extended synthesis: A response to Müller and Pigliucci. *Biological Theory* 5: 395–396.

Crews, David, Andrea C. Gore, Timothy S. Hsu, Nygerma L. Dangleben, Michael Spinetta, Timothy Schallert, Matthew D. Anway, and Michael K. Skinner. 2007. Transgenerational epigenetic imprints on mate preference. *Proceedings of the National Academy of Sciences of the United States of America* 104: 5942–5946.

Cropley, Jennifer E., Catherine M. Suter, Kenneth B. Beckman, and David I.K. Martin. 2006. Germ-line epigenetic modification of the murine Avy allele by nutritional supplementation. *Proceedings of the National Academy of Sciences of the United States of America* 103: 17308–17312.

Cuzin, François, Valérie Grandjean, and Minoo Rassoulzadegan. 2008. Inherited variation at the epigenetic level: Paramutation from the plant to the mouse. *Current Opinion in Genetics & Development* 18(2): 193–196.

Dawkins, Richard. 1976. *The selfish gene*. Oxford: Oxford University Press.

Dawkins, Richard. 2004. Extended phenotype–but not too extended. A reply to Laland, Turner and Jablonka. *Biology and Philosophy* 19: 377–396.

Daxinger, Lucia, and Emma Whitelaw. 2010. Transgenerational epigenetic inheritance: More questions than answers. *Genome Research* 20: 1623–1628.

Deecke, Volker B., John K.B. Ford, and Paul Spong. 2000. Dialect change in resident killer whales: Implications for vocal learning and cultural transmission. *Animal Behaviour* 60: 629–638.

Dewsbury, Donald A. 1999. The proximate and the ultimate: Past, present, and future. *Behavioural Processes* 46: 189–199.

Dingemanse, Niels J., J.N. Anahita, Denis Réale Kazem, and Jonathan Wright. 2010. Behavioural reaction norms: Animal personality meets individual plasticity. *Trends in Ecology & Evolution* 25: 81–89.

Dobzhansky, Theodosius G. 1937. *Genetics and the origin of species*. New York: Columbia University Press.

Fisher, Ronald A., and Henry J. Bennett. 1999. *The genetical theory of natural selection: A complete variorum edition*. New York: Oxford University Press.

Galef, Bennett G. 1992. The question of animal culture. *Human Nature* 3: 157–178.

Galef Jr. Bennett G., and Kevin N. Laland. 2005. Social learning in animals: Empirical studies and theoretical models. *BioScience* 55: 489–499.

Ghalambor, Cameron K., Lisa M. Angeloni, and Scott P. Carroll. 2010. Behaviour as phenotypic plasticity. In *Evolutionary behavioral ecology*, ed. David F. Westneat and Charles W. Fox, 90–107. New York: Oxford University Press.

Gilbert, Scott F., and David Epel. 2009. *Ecological developmental biology: Integrating epigenetics, medicine, and evolution*. Sunderland: Sinauer.

Gilbert, Scott F., John M. Opitz, and Rudolf A. Raff. 1996. Resynthesizing developmental and evolutionary biology. *Developmental Biology* 173: 357–372.

Godfrey-Smith, Peter. 2009. *Darwinian populations and natural selection*. Oxford: Oxford University Press.

Gottlieb, Gilbert. 2001. *Individual development and evolution: The genesis of novel behaviour*. London: Lawrence Erlbaum.

Grant-Downton, Robert T., and H.G. Dickinson. 2006. Epigenetics and its implications for plant biology 2. The 'epigenetic epiphany': Epigenetics, evolution and beyond. *Annals of Botany* 97: 11–27.

Griffiths, Paul E. 2008. Ethology, sociobiology, and evolutionary psychology. In *A companion to the philosophy of biology*, ed. Sahorta Sarkar and Anya Plutynski, 393–414. Oxford: Wiley-Blackwell.

Hall, Brian K. 1999. *Evolutionary developmental biology*. Dordrecht: Kluwer.

Hall, Brian K. 2000. Guest editorial: Evo-devo or devo-evo—Does it matter? *Evolution & Development* 2: 177–178.

Hall, Brian K. 2003. Evo-devo: Evolutionary developmental mechanisms. *International Journal of Developmental Biology* 47: 491–496.

Hall, Brian K. 2011. A brief history of the term and concept epigenetics. In *Epigenetics: Linking genotype and phenotype in development and evolution*, ed. Benedikt Hallgrímsson and Brian K. Hall, 9–13. Berkeley: University of California Press.

Hauser, Marie-Theres, Werner Aufsatz, Claudia Jonak, and Christian Luschnig. 2011. Transgenerational epigenetic inheritance in plants. *Biochimica et Biophysica Acta* 1809: 459–468.

Heyes, C.M. 1993. Imitation, culture and cognition. *Animal Behaviour* 46: 999–1010.

Holzhaider, Jennifer C., Gavin R. Hunt, and Russell D. Gray. 2010. Social learning in New Caledonian crows. *Learning & Behavior* 38: 206–219.

Houck, Lynne D., and Lee C. Drickamer. 1996. *Foundations of animal behavior: Classic papers with commentaries*. Chicago: University of Chicago Press.

Hunt, Gavin R., and Russell D. Gray. 2003. Diversification and cumulative evolution in New Caledonian crow tool manufacture. *Proceedings of the Royal Society of London B* 270: 867–874.

Hunt, Gavin R., and Russell D. Gray. 2007. Parallel tool industries in New Caledonian crows. *Biology Letters* 3: 173–175.

Huxley, Julian S. 2010. *Evolution: The modern synthesis*. Cambridge, MA: MIT Press.

Jablonka, Eva, and Marion J. Lamb. 2005. *Evolution in four dimensions: Genetic, epigenetic, behavioral, and symbolic variation in the history of life*. Cambridge, MA: MIT Press.

Jablonka, Eva, and Marion J. Lamb. 2008. Soft inheritance: Challenging the modern synthesis. *Genetics and Molecular Biology* 31: 389–395.

Jablonka, Eva, and Gal Raz. 2009. Transgenerational epigenetic inheritance: Prevalence, mechanisms, and implications for the study of heredity and evolution. *The Quarterly Review of Biology* 84: 131–176.

Kenward, Ben, Christian Rutz, Alex A.S. Weir, and Alex Kacelnik. 2006. Development of tool use in New Caledonian crows: Inherited action patterns and social influences. *Animal Behaviour* 72: 1329–1343.

Klopfer, P.H., and J.P. Hailman. 1972. *Function and evolution of behavior*. Reading: Addison-Wesley.

Krebs, John R., and Nicholas B. Davies. 1997. *Behavioral ecology*. Oxford: Wiley-Blackwell.

Kroodsma, Donald E., and John R. Krebs. 1980. Repertoires and geographical variation in bird song. *Advances in the Study of Behavior* 11: 143–177.

Laland, Kevin N., and Vincent M. Janik. 2006. The animal cultures debate. *Trends in Ecology & Evolution* 21: 542–547.

Laland, Kevin N., Culum Brown, and Jens Krause. 2003. Learning in fishes: From three-second memory to culture. *Fish and Fisheries* 4: 199–202.

Laland, Kevin N., John F. Odling-Smee, and Scott F. Gilbert. 2008. EvoDevo and niche construction: Building bridges. *Journal of Experimental Zoology B* 310: 549–566.

Laland, Kevin N., Kim Sterelny, John F. Odling-Smee, William Hoppitt, and Tobias Uller. 2011. Cause and effect in biology revisited: Is Mayr's proximate-ultimate dichotomy still useful? *Science* 334: 1512–1516.

Langergraber, Kevin E., Christophe Boesch, Eiji Inoue, Miho Inoue-Murayama, John C. Mitani, Toshisada Nishida, Anne Pusey, Vernon Reynolds, Grit Schubert, Richard W. Wrangham, Emily Wroblewski, and Linda Vigilant. 2011. Genetic and 'cultural' similarity in wild chimpanzees. *Proceedings of the Royal Society B* 278: 408–416.

Laubichler, Manfred D. 2009. Evolutionary developmental biology offers a significant challenge to the neo-Darwinian paradigm. In *Contemporary debates in philosophy of biology*, ed. Francisco J. Ayala and Robert Arp, 199–212. Oxford: Wiley-Blackwell.

Leadbeater, Ellouise, and Lars Chittka. 2007. Social learning in insects – From miniature brains to consensus building. *Current Biology* 17: R703–R713.

Lefebvre, Louis, and Julie Bouchard. 2003. Social learning about food in birds. In *The biology of traditions: Models and evidence*, ed. Dorothy M. Fragaszy and Susan Perry, 94–126. Cambridge: Cambridge University Press.

Lehner, Philip N. 1998. *Handbook of ethological methods*. Cambridge: Cambridge University Press.

Lewontin, Richard C. 1970. The units of selection. *Annual Review of Ecology and Systematics* 1: 1–18.

Manning, Aubrey. 2005. Four decades on from the 'four questions'. *Animal Biology* 55: 287–296.

Manning, Aubrey, and Marian S. Dawkins. 1998. *An introduction to animal behaviour*. Cambridge: Cambridge University Press.

Martin, Paul, and Patrick Bateson. 2007. *Measuring behaviour: An introductory guide*. Cambridge: Cambridge University Press.

Mayr, Ernst. 1961. Cause and effect in biology. *Science* 134: 1501–1506.

Mayr, Ernst. 1982. *The growth of biological thought: Diversity, evolution, and inheritance*. Cambridge, MA: Belknap Press.

Mayr, Ernst. 1993. What was the evolutionary synthesis? *Trends in Ecology & Evolution* 8: 31–34.

McGrew, William C. 1998. Behavioral diversity in populations of free-ranging chimpanzees in Africa: Is it culture? *Human Evolution* 13: 209–220.

Mery, Frédéric, and James G. Burns. 2010. Behavioural plasticity: An interaction between evolution and experience. *Evolutionary Ecology* 24: 571–583.

Mery, Frédéric, Susana A.M. Varela, Étienne Danchin, Simon Blanchet, Deseada Parejo, Isabelle Coolen, and Richard H. Wagner. 2009. Public versus personal information for mate copying in an invertebrate. *Current Biology* 19: 730–734.

Minelli, Alessandro. 2009. Evolutionary developmental biology does not offer a significant challenge to the neo-Darwinian paradigm. In *Contemporary debates in philosophy of biology*, ed. Francisco J. Ayala and Robert Arp, 213–226. Oxford: Wiley-Blackwell.

Morgan, Hugh D., Heidi G.E. Sutherland, David I.K. Martin, and Emma Whitelaw. 1999. Epigenetic inheritance at the agouti locus in the mouse. *Nature Genetics* 23: 314–318.

Müller, Gerd B. 2007. Evo–devo: Extending the evolutionary synthesis. *Nature Reviews Genetics* 8: 943–949.

Müller, Gerd B. 2008. Evo-devo as a discipline. In *Evolving pathways: Key themes in evolutionary developmental biology*, ed. Alessandro Fusco and Giuseppe Fusco, 5–30. Cambridge: Cambridge University Press.

Müller, Gerd B. 2010. Epigenetic innovation. In *Evolution: The extended synthesis*, ed. Massimo Pigliucci and Gerd B. Müller, 307–332. Cambridge, MA: MIT Press.

Müller, Gerd B., and Stuart A. Newman. 2005. The innovation triad: An EvoDevo agenda. *Journal of Experimental Zoology B* 304: 487–503.

Müller, Gerd B., and Massimo Pigliucci. 2010. Extended synthesis: Theory expansion or alternative? *Biological Theory* 5: 275–276.

Mundinger, P.C. 1982. Microgeographic and macrogeographic variation in the acquired vocalizations of birds. In *Acoustic communication in birds*, vol. 2, ed. Donald Kroodsma, Edward H. Miller, and Henri Ouellet, 147–208. New York: Academic.

Odling-Smee, F. John, Kevin N. Laland, and Marcus W. Feldman. 2003. *Niche construction: The neglected process in evolution*. Princeton: Princeton University Press.

Pigliucci, Massimo. 2007. Do we need an extended evolutionary synthesis? *Evolution* 61: 2743–2749.

Pigliucci, Massimo. 2008. The proper role of population genetics in modern evolutionary theory. *Biological Theory* 3: 316–324.

Pigliucci, Massimo. 2009. An extended synthesis for evolutionary biology. *Annals of the New York Academy of Sciences* 1168: 218–228.

Pigliucci, Massimo, and Gerd B. Muller. 2010. *Evolution – The extended synthesis*. Cambridge, MA: MIT Press.

Podos, Jeffrey, and Paige S. Warren. 2007. The evolution of geographic variation in birdsong. *Advances in the Study of Behavior* 37: 403–458.

Raff, Rudolf A. 2000. Evo-devo: The evolution of a new discipline. *Nature Reviews Genetics* 1: 74–79.

Rakyan, Vardhman K., Jost Preis, Hugh D. Morgan, and Emma Whitelaw. 2001. The marks, mechanisms and memory of epigenetic states in mammals. *Biochemical Journal* 356: 1–10.

Reik, Wolf, Wendy Dean, and Jörn Walter. 2001. Epigenetic reprogramming in mammalian development. *Science* 293: 1089–1093.

Richards, Eric J. 2006. Inherited epigenetic variation—Revisiting soft inheritance. *Nature Reviews Genetics* 7: 395–401.

Richards, Christina L., Oliver Bossdorf, and Massimo Pigliucci. 2010. What role does heritable epigenetic variation play in phenotypic evolution? *BioScience* 60: 232–237.

Sherry, David F., and Bennett G. Galef. 1984. Cultural transmission without imitation: Milk bottle opening by birds. *Animal Behaviour* 32: 937–938.

Sherry, David F., and Bennett G. Galef. 1990. Social learning without imitation: More about milk bottle opening by birds. *Animal Behaviour* 40: 987–989.

Shettleworth, Sarah J. 2010. *Cognition, evolution, and behaviour*. Oxford: Oxford University Press.

Sih, Andrew, Alison M. Bell, J. Chadwick Johnson, and Robert E. Ziemba. 2004a. Behavioral syndromes: An integrative overview. *The Quarterly Review of Biology* 79: 241–277.

Sih, Andrew, Alison Bell, and J. Chadwick Johnson. 2004b. Behavioral syndromes: An ecological and evolutionary overview. *Trends in Ecology & Evolution* 19: 372–378.

Slater, Peter J.B., and T.R. Halliday. 1994. *Behaviour and evolution.* Cambridge: Cambridge University Press.

Stebbins, G. Ledyard, and Francisco J. Ayala. 1981. Is a new evolutionary synthesis necessary? *Science* 213: 967–971.

Sterelny, Kim. 2000. Development, evolution, and adaptation. *Philosophy of Science* 67: S369–S387.

Sterelny, Kim, and Paul E. Griffiths. 1999. *Sex and death: An introduction to philosophy of biology.* Chicago: University of Chicago Press.

Thornton, Alex, and Katherine McAuliffe. 2006. Teaching in wild meerkats. *Science* 313: 227–229.

Tinbergen, Niko. 1951. *The study of instinct.* Oxford: Clarendon Press.

Tinbergen, Niko. 1963. On aims and methods of ethology. *Zeitschrift für Tierpsychologie* 20: 410–433.

Tomasello, Michael. 1994. Cultural transmission in the tool use and communicatory signaling of chimpanzees? In *"Language" and intelligence in monkeys and apes: Comparative developmental perspectives*, ed. Sue Taylor Parker and Kathleen R. Gibson, 274–311. Cambridge: Cambridge University Press.

Tomasello, Michael. 1999a. The human adaptation for culture. *Annual Review of Anthropology* 28: 509–529.

Tomasello, Micahel. 1999b. *The cultural origins of human cognition.* Cambridge, MA: Harvard University Press.

Waterland, Robert A., Michael Travisano, and Kajal G. Tahiliani. 2007. Diet-induced hypermethylation at agouti viable yellow is not inherited transgenerationally through the female. *The FASEB Journal* 21: 3380–3385.

Whiten, Andrew. 2005. The second inheritance system of chimpanzees and humans. *Nature* 437: 52–55.

Whiten, A., J. Goodall, W.C. McGrew, T. Nishida, V. Reynolds, Y. Sugiyama, C.E.G. Tutin, R.W. Wrangham, and C. Boesch. 1999. Cultures in chimpanzees. *Nature* 399: 682–685.

Wolff, George L., Ralph L. Kodell, Stephen R. Moore, and Craig A. Cooney. 1998. Maternal epigenetics and methyl supplements affect agouti gene expression in Avy/a mice. *The FASEB Journal* 12: 949–957.

Youngson, Neil A., and Emma Whitelaw. 2008. Transgenerational epigenetic effects. *Annual Review of Genomics and Human Genetics* 9: 233–257.

Yurk, H., L. Barrett-Lennard, J.K.B. Ford, and C.O. Matkin. 2002. Cultural transmission within maternal lineages: Vocal clans in resident killer whales in southern Alaska. *Animal Behaviour* 63: 1103–1119.

Constructing the Cooperative Niche

Kim Sterelny

Abstract Humans contrast with their great ape relatives in many ways, but one of the most striking is our richly cooperative social lives. The explanation of this difference is complex and multi-factorial. But this paper argues that one central element is niche construction. Hominins are inveterate and extensive niche constructors. Individually and collectively, we have deeply affected our physical and biological environment, and have used technology to filter and transform the selective effects of the changed physical and biology worlds in which we have lived. But members of our lineage have not just acted on physical and biological environments; they have organised their informational environment too. Not just their own, but that of the next generation. While intensive and active teaching is probably a recent phenomenon, teaching itself is not. Furthermore, adults structure the learning environment of the next generation in many other ways: by acting as models of adult life; by providing supervised, safer environments; by providing toys, tools and props that structure and support trial and error learning. So the skills, values, ideas, information, and expected modes of social interaction and behaviour are made accessible to the next generation. This happens in circumstances which have often been adapted to enhance learning. The main theme of this paper is to show that humans cooperate more than other great apes largely because they reconstruct their environment more than other great apes, and one aspect of that reconstruction has been to make a world in which cooperation could survive and expand.

K. Sterelny (✉)
School of Philosophy, Australian National University, Acton, Australia
e-mail: kim.sterelny@anu.edu.au

G. Barker et al. (eds.), *Entangled Life*, History, Philosophy and Theory
of the Life Sciences 4, DOI 10.1007/978-94-007-7067-6_13,
© Springer Science+Business Media Dordrecht 2014

1 A Puzzling Pattern

The hominin lineage diverged from its *pan* sister group six million years or so ago (Klein 2009a), and for the first half of this period there is little evidence that hominin social life and cognitive capacity varied greatly from great ape sister clades. But by about four million years ago, some hominin species were obligately bipedal (Klein 2009b), and by 3.5 million years ago, we begin to see evidence of increased tool use and a shift to a richer and more meat-based diet (McPherron et al. 2010). The Oldowan lithic industry dates from about 2.3 million years ago (Foley and Lahr 2003), and by then there are clear signs of a major change, with habilenes showing a modest increases in relative brain size, reduced teeth, jaws and guts. Our ancestors were changing. Over the next two million years the basic parameters that characterise late hominins evolved: larger brains (especially the neocortex and cerebellum); a longer life span, including both a long period of juvenile dependency and (with women) an intriguingly long period of active competence after menopause; large body size with quite modest sexual dimorphism. Hominins became increasingly technologically adept, with the Acheulian stone industry establishing at about 1.7 mya; fire from perhaps 800 kya; and complex, multi-stage stone tool techniques from perhaps 300 kya (Foley and Gamble 2009). Hominins spread ecologically and geographically. As these ecological and technological changes emerged, hominins also evolved minds and social lives that were very different from those of other great apes.

It is very difficult to specify the changes that lead to the complex and cooperative social lives of late hominins. There is, for example, nothing like consensus on the dates of emergence of language, or of paternal investment in children. That said, it is clear that one change was the evolution of a heavy (ultimately, very heavy) dependence on cooperation, including informational cooperation in social learning and teaching. Great apes will cooperate, and will act prosocially in minor ways without any (obvious) expectation of return benefit. But unlike humans, they appear to be largely blind to one another's informational needs (Warneken et al. 2007; Warneken and Tomasello 2009). Despite these uncertainties, I think it's clear that expanded hominin cooperation has deep roots, dating at least to the erectines (about 1.7 mya) and quite likely to the habilenes 2.3 mya. Sarah Hrdy and Kristin Hawkes have argued convincingly that erectine babies and infants were so expensive that erectine life history would not be viable without significant reproductive cooperation (Burkart et al. 2009; Hawkes 2003; Hrdy 2009). It is equally clear that the shift to a meat-based diet depended on ecological cooperation. Hunting large game with short-range, low velocity weapons at acceptable levels of risk depends on cooperation; and hominins have been hunting large game at least since 400 kya and almost certainly much longer (Stiner 2002).

In other work, I have defended a three-stage model of the rise of cooperation in hominin social life (Sterelny 2013a, b; Sterelny forthcoming). The first stage is the transition from the hierarchical and individual world of great apes (for the most part, chimps and other great apes forage individualistically, consuming as they go) to a world of mutualistic, egalitarian, cooperative foragers. Collective defence, collective hunting, and power scavenging are profitable mutualisms: the rewards of cooperation are produced collectively, but they are divided and enjoyed at the point of production (Tomasello et al. 2012). Mutualism does not depend on mutual tracking and policing over time, and so it is less cognitively and socially demanding than forms of cooperation that depend on reciprocation over time. Erectines were probably mutualist foragers; their larger-brained, large-game hunting successors, the heidelbergensians, almost certainly were. Reciprocation-based cooperation ultimately became a central form of forager cooperation, as some mix of technological advance and ecological necessity drove an expansion of the range of resources harvested, an increase in specialization, and a shift to a more dispersed social world. This is my second stage of the evolution of cooperation, and I argue that the emergence of archaeological signals of ritual and ideology between about 120 and 75 kya is an indirect signal of forager economies dependent on reciprocation-based cooperation—it is a sign of the increased conflict risk and cognitive load of that form of cooperation.[1] The final stage is much more recent. Beginning with the Pleistocene-Holocene transition, about 10 kya, many human societies became more sedentary, more complex, and less egalitarian. There was a significant spike in intergroup violence (Seabright 2010). Despite the erosion of face-to-face mechanisms of trust, in many of these societies the social contract survived, and major problems of collective action were solved (Bogucki 1999; Flannery and Marcus 2012). Large scale, perhaps even cross-generation, cooperation had arrived.

I have argued that this trajectory has been driven by positive feedback, for at each stage in the evolution of hominin social life there were cooperation profits to be had, just an innovation or two away. The human career is very largely a case study in the profit of cooperation. Hominins began as a minor player in a very rich East African fauna. But our lineage speciated, invaded most terrestrial ecologies, dispersed (even in some earlier forms) over much of the old world, and vastly expanded its total population. Perhaps the hominin cooperation profile is not the whole explanation of that difference, but it is surely central to it. I do not intend to repeat my account of this grand narrative in this paper. Rather, my project here is to identify the resources we need to explain this social revolution and its consequences, and to make explicit the evolutionary mechanisms that underlie it. Like other evolutionary theorists, I see cooperation as puzzling and problematic, but not just because it is hard to understand why cooperation is not undermined by free-riding. Rather, it is because we cannot see cooperation as a feature of individual phenotypes evolving in response to environmental pressure; it is not like salt tolerance in

---

[1]There is a rich archaeological literature on "behavioral modernity," i.e., on the archaeological signs that ancient foragers resemble those known from ethnography. An important issue is the fact that these signs are about 100,000 years younger than our species (d'Errico 2003; Henshilwood and Marean 2003; Nowell 2010; Sterelny 2011). I interpret many of these signs of modernity as signs that a reciprocation-based economy is replacing a mutualistic economy.

Australian plants, or the sexual dimorphism of many Australian parrots. To explain cooperation, we need to explain the co-construction of individual phenotypes, social structures, and selective environments. In the next section, I identify some of the core cognitive, technological, morphological, and environmental preconditions of the cooperation transition, and sketch some of the ways in which the evolution of cooperation interacts with these enabling factors, leveraging further change. These levers are center stage in Sect. 3. In Sect. 4, I discuss the upshot, showing how these interactions illuminate some puzzling features of human cognition and social life.

2   Cooperation: The Business Plan

In 2003, building on the work of Richard Lewontin, John Odling-Smee, Kevin Laland, and Marcus Feldman published their manifesto on niche construction (Odling-Smee et al. 2003). That work had many threads, but one was to insist that the adaptive fit between organism and environment often depends on organisms, individually or collectively, modifying their environment, not just organisms responding to their environment. This chapter sees hominin cooperation as a particularly important example of this two-way interplay between organisms and environments. Cooperation powers niche construction. But niche construction powers cooperation too, and identifying those links will be a major theme of this chapter. More specifically, to explain cooperation in the hominin lineage, and to explain the contrast between the hominins and the great apes, we have to identify:

(a) the potential profits cooperation can generate;
(b) the investment needed to realize those profits; in this case, the cognitive, technical, and social prerequisites of specific forms of cooperation, and the processes by which those prerequisites came to be in place;
(c) the policing mechanisms that ensure that the profits are not distributed in ways that destabilize cooperation.

These factors will not be constant over the three pulses that characterise the hominin experiment in cooperation: the profit-investment-policing profile of the very large brained, highly social hominins of the last half million years is likely to be very different from that of habilenes or early erectines.

2.1   Profits

As noted above, hominins diverged from the great apes about six million years ago. Orangutans and gorillas, with their dietary specialization, may not be just a social or technical innovation or two away from profitable collective foraging or defence. There are few economies of scale in browsing and fruit-eating, and these are formidable animals living in deep cover. Their ancestors were probably not at high predation risk. In the typical habitats of these great apes, the seasonal footprint is muted, and the resources they target are abundant, though individually of low value. So these animals do not have to manage risk in the way those that target high value, but rare and scattered resources do. As the classic vampire bat studies show, risk management often rewards cooperation (Wilkinson 1990).

For similar reasons, there is relatively little pressure on these vegetarian great apes for active information-sharing. Gorillas and orangutans experience great botanical diversity, and their young need to learn to identify and exploit edible plants (plants often do not want to be eaten, and protect themselves with spines and thorns) (Byrne 1995, 2000, 2004). But simple, passive social learning probably suffices: young orangutans travel with their mothers, and if her interests and acts are especially salient and interesting to her young, that adaptive bias in trial and error learning probably suffices. There is not much need for active teaching, nor for information sharing and coordination between adults. There are no fruit in Borneo jungles that can be harvested only by orangutans acting together. So perhaps it is no surprise that sustained and active cooperation is not the default for great apes: it needs special explanation.

By contrast, early hominins were probably generalist omnivores living in variable environments (Klein 2009b). In most circumstances, for such animals, in such environments, ecological cooperation is potentially profitable, because collective action, specialization and the division of labor, and cooperative risk management, will all deliver to agents more usable resources at less risk—so long as they can coordinate their activities effectively. So the basic ecological difference between woodland omnivores and forest vegetarians explains why the forest vegetarians never experienced cooperation take-off. However, the *pan* species, like our ancestors, are generalist omnivores. Yet in chimps and bonobos cooperation is limited and (at least in chimps) not very stable. Males form coalitions both to promote their political interest and in defending and extending their territory in conflict with other bands. Males also hunt collectively, though perhaps not cooperatively. So there is some limited ecological cooperation, but almost no informational or reproductive cooperation (de Waal 2008; de Waal and Suchak 2010). So while the basic ecological difference between deep forest herbivores and early hominins might explain the cooperation differences in those lineages, we need an explanation of why the *pan* lineage did not experience cooperation lift-off.

There are some plausible suggestions: (i) chimps, especially males, are extremely strong and dangerous, and hence predation pressure did not select for collective defence, an important early form of cooperation; (ii) early hominins lived in more open habitat, and for that reason, were under stronger selection for collective defense (Foley 1995); (iii) cooperation is linked to bipedality, a morphological and ecological change with profound social and cognitive consequences. Bipedality is important for at least three reasons: first, it frees the hands from the demands of locomotion; second, it indicates a shift to larger and less tree-dominated territories, with their opportunities and dangers; third, it forced our ancestors to sleep on the ground, thus making them more vulnerable to predators at night. If bipedal locomotion was linked to a shift to woodland and grassland environments, it also exposed hominins to predation, especially if that was coupled with less capacity to shelter in those trees that were available. So this change is a potential cooperation trigger both because fully bipedal apes must sleep on the ground (the predation-cooperation connection again) and because bipedal apes can evolve the capacity to use very simple weapons, which as we will see, may have reduced the costs of controlling defection and free-riding. The problem of understanding the initial divergence of the hominins is real, but because there are a number of plausible scenarios, not because there are none.

So in summary, and focusing on the earliest stages of hominin cooperation take-off, cooperation profits were probably more readily available to our lineage; habitat shifts and phenotype changes brought the first of those profits within, or close to, existing phenotypic variation. As soon as hominins acquired very rudimentary weapons (sticks as clubs or for jabbing; stones as missiles), collective defense and power scavenging opportunities open up. These forms of collective action depend only on the ability to monitor one another's behavior: mobwork is enough to secure some important benefits of cooperation. However, once mobwork became an enduring and important feature of hominin lives, that cooperative foraging practice selected for the expansion of capacities to communicate and coordinate. Hunting and power scavenging required some capacities to share information: in coordination; perhaps in planning; perhaps in recruitment.[2]

Changes in hominin phenotypes then open up new potential cooperation profits. Sarah Hrdy, Kristin Hawkes, and their colleagues have argued that reproductive cooperation is an important, foundational form of human cooperation, and its profit depends largely on human life history (Hawkes 2003; Hawkes et al. 1998; Hrdy 2009). Cooperation is most likely to evolve if help is cheap to give but very valuable to receive, and with humans, reproductive cooperation sometimes has that fortunate asymmetry. Hominins have long been bipedal, and perhaps by the habilenes (about 2.3 mya) and certainly by the erectines (1.7 mya), the characteristics that make sapiens children so expensive had begun to emerge. Ancient human mothers were delivering large-brained, hyper-dependent babies, babies that could not even cling on to their mother. Birth itself was physically challenging; infants were immobile compared to adult range sizes; children were dependent significantly longer than their great ape equivalents. Some forms of aid are quite expensive—actively provisioning children; carrying them significant distances—and help of this kind probably initially depended (and largely continues to depend) on kin selection. But many important forms of help are quite cheap: it costs little to offer a birthing mother help, simple care, and protection; to keep an eye on children for an hour or two at a base camp while she forages (especially if you are already keeping an eye on your own); to carry an infant for a few minutes while the mother attends to some urgent task. If this help comes from near-adult girls in the group, those girls gain valuable experience in return.

---

[2]If the group did not forage as a single unit, those who found a major kill or killing opportunity would need to recruit others; power scavenging offers large economies of scale.

Let's briefly step back from detailed conjecture to expose the explanatory strategy: identify significant cooperation profits available to an omnivorous, open-range, bipedal great ape, and show how the exploitation of these profits leads to further changes, both in individual phenotype and environment. These changes establish pull-factors that select for further cooperation, and further phenotypic and environmental change. The strategy, then, is to show how a positive feedback loop is established, and why it was stable over three million years or so.

2.2 Investment

It is one thing for there to be cooperation profits in the vicinity, another to be able to exploit them. The opportunities must be recognized; they must be salient to the agents. Cooperative action must be coordinated. Even when agents cooperate to lower risk, by sharing their individual successes to reduce the footprint of luck, or to exploit complementary resources, they depend not just on trust but on coordination: minimally, where and when they will meet to share and exchange. Some forms of profitable collective action require only minimal coordination—that agents assemble at the same time and place, and are focused on the same task. But others demand some division of labor and role specialization. It is for this reason that power scavenging very likely evolved before systematic hunting. For power scavenging requires only minimal coordination: a noisy mob can drive a solitary carnivore from a kill, but systematic hunting often requires planning, coordination, and a division of labor: for example, if prey are spooked into an ambush site, or when prey is exhausted by relays of hunters in endurance hunting.

Hunting is often teamwork, not mobwork, and great apes probably do not have a baseline capacity for teamwork. We owe this insight to Michael Tomasello, who has emphasised the cognitive demands of this more nuanced form of collective action (Tomasello 2009; Tomasello and Carpenter 2007; Tomasello et al. 2005). Teamwork requires both sophisticated theory of mind capacities and the ability to represent the structure of a collective task in an agent-neutral way. Great apes probably have at best a limited form of this capacity, and that is why they struggle with role reversal tasks. Much of the social intelligence literature has focused on the cognitive and informational demands of policing cooperation (Humphrey 1976; Whiten and Byrne 1997). But once we move beyond the profits of mob activity in responding to unplanned opportunities or risks, the cognitive-informational demands of cooperation are very significant. If Tomasello is right, chimps have the cognitive capacity to act in a mob but not in a team.

Collective action depends on motivational preconditions, too. Most obviously, agents need enough tolerance and trust to act in close association with others. That is not trivial: tolerance and trust must be gained despite aversive interactions with some in the group, for no band is free of conflict. A little less obviously, impulse control is necessary.[3] A stag hunt will fail if one of the team is distracted by a passing rabbit, and chases off after it, leaving a hole in the human trap. This is not defection; the distracted hunter would have done better with his share of the stag, even if he caught the rabbit. It is a triumph of the now over the future. The profits of cooperation often require persistence, maintaining focus over time. Once again, we see the same explanatory strategy at work: early forms of hominin cooperation depend only on cognitive, communicative, and motivational capacities at or near great ape baselines. But once these early forms of cooperation become a default form of life, that changes both the selective and developmental environment. Young hominins grow up in a more tolerant and cooperative world, and selection favors those that develop adaptively in that world. Moreover, further cooperation profits are available to those with enough cooperation-coordination-impulse control capacities to inch their way to coordinated collective action and thence to collective action supported by planning and/or role specialisation. Teamwork evolves incrementally, with each increment bringing new activities within reach or reducing the costs and risks of existing activities.

2.3 Policing

Much of the literature on the evolution of cooperation is focused on policing. That focus is driven by an important insight: the profits of cooperation often do not depend on a full contribution from each of those that stand to profit, and this creates a temptation to free-ride. Now suppose, as is plausible, that:

(a) if free-riding invades, it destabilizes cooperation;
(b) free-riding will invade, unless active measures are taken to block or deter it;
(c) active anti-free-rider measures are not cost free.

We then have a puzzle: who pays these costs, and why are they worth paying (see, for example, Okasha 2006). This puzzle is so pressing that many have taken the problem of human cooperation to largely reduce to that of explaining why active measures—punishment costs—are cheap enough to make stable cooperation possible. To take three examples, Don Ross argues that punishment is cheap because of human motivational sensitivity to social rewards and punishments (Ross 2006b); Bowles and Gintis argue that punishment is cheap because it invades as a conditional strategy, triggered only in the presence of sufficient punishers to divide the cost between them (Bowles and Gintis 2011); Paul Bingham argues that the invention of weapons, especially projectile weapons, made punishment cheap by allowing a larger coalition to simultaneously attack a recalcitrant cheat, thus reducing the risk to each individual in the coalition (Bingham 1999, 2000).

---

[3]Wynn and Coolidge have long argued that hominin cognitive evolution is largely an expansion of working memory. For them, working memory seems to include executive function skills: planned behaviour and the capacity to resist distraction (Wynn and Coolidge 2004, 2010).

If the three-phase model of Sect. 1 is right, the nature and cost basis of deterrence—of policing free-riding—will almost certainly have changed with changes in the role of cooperation in human social life. Late evolved hominins[4] do typically care profoundly about how others see them, and they do have formidable capacities to communicate and coordinate. So the mechanisms identified by Ross, Bowles, and Gintis certainly help explain the stability of reciprocation-based forager economies (and cooperation in complex Holocene groups). But the growth of cooperation in the transition from great ape society to egalitarian mutualist foragers cannot depend on such sophisticated cognitive and communicative tools, for these evolved only because hominins had long lived in a cooperative milieu. Bingham's proposal is a more plausible explanation of cheap deterrence (or cheap enough deterrence) in the initial phase of divergence from great ape patterns. But he underestimates the cognitive and motivational prerequisites of coalitional enforcement, even with weapons. For in earlier forms of hominin social life, cooperation was threatened not so much by lazy shirkers but by active and dangerous bullies, by alpha males who simply seize what they want (Boehm 1999, 2012). Great ape societies show that such alphas are aware of the threat posed by coalitions, and attempt to break them up (de Waal 2008). So early forms of coalitional enforcement probably do depend on the extra threat posed by armed coalitions, but they also depend on an enhanced motivational and cognitive platform built by joint activity that is somewhat profitable for all despite a less-than-equitable distribution of profits. Collective defence is one such activity; mob hunting, as in chimpanzee monkey hunting, is another. Monkey hunts do not result in all the chimps getting roughly equal portions of monkey. It is important to recall that coalitional enforcement probably did not evolve from scratch in the hominin lineage; chimps do occasionally lose patience with alphas and collectively hound them (de Waal 2008). What was novel in the hominin lineage was not the existence of coalitions from below, but such coalitions exerting sustained, long-term pressure, flattening social hierarchies in the hominin lineage, perhaps for a couple of million years or more (Boehm 2012). Early hominins learned to tolerate and trust one another more, to act together in more coordinated ways, and to stay on the job. Those added capacities (plus the use of weapons) emerged in ecological contexts, and were then coopted into social ones. The result was a cooperative milieu stable enough to allow selection to build the more complex deterrence mechanisms identified by Ross and others.

3 Trigger and Feedback

I noted above that one of the puzzles in understanding hominin cooperation is identifying the trigger of the initial hominin-pan divergence. My best guess is that that initial trigger was increased predation risk, selecting for shared vigilance

---

[4]By this I mean sapiens, Neanderthals, and their immediate predecessor: the very large-brained hominins of the last half a million years.

and active collective defence. There is some evidence of stone tool use at about 3.5 million years ago. A group of hominins armed with sticks, throwing stones, and making a loud racket might well deter attack by medium-size predators that would certainly be a threat to an isolated australopithecine; the hominins of three to four million years ago were not especially formidable individuals. Likewise, unless they slept close together and were prepared to respond to danger collectively, they would have been very vulnerable to leopards and other nocturnal predators. If these conjectures are roughly right, we would expect four consequences:

(i) Tolerance and Coordination. There would be selection for motivational and cognitive changes in the hominin lineage, probably initially just increased tolerance for the proximity of others, and hence somewhat improved impulse control. This would be scaffolded by phylogenetically ancient mechanisms of association and affiliation: the more a group of hominins were successful in the company of one another (in, for example, deterring hyena attack), the more they would like being in one another's presence. Once somewhat cooperative group life became the default hominin experience, there would be positive selection for coordination and the capacity to anticipate others' actions.

(ii) Developmental Environment. There would be changes in the developmental environment. Sarah Hrdy has emphasized the fact that the mother forms the whole social environment of infant apes. In hominins, that changed. If the australopithecines of three million years ago slept and foraged in the immediate company of others, tolerating one another's close presence, even very young hominins would have experienced a social environment of other adults and older juveniles, training the infant from a young age for life in company and giving that infant more opportunity for social learning (Burkart et al. 2009; Hrdy 2009). Moreover, once life in company has become a routine aspect of hominin experience, this is likely to be reinforced by selection on mothers to seek help. Upright mothers have a problem with young infants; they have to be carried, as they cannot ride on their mother's back, holding on themselves, as many infant primates do. Furthermore bipedalism exacerbates the costs of immobile infants, for it almost certainly signals increased range size. Thus there is selection on mothers to seek and give help in carrying and/or creching infants. Helping protect and carry the young is a low-cost, high-value form of aid, and subadult females can learn crucial caring skills by helping mothers. It is also true that the overall mobility of the group is improved if mothers do not have to carry infants the whole time: group members can chase resources more effectively without having to leave exposed the most vulnerable individuals in their cohort. As above, once tolerance is established—once mothers do not fear harm to their infants at the hands of males, and once adults and juveniles tolerate the young and the curious amongst them—a platform is available for a further elaboration of reproductive cooperation.

(iii) Ecological Opportunity. Intimidating predators by cooperating as a crudely armed mob opens up an ecological opportunity: power scavenging. Initially, armed mobs of australopithecines could probably merely have driven the less formidable predators and scavengers from their kills—perhaps initially only to get marrow from the large bones of herbivores, or they could use their early skills with stone to break open these bones. But once this opportunity is first seized, it can expand incrementally. For example, hominin bands can learn to recognise the natural signs of a kill: vulture behaviour; the drag marks leopards make when storing a kill in trees, away from hyenas. They can learn to communicate—one foraging group recruiting others, if a really major prize is available. They can become more adept at driving predators from kills by volleys of thrown stones. Leverage and dexterity is one of the payoffs of bipedality, and predators cannot afford serious injury. They must be risk averse.

(iv) Morphology, Life History and Social Learning. A shift to power scavenging (perhaps with opportunistic small game hunting) adds more meat and fat to the diet, easing energetic constraints on brain size, reducing the mechanical demands on teeth and jaws, and allowing gut mass (also expensive tissue) to shrink (Roebroeks 2007). By two million years ago, there had clearly been a major change in hominin diets, for hominin skeletons show a marked reduction in tooth and jaw size: Richard Wrangham hangs his hypothesis of the early evolution of cooking on this morphological transformation (Carmody and Wrangham 2009; Wrangham 2001, 2009). I noted above that great apes are extractive foragers, using resources that require skill and knowledge to harvest (Byrne 2002, 2004). Add three ingredients from above to this extractive foraging basis, and we see how positive feedback can drive the expansion of hominin cooperation. First: bipedalism opens potentials for cognitive, behavioural, and morphological specializations, supercharging extractive foraging. Second: hominins develop in an intimate environment that makes social learning more reliable. Third: cooperation adds the potentials of collective action, teamwork, and specialization to the existing baseline of skilled extractive foraging. There is positive feedback between information sharing, ecological cooperation, and reproductive cooperation.

Social learning and information sharing support effective foraging by giving hominins access to new resources and by helping them extract more from their existing resource base. Power scavenging depends on understanding the local environment and animal behavior, both in locating scavengers' kills, and knowing how and when to drive a dangerous animal from its own kill. Hunting, once it moves beyond opportunistically seizing anything small or vulnerable a party might by chance come across, even more obviously depends on the ability to read the landscape and to understand animal behaviour. Kim Shaw-Williams has shown that one consequence of going bipedal is that, first, physical tracks become more perceptually salient, and scent trails less salient, and that (second) physical tracks are much more information-rich than scent trails. Hominins became the first primate, almost certainly the first animal, to exploit this rich source of information (Shaw-Williams 2011).

Power scavenging and hunting were supported by technology, though to the extent that we can tell this from the physical record, the human toolkit remained fairly simple until the last few hundred thousand years. Even so, almost certainly its manufacture and use depended on social transmission, perhaps including active teaching (Csibra and Gergely 2011; Stout 2011). But technology and technique was important not just in capturing resources but in preparing them. For the last decade, Richard Wrangham has shown the importance of cooking, and of food processing more generally. Cooking increases food value, and reduces the time and effort of eating. Chimps spend three or four times longer in eating than do typical humans, because they have to chew their food intensively, just to make it edible. Cooking improves our time budget, not just our energy budget (Wrangham 2009). In addition, and perhaps still more importantly, food preparation makes a whole new source available. Many plants that live in seasonal environments develop underground storage organs ("USOs"). These are rich sources of starch, but they are often difficult to find and dig out, so good botanical skills are needed. Moreover, most are protected chemically, and they cannot be eaten until they are processed, by one or more of soaking, washing, and cooking (O'Connell et al. 1999).

In all probability, early hominin social learning was richer and more reliable than great ape social learning only as a passive by-product of changes in adult activity patterns. If adults stay together in cohesive bands, while making simple tools or using them to process challenging resources in their environment, juvenile learning environments change. They are exposed to more adult models, and adult ecological choices shape their environment of exploration learning. Once social learning and information flow becomes more deliberate and bi-directional, further opportunities open up. Specialist tool kits and an expanded material technology are relatively recent, dating to perhaps 100 kya (McBrearty 2007; McBrearty and Brooks 2000). On the other hand, large game hunting requires coordination and communication, not just cooperative intent in an armed mob, and there is clear evidence of systematic large game hunting much earlier—perhaps as early as 1.7 mya, and certainly at about 400 kya (Boehm 2012; Jones 2007). And while the technological base of the Middle Stone Age of 200 kya is not varied, there is evidence of compound tools and the use of adhesives (Wadley 2010). So although the flow of technical skill between the generations might not have required rich communicative capacities until perhaps the last 200,000 years or so, large game hunting shows that active communication and collaboration—a deliberate, two way flow of information exchange—is half a million years old, perhaps much older.

In short, then, more reliable and more extensive social learning makes foraging more profitable. But equally, profitable foraging supports more reliable and extensive social learning. Successful cooperative foraging supports longer childhoods—one very important life history difference between the hominins and the great apes—thus giving young hominins longer to acquire critical skills. Kaplan, Gurven and their colleagues place great weight on this factor, providing data suggesting that foragers do not become fully self-sufficient until they are almost 20 but, once they are competent, produce more than they consume for decades (Gurven et al. 2006; Kaplan et al. 2009). So an intergenerational subsidy supports social learning, but that social learning supports skills which in turn makes the subsidy possible. Profitable foraging supports larger groups, which makes social learning more reliable in the short run, by giving the young more models to learn from, and in the medium run, by making it less likely that skills will be lost through the unlucky death of a few key individuals. Recent modelling has revealed that small populations are surprisingly vulnerable to the loss of information by unlucky accident (Powell et al. 2009). And it also helps explain longer hominin life expectancies—sapiens life expectancy is a couple of decades longer than that of chimps. Cooperation reduces the risk of predation, and even very simple care makes many illnesses and accidents survivable; animals without social support are desperately at risk if seriously hurt or ill. The take-home message remains the same: an initial ecological trigger builds an adaptive platform, which is then elaborated through positive feedback.

4 The Peculiarities of the Beast

So far in this chapter I have outlined a picture of the incremental evolution of human cooperation, and provided a framework for that trajectory. I have attempted to make explicit some of the preconditions—both internal and external preconditions—of cooperation take-off, and sketched some of the feedback loops through which those enabling factors themselves changed with changes in hominin cooperation. In this final section, I focus on the upshot. In understanding this trajectory we understand some of the very strange features of human life and cognition.

4.1 Individuals and Groups

Great apes are typically social, and hominins inherited this trait; we are social too. But because humans cooperatively modify their environment, including their social environment, human social life is very different from great ape social life. Human groups are not mere aggregates or heaps. At least since the evolution of reciprocation-based forager bands, and obviously since the emergence of the much larger and more complex societies of the Holocene, human groups are more like systems than populations. (i) Humans do not just belong to bands (and the like); they identify with the groups of which they are a part; they and others recognise their membership, especially as individuals often display insignias of group identity (as in gang patches and the like); individuals often have strong emotions of affiliation and loyalty to the groups to which they belong. (ii) Individuals within groups often have stable, distinctive roles, roles that shape their actions in predictable ways. (iii) Groups often have significant internal structure. In small traditional societies, this structure often takes the form of genealogical groupings: families, clans, moieties (Barnard 2011). But economic units exist as well, in stable hunting partnerships. In the Holocene, with its larger cultures, economic and other institutions became increasingly important (Seabright 2010). So even in traditional small-scale societies, human groups are complex, with vertical complexity and horizontal differentiation.

(iv) Groups sometimes engage in planned, coordinated activity with a collective product. Quite often, that product is not physical or biological but informational or representational. The legal system of a culture is clearly the collective product of that culture, and is a characteristic of the culture as a whole rather than of the individuals within the culture. The same is true of much of the normative and ideological life of a group: its norms, customs, religious rituals, and representations.

These collective products challenge the project of developing an evolutionary account of hominin social life. To a reasonable approximation, great ape social life can be explained by explaining the cognition and behaviour of individual apes; patterns in great ape social life are mostly simple statistical patterns in the summed behavior of individual agents. It is far from obvious that the same is true of human social life, in even the simplest of human societies. One response is to treat the groups themselves as units of selection (Bowles and Gintis 2011). Cultural group selection models probably do explain some features of human social life, but the conditions under which groups themselves are selected are quite restrictive: only a few features of groups are selectable, and that only in a rather narrow range of circumstances.

An alternative and more general approach is to see human social life as a more elaborated version of something seen quite often in the animal world. Think, for example, of swarming or flocking behaviour: functionally co-ordinated collective behavior that is the result of individual agents following simple local rules, typically in response not to any perception of the group as a whole but to the actions of their immediate neighbours. Just as flocks and swarms are the collective product of individual decision for individual benefit, human social life is the collective product of individual decision for individual benefit, but with the following important added features. (i) The collective phenomenon is not just an aggregate of individual decision, because of the ways human groups are structured—with stable individual roles, and persisting levels of organization between the individual and the group as a whole. (ii) In part because human groups are highly structured, these collective products are stable and persisting. Human social life is characterised by repeated patterns of interaction and a stable, organized informational environment. Local skills, customs, norms, and habitual patterns of interaction are on display, and this makes both coordination between adults and the enculturation of the next generation more reliable and predictable.[5] Human groups do not hide their norms, expectations, and customs from each other or the next generation. (iii) As a consequence of the stability of the collective phenomena, the collective character of the group influences individual phenotype, in both ontogeny and development. (iv) This mutual causal influence can result in positive feedback; we have already seen the example of

---

[5]Don Ross goes further, arguing that humans shape one another's psychology and habits, creating in one another stable and relatively public intentional profiles, making longer term collaboration and coordination possible. Our unshaped brains would leave us with much less stable world views and preference functions, and hence make our moment-by-moment decision making far less predictable (Ross 2006a).

the connection between effective group size and innovation. There is persuasive modelling (with some archaeological support) to indicate that once a threshold is reached, humans are able to retain informational resources much more reliably, and innovate more frequently. Size gives redundancy in retaining critical information (small groups can easily lose rare skills through unlucky accident), offers the next generation more and more diverse models in social learning, and affords more opportunities for specialization. Once the retention-innovation cycle takes off, groups extract more resources more efficiently from their environment, thus making it more likely that they can support or expand their population base. In sum, seeing humans as collective niche constructors helps explain the fact that the collective features of human groups can drive an evolutionary trajectory, transforming individual phenotypes, without the group itself being a unit of selection.

4.2 Cognitive Perversity

It is obvious that humans are far more intelligent (admittedly, in ways that are difficult to make precise) than other great apes. Individually and collectively we understand far more about our environment—physical, biological, social, psychological—than do other great apes. Individually and collectively, we are good at putting this information to use in making long-term plans (we do not just live in the present); in organizing collective action; and in making and using physical and social tools. Our capacities for efficient reason are far more highly developed than those of the great apes. But a strikingly large fraction of the representational and informational resources of the human mind is not devoted to efficient reason— to representing our environments and their latent possibilities, opportunities, and risks. A sizable fraction is devoted to (i) fictions, narratives (and other depictions known not to be veridical), and myths, quasi-fictional, quasi-historical narratives of special importance and affective power; (ii) religions, which typically involve stories and claims about the history and workings of the world that are presented as true, and which seem to be taken as true, despite the fact that efficient cognition would show them to be profoundly implausible and without evidential support; (iii) norms— humans typically represent themselves as living in a world of prohibitions and obligations, not just in a world of natural facts.

In sum: we think normatively; we represent the world in religious and magical terms; we consume and produce stories and other fictions, often knowing that they are fictions, perhaps enjoying them because they are fictions. There is much more to human minds than information-gathering and efficient instrumental reasoning about our actual environment. It is has often been argued that there is a crucial connection between these apparently perverse features of the human mind and our propensity to cooperate (see for example (Joyce 2006; Kitcher 2011; Wilson 2002). I think this idea is right, and that it is an increasingly important factor stabilizing human cooperation over the last 100,000 years: that is, cooperation in reciprocation-based forager economies and the farming economies that succeeded them. Sapiens hominins (and perhaps earlier ones) have invented a set of social or cultural tools—ways of organizing their social environment to enhance cooperation. Those tools are an utterly pervasive feature of contemporary environments: legal and institutional frameworks coordinate markets and other systems of social interaction to minimise conflict costs and increase their predictability and efficiency.[6] In my view, these social tools have a deep history: they date back in time to the mid-Pleistocene, 100 kya, and perhaps earlier. (I have argued for this elsewhere (Sterelny 2012b); since my main focus in this chapter has been on early phases of hominin cooperation, it is not front and center here.)

4.3 Niche Construction

Let me end by returning to the overarching theme of niche construction. Many organisms (and groups of organisms) act on their environment in ways that significantly alter that environment; significantly enough to affect the intensity and direction of selection; significantly enough so that organism-environment effects help explain the adaptive fit between the organism's phenotype and its environment. Hominins, and especially recent hominins, are major league niche constructors. Holocene humans have transformed their physical and biological environment: over the last 10,000 years, an increasing fraction of our species have lived in built environments and eaten from intensively managed biological resources. In doing so, we have almost certainly exerted transforming effects on our symbionts, parasites, pathogens, and commensals, and living in these constructed environments has had effects on humans, too. While Pleistocene humans did not transform their physical and biological environment quite so profoundly, shelters, clothes, fire, and resource management have deep histories. But humans do not just intervene on their physical and biological environment; they organize their informational environment too, and not just their own, but that of the next generation. Intensive, prolonged active teaching is probably an artefact of contemporary environments. This is not the case for teaching more generally: while social learning is found in many species, active teaching is rare (Hewlett et al. 2011). Moreover, adults structure the learning environment of the next generation in many other ways: by providing supervised, safer environments; by providing toys, tools, and props that structure and support trial-and-error learning; by merely being tolerant of the curiosity of the young. So ideas, information, and skills are made available to the next generation in a physical environment that is often physically and biologically modified to enhance learning (Sterelny 2012a). The extent and transforming character of human niche

---

[6]That is not all they do, of course, and while these contemporary social mechanisms do coordinate and regularize interaction, making forms of cooperation possible that would be otherwise inconceivable, they also distribute the profits of those interactions very unequally, while often at the same time entrenching those inequalities.

construction is far from unique. Termites, for example, live in worlds that are almost entirely termite-constructed, and their entire phenotypes are adapted to a world they have made (Turner 2000). But human niche construction is unique for a primate, for a great ape. Other great apes modify their environments: for example, chimps build nests in which they sleep. But these modifications are fairly minor: to a reasonable approximation, great apes live in the world as they find it, rather than the world as they reconstruct it. Humans do not. Part of the reconstruction that humans have engaged in has been the making of a world that enabled cooperation to survive and expand.

References

Barnard, Alan. 2011. *Social anthropology and human origins*. Cambridge: Cambridge University Press.

Bingham, Paul. 1999. Human uniqueness: A general theory. *Quarterly Review of Biology* 74: 133–169.

Bingham, Paul. 2000. Human evolution and human history: A complete theory. *Evolutionary Anthropology* 9: 248–257.

Boehm, Chris. 1999. *Hierarchy in the forest*. Cambridge, MA: Harvard University Press.

Boehm, Christopher. 2012. *Moral origins: The evolution of virtue, altruism and shame*. New York: Basic Books.

Bogucki, Peter. 1999. *The origins of human society*. Oxford: Blackwell.

Bowles, Sam, and Herbert Gintis. 2011. *A cooperative species: Human reciprocity and its evolution*. Princeton: Princeton University Press.

Burkart, Judith, Sarah Blaffer Hrdy, and Carel van Schaik. 2009. Cooperative breeding and human cognitive evolution. *Evolutionary Anthropology* 18: 175–186.

Byrne, R. 1995. *The thinking ape: Evolutionary origins of intelligence*. Oxford: Oxford University Press.

Byrne, Richard. 2000. Evolution of primate cognition. *Cognitive Science* 24: 543–570.

Byrne, Richard. 2002. Seeing actions as hierarchically organized structures: Great ape manual skills. In *The imitative mind*, ed. Andrew Meltzoff and William Prinz, 122–140. Cambridge: Cambridge University Press.

Byrne, Richard. 2004. The manual skills and cognition that lie behind hominid tool use. In *Evolutionary origins of great ape intelligence*, ed. Anne Russon and David R. Begun, 31–44. Cambridge: Cambridge University Press.

Carmody, Rachael, and Richard Wrangham. 2009. The energetic significance of cooking. *Journal of Human Evolution* 57: 379–391.

Csibra, Gergely, and György Gergely. 2011. Natural pedagogy as evolutionary adaptation. *Philosophical Transactions of the Royal Society B* 366: 1149–1157.

d'Errico, F. 2003. The invisible frontier: A multiple species model for the origin of behavioural modernity. *Evolutionary Anthropology* 12: 188–202.

de Waal, Franz. 2008. *Chimpanzee politics: Power and sex among apes*, 25th anniversary ed. Baltimore: John Hopkins University Press.

de Waal, Franz, and Malini Suchak. 2010. Prosocial primates: Selfish and unselfish motivations. *Philosophical Transactions of the Royal Society B* 365: 2711–2722.

Flannery, Kent, and Joyce Marcus. 2012. *The creation of inequality*. Cambridge, MA: Harvard University Press.

Foley, Robert. 1995. *Humans before humanity*. Oxford: Blackwell.

Foley, Robert, and Clive Gamble. 2009. The ecology of social transitions in human evolution. *Philosophical Transactions of the Royal Society B* 364: 3267–3279.

Foley, Robert, and Marta Mirazon Lahr. 2003. On stony ground: Lithic technology, human evolution, and the emergence of culture. *Evolutionary Anthropology* 12: 109–122.

Gurven, Michael, Hillard Kaplan, and Maguin Gutierrez. 2006. How long does it take to become a proficient hunter? Implications for the evolution of extended development and long life span. *Journal of Human Evolution* 51: 454–470.

Hawkes, Kristin. 2003. Grandmothers and the evolution of human longevity. *American Journal of Human Biology* 15: 380–400.

Hawkes, K., J.F. O'Connell, N.G. Blurton Jones, H. Alvarez, and E. Charnov. 1998. Grandmothering, menopause, and the evolution of human life histories. *Proceedings of the National Academy of Sciences of the United States of America* 95: 1336–1339.

Henshilwood, Christopher, and Curtis Marean. 2003. The origin of modern behavior. *Current Anthropology* 44: 627–651 (includes peer commentary and author's responses).

Hewlett, Barry, Hillary Fouts, Adam Boyette, and Bonnie Hewlett. 2011. Social learning among Congo Basin hunter-gatherers. *Philosophical Transactions of the Royal Society B* 366: 1168–1178.

Hrdy, Sarah Blaffer. 2009. *Mothers and others: The evolutionary origins of mutual understanding*. Cambridge, MA: Harvard University Press.

Humphrey, Nicholas. 1976. The social function of intellect. In *Growing points in ethology*, ed. P.P.G. Bateson and R.A. Hinde, 303–317. Cambridge: Cambridge University Press.

Jones, Martin. 2007. *Feast: Why humans share food*. Oxford: Oxford University Press.

Joyce, Richard. 2006. *Evolution of morality*. Cambridge, MA: MIT Press.

Kaplan, Hilliard, Paul Hooper, and Michael Gurven. 2009. The evolutionary and ecological roots of human social organization. *Philosophical Transactions of the Royal Society B* 364: 3289–3299.

Kitcher, Philip. 2011. *The ethical project*. Harvard: Harvard University Press.

Klein, Richard G. 2009a. Darwin and the recent African origin of modern humans. *Proceedings of the National Academy of Sciences of the United States of America* 106: 16007–16009.

Klein, Richard G. 2009b. *The human career: Human biological and cultural origins*, 3rd ed. Chicago: University of Chicago Press.

McBrearty, Sally. 2007. Down with the revolution. In *Rethinking the human revolution: new behavioural and biological perspectives on the origin and dispersal of modern humans*, ed. K. Paul Mellars, Bar-Yosef O. Boyle, and C. Stringer, 133–151. Cambridge: McDonald Institute Archaeological Publications.

McBrearty, S., and A. Brooks. 2000. The revolution that wasn't: A new interpretation of the origin of modern human behavior. *Journal of Human Evolution* 39: 453–563.

McPherron, Shannon, Zeresenay Alemseged, Curtis Marean, Jonathan Wynn, Denné Reed, Denis Geraads, René Bobe, and Hamdallah Béarat. 2010. Evidence for stone-tool-assisted consumption of animal tissues before 3.39 million years ago at Dikika, Ethiopia. *Nature* 466: 857–860.

Nowell, April. 2010. Defining behavioral modernity in the context of Neandertal and anatomically modern human populations. *Annual Review of Anthropology* 39: 437–452.

O'Connell, J.F., K. Hawkes, and N.G. Blurton Jones. 1999. Grandmothering and the evolution of *Homo erectus*. *Journal of Human Evolution* 36: 461–485.

Odling-Smee, F. John, Kevin Laland, and Marcus Feldman. 2003. *Niche construction: The neglected process in evolution*, Monographs in population biology. Princeton: Princeton University Press.

Okasha, Samir. 2006. *Evolution and the units of selection*. Oxford: Oxford University Press.

Powell, Adam, Stephen Shennan, and Mark Thomas. 2009. Late Pleistocene demography and the appearance of modern human behavior. *Science* 324: 1298–1301.

Roebroeks, Will (ed.). 2007. *Guts and brains: An integrative approach to the hominin record*. Leiden: Leiden University Press.

Ross, Don. 2006a. The economic and evolutionary basis of selves. *Cognitive Systems Research* 7: 246–258.

Ross, Don. 2006b. Evolutionary game theory and the normative theory of institutional design: Binmore and behavioral economics. *Politics, Philosophy and Economics* 5: 51–80.

Seabright, Paul. 2010. *The company of strangers: A natural history of economic life*. Princeton: Princeton University Press.

Shaw-Williams, Kim. 2011. *The triggering track-ways theory*. M.A. thesis, Victoria University of Wellington.

Sterelny, Kim. 2011. From hominins to humans: How sapiens became behaviourally modern. *Philosophical Transactions of the Royal Society B* 366: 809–822.

Sterelny, Kim. 2012a. *The evolved apprentice*. Cambridge, MA: MIT Press.

Sterelny, Kim. 2012b. Morality's dark past. *Analyse and Kritik* 34: 95–116.

Sterelny, Kim. 2013a. Cooperation in a complex world: The role of proximate factors in ultimate explanations. *Biological Theory* 7: 358–367.

Sterelny, Kim. 2013b. Life in interesting times: Cooperation and collective action in the Holocene. In Cooperation and its evolution, eds. Kim Sterelny, Richard Joyce, Brett Calcott, and Ben Fraser, 89–108. Cambridge, MA: MIT Press.

Sterelny, Kim. Forthcoming. Signals, symbols and norms. *Biological Theory*.

Stiner, Mary C. 2002. Carnivory, coevolution, and the geographic spread of the genus Homo. *Journal of Archaeological Research* 10: 1–63.

Stout, Dietrich. 2011. Stone toolmaking and the evolution of human culture and cognition. *Philosophical Transactions of the Royal Society Series B* 366: 1050–1059.

Tomasello, Michael. 2009. *Why we cooperate*, Cambridge, MA: MIT Press.

Tomasello, Michael, and Malinda Carpenter. 2007. Shared intentionality. *Developmental Science* 10: 121–125.

Tomasello, Michael, Malinda Carpenter, Josep Call, Tanya Behne, and Henrike Moll. 2005. Understanding and sharing intentions: The origins of cultural cognition. *Behavioral and Brain Sciences* 28: 675–691.

Tomasello, Michael, Alicia P. Melis, Claudio Tennie, Emily Wyman, and Esther Herrmann. 2012. Two key steps in the evolution of human cooperation: The interdependence hypothesis. *Current Anthropology* 53: 673–692.

Turner, J. Scott. 2000. *The extended organism: The physiology of animal-built structures*. Cambridge, MA: Harvard University Press.

Wadley, Lyn. 2010. Compound-adhesive manufacture as a behavioral proxy for complex cognition in the Middle Stone Age. *Current Anthropology* 51: S111–S119.

Warneken, Felix, and Michael Tomasello. 2009. Varieties of altruism in children and chimpanzees. *Trends in Cognitive Science* 13: 397–402.

Warneken, Felix, Brian Hare, Alicia Melis, Daniel Hanus, and Michael Tomasello. 2007. Spontaneous altruism by chimpanzees and young children. *PLoS Biology* 5: e184. doi:110.1371.

Whiten, A., and R. Byrne (eds.). 1997. *Machiavellian intelligence II: Extensions and evaluations*. Cambridge: Cambridge University Press.

Wilkinson, Gerald S. 1990. Food sharing in vampire bats. *Scientific American* 262: 64–70.

Wilson, David Sloan. 2002. *Darwin's cathedral: Evolution, religion and the nature of society*. Chicago: University of Chicago Press.

Wrangham, Richard. 2001. Out of the pan, into the fire: How our ancestors' evolution depended on what they ate. In *Tree of life*, ed. Franz de Waal, 121–143. Cambridge, MA: Harvard University Press.

Wrangham, Richard. 2009. *Catching fire: How cooking made us human*. London: Profile Books.

Wynn, Thomas, and Frederick Coolidge. 2004. The expert Neanderthal mind. *Journal of Human Evolution* 46: 467–487.

Wynn, Thomas, and Frederick Coolidge. 2010. How Levallois reduction is similar to, and not similar to, playing chess. In *Stone tools and the evolution of human cognition*, ed. April Nowell and Iain Davidson, 83–103. Boulder: University of Colorado Press.

Printed by Printforce, the Netherlands